ISDB-T
A Handbook for Broadcast Engineers

Masayuki Ito
Yasuo Takahashi
James Rodney Santiago

Copyright © 2017 Masayuki Ito, Yasuo Takahashi, James Rodney Santiago
All rights reserved.
ISBN: 1544718144
ISBN-13:   978-1544718149

TABLE OF CONTENTS

TABLE OF CONTENTS i
PREFACE vii
Chapter I.   Introduction of Digital Terrestrial Broadcasting System 1
   I.1   History of Digitization in the Broadcasting Service 1
   I.2   Comparison between Analog and Digital Broadcasting 2
Chapter II.   Structure of Digital Broadcast Systems 5
   II.1   Structure of Digital Broadcast Systems 5
      II.1.1   Source coding technology 6
      II.1.2   Multiplexing technology 6
   II.2   Transmission Technology 8
      II.2.1   Signal modulation in digital broadcasting 8
      II.2.2   Various transmission technology in different media 10
   II.3   Multi Carrier or Single Carrier in Terrestrial Broadcast 11
      II.3.1   Transmission capacity of OFDM and single carrier 12
      II.3.2   Robustness against multi-path interference 13
      II.3.3   Introduction of the "Guard Interval" 15
   II.4   Hierarchical Transmission 19
      II.4.1   What is "Hierarchical transmission"? 19
      II.4.2   Non uniform mapping (adopted in DVB-T) 20
      II.4.3   Segmented OFDM transmission in ISDB-T 23
      II.4.4   Mobile reception service in terrestrial broadcast 24
   II.5   Error Correction System (Channel Coding) 25
   II.6   Digital TV Standards 26
      II.6.1   ISDB-T 27
      II.6.2   DVB-T 28
      II.6.3   ATSC 30
      II.6.4   DTMB 32
   II.7   Structure of the ISDB-T Standards 35
      II.7.1   Requirement for ISDB-T and solution 35
      II.7.2   Structure of ISDB-T technical standards (Japan) 36
      II.7.3   Structure of ISDB-T technical standard (Brazil) 37
      II.7.4   Commonalities and differences between ARIB/ABNT standards 38
Chapter III.   ISDB-T Transmission System 41
   III.1   Technical Overview 41
      III.1.1   Transmission parameters 41
      III.1.2   Functional block diagram of an ISDB-T transmission system 48
   III.2   Hierarchical Transmission in ISDB-T 50
      III.2.1   Concept and rules of hierarchical transmission in ISDB-T 51
      III.2.2   TS re-multiplexer 57
      III.2.3   Division of TS into hierarchical layers 60
      III.2.4   "Delay adjustment" before the byte interleaver 61
      III.2.5   "Delay adjustment" before carrier modulation 62
      III.2.6   Combining hierarchical layers 63

- III.3 Channel Coding 63
  - III.3.1 Overview of channel coding 63
  - III.3.2 Outer coder... 64
  - III.3.3 Energy dispersal 65
  - III.3.4 Byte interleaving.. 66
  - III.3.5 Inner coder 67
  - III.3.6 Mapping and bit interleaving 69
  - III.3.7 Time interleaving 72
  - III.3.8 Frequency interleaving 75
- III.4 OFDM Framing, Modulation... 76
  - III.4.1 OFDM framing.. 77
  - III.4.2 Scattered Pilot (SP) 78
  - III.4.3 Auxiliary Channel (AC) 80
  - III.4.4 Transmission and Multiplexing Configuration Control (TMCC) 80
  - III.4.5 OFDM modulation and spectrum 82
  - III.4.6 Guard interval 83
- III.5 Calculation of Data Rates... 83
  - III.5.1 Calculation examples of 6MHz system.. 85
  - III.5.2 Calculation examples of 8MHz system.. 86

Chapter IV. Multiplexing and PSI/SI... 89
- IV.1 Structure of multiplexed signal... 89
  - IV.1.1 Signal processing for TS generation 89
  - IV.1.2 PES packet 90
  - IV.1.3 TS packet.. 91
  - IV.1.4 Section data... 94
  - IV.1.5 Synchronization control... 96
- IV.2 Program Specific Information (PSI) 97
  - IV.2.1 The PAT and PMT 97
  - IV.2.2 Network Information Table (NIT) 99
  - IV.2.3 Conditional Access Table (CAT).. 101
- IV.3 Service Information (SI).. 103
  - IV.3.1 Service Description Table (SDT).. 104
  - IV.3.2 Event Information Table (EIT) 104
  - IV.3.3 Other tables (BIT, TOT) 106

Chapter V. Video and Audio Coding System 107
- V.1 Video Coding Technology 107
  - V.1.1 Input video format 107
  - V.1.2 Video resolution and aspect ratio 110
- V.2 MPEG-2 Video Compression 111
  - V.2.1 Signal configuration... 112
  - V.2.2 Motion estimation and motion compensation 114
  - V.2.3 Discrete Cosine Transform (DCT)... 114
  - V.2.4 Quantization and variable length coding 116
- V.3 H.264 Coding System 118
- V.4 Audio Coding Technology 121
  - V.4.1 Sampling and quantization of audio signals.. 121
  - V.4.2 Audio coding algorithm 122
  - V.4.3 Specification of the audio coding in the standard... 124
  - V.4.4 Comparison of the Japanese and Brazilian audio coding system 126

Chapter VI. Datacasting System 127
- VI.1 Data Broadcasting Service and Classification 127
  - VI.1.1 Profiles 128
- VI.2 Broadcast Markup Language (BML) 129
  - VI.2.1 Language structure of BML 129
  - VI.2.2 Monomedia 132
  - VI.2.3 Remote controller operation 135
  - VI.2.4 Layout design with CSS 135
  - VI.2.5 Coordinate system and z-index 137
  - VI.2.6 Plane model 138
  - VI.2.7 Scripting 139
- VI.3 Ginga 141
  - VI.3.1 Ginga-J 142
  - VI.3.2 Ginga-NCL 145
- VI.4 Caption and Superimpose 149
- VI.5 Transmission of Data Broadcasting 151
  - VI.5.1 Independent PES transmission protocol 151
  - VI.5.2 Data carousel transmission protocol 151
  - VI.5.3 Event message transmission protocol 152

Chapter VII. ISDB-T receiver 155
- VII.1 ISDB-T receiver overview 155
- VII.2 Structure of digital receivers 156
- VII.3 Synchronization Technology in Digital Receiver 157
  - VII.3.1 Block diagram of tuner/demodulator 157
  - VII.3.2 Synchronization techniques in ISDB-T receiver 159
- VII.4 Composition and Specification of ISDB-T receiver 164
  - VII.4.1 Specifications of tuner/demodulator 164
  - VII.4.2 Functions and specifications of back end 165
  - VII.4.3 Software of digital receiver 168

Chapter VIII. One-Seg Broadcasting Service 173
- VIII.1 Transmission Technology for One-Seg 174
  - VIII.1.1 The relationship between transmission bandwidth and receiving bandwidth 174
  - VIII.1.2 Reduction of power consumption of portable receiver and transmission rate 175
  - VIII.1.3 Operational guideline for hierarchical transmission 177
  - VIII.1.4 Block diagram of One-Seg receiver 180
  - VIII.1.5 Adaptive reception in mobile reception 181
- VIII.2 Video Coding for One-Seg Service 182
  - VIII.2.1 Video coding definition in ARIB STD-B24 183
  - VIII.2.2 Video coding definition in ARIB TR-B14 183
- VIII.3 Audio Coding for One-Seg Service 184
  - VIII.3.1 Audio coding definition in TR-B14 185
- VIII.4 Data Broadcasting for One-Seg Service 185
- VIII.5 Presentation on One-Seg 187
- VIII.6 Basic receivers 188

Chapter IX. Emergency Warning Broadcasting System 191
- IX.1 Overview of EWBS 191
- IX.2 Signaling of EWBS 192
  - IX.2.1 Emergency alarm broadcasting signal in the TMCC 192
  - IX.2.2 Emergency Information Descriptor in PMTs 193

IX.2.3	Operation of EWBS	195
IX.3	Emergency Earthquake Warning	195
Chapter X.	Deployment of Digital Terrestrial Transmission Network	199
X.1	The Digital Transmission Network	199
X.1.1	Differences between analog and digital transmission network	199
X.1.2	Required signal to noise ratio	200
X.1.3	Transmission network configuration	201
X.2	SFN and Signal Transmission in a Network	201
X.2.1	Types of transmission network	201
X.2.2	Synchronization for SFN	202
X.2.3	Types of interface point in transmission network	202
X.2.4	Types of synchronization scheme for SFN	203
X.2.5	Examples of SFN construction	204
X.2.6	Additional information multiplexed on a broadcast TS for network control	208
X.3	Degradation Factors in a Transmission Network and the Methods of Evaluation	217
X.3.1	Network model	217
X.3.2	Individual degradation factors	219
X.3.3	Signal degradation in the transmission link	221
X.3.4	Degradations caused by the transmitting equipment	223
X.4	Examples of Signal Improvement Technologies in a Transmission Network	228
X.4.1	Feedback pre-distortion compensation type amplifier	228
X.4.2	Feed forward type amplifier	228
X.4.3	Coupling loop interference canceller	229
Chapter XI.	Broadcaster's Infrastructure for Digital Terrestrial Broadcast	233
XI.1	Construction of Broadcaster's Facility and Migration to Digital broadcasting	233
XI.1.1	Transition to digital system	233
XI.2	Master Studio Subsystem	235
XI.2.1	Encoder	237
XI.2.2	Multiplexer	237
XI.2.3	PSI/SI	237
XI.2.4	Data-casting subsystem	238
XI.3	Transmitter and Transmission Network	239
XI.3.1	Important points on digital TV transmission networks design	240
XI.3.2	Network relay link	240
XI.3.3	Technical topics for digital TV transmission network	242
Chapter XII.	Measurement and Evaluation of Digital Broadcasting	245
XII.1	Overview of Measurement for Digital Broadcasting System	246
XII.1.1	Purpose of measurement for digital broadcasting signal	246
XII.1.2	Classification based on measurement configuration	246
XII.2	Overview of Measurement/Evaluation Items for System and Receiver	247
XII.2.1	Measurement/evaluation items for system and receiver	247
XII.2.2	Overview of each measurement items for system/receiver evaluation	248
XII.3	Overview of Measurement/Evaluation Items for Transmitter & Transmission Network	252
XII.3.1	Measurement/evaluation items for transmitter & transmission network	252
XII.3.2	Overview of each measurement item	253
XII.3.3	Intermodulation (Shoulder)	254
XII.3.4	Modulation Error Ratio (MER)	254
XII.3.5	Bit Error Rate (BER) (PN Method)	255
XII.3.6	BER (Simple measurement method using broadcast signal)	255

XII.3.7	Equivalent Noise Degradation (END) and Equivalent Noise Floor (ENF)	257
XII.3.8	Delay profile	258
XII.4	Some Examples of Measurement Data	259

Chapter XIII. Frequency Allocation and Channel Planning 263
- XIII.1 International Frequency Allocation 263
- XIII.2 National Frequency Allocation 265
- XIII.3 Parameters and Process of the Channel Planning 267
  - XIII.3.1 Minimum usable field strength 268
  - XIII.3.2 Field strength calculation 269
  - XIII.3.3 Protection ratio 270

INDEX 273

ABOUT THE AUTHOR 277
- Masayuki ITO 277
- Yasuo TAKAHASHI 277
- James Rodney P. SANTIAGO 277

(This page is left intentionally blank.)

PREFACE

This book is intended to serve as a compendium of practical knowledge on ISDB-T product design and operation. Although the book will prove most useful to the engineer engaged in ISDB-T product design, it will also be of considerable use to the broadcast operator in understanding the capabilities and limitations of the ISDB-T system. The treatment is directed toward understanding the engineering of the ISDB-T system in addition to its design concepts. The contents in this book cover the subject up to 2012. For updates on recent developments such as 4K/8K, please wait till the next edition.

This book is organized into four parts. The first part, consisting of Chapters 1 through 6, delineates the technology elements of an ISDB-T system. It also covers some principles and fundamental technologies that are common to the digital terrestrial broadcast systems. The second part, consisting of Chapters 7 and 8, deals with basic receiver types and design requirements of core technology elements. The third part, Chapters 9 through 12, treats a number of areas of deployment of broadcast networks. The final part, consisting of Chapter 13, deals with topics associated with regulatory matters. This chapter contains useful information that is essential to broadcasters and government officers.

During the process of writing this book, many individuals from the community took time out to help us. We would like to give a special thanks to all the experts who took part in DiBEG activity and for giving valuable feedback and contributions to this book.

We also thank the ISDB-T developers and engineers, especially those who participated in ARIB standardization activity and those from NHK STRL for inventing this great broadcasting technology. So much hard work has gone into making ISDB-T one of the most widely used terrestrial broadcasting standards today.

The authors would like to offer an extra special thanks to many other individuals without whose contributions this book would not have been possible.

Masayuki Ito
Yasuo Takahashi
James Rodney Santiago

(This page is left intentionally blank.)

Chapter I. Introduction of Digital Terrestrial Broadcasting System

I.1 History of Digitization in the Broadcasting Service

Analog TV broadcasting service has been in existence for more than 50 years. The use of digital technology in the broadcasting area started around 1980's especially in studio recording and processing use. Comparing to analog, digital technology is more efficient since the nature of the information does not change despite of the multi-generation transfer. In addition, the storage and archiving of audio/video are more efficient with the digital technology since compression techniques can be used to minimize the required physical size and data storage capacity. Another main factor is that the digital technology enables random and near real-time access to the stored audio and video very easily.

During this period in the 1980's, compression techniques can only be applied effectively to the processing of audio and video materials within the production area and cannot be extended up to the transmission part due to the fact that the technology at that time is not yet mature and sufficient to support the required channel capacity, which led to further research in compression. Electronics manufacturers paved the way for significant progress in LSI technology that allowed its practical use in the transmission part in the early 1990's.

During the development of digital transmission technologies for communication networks, it was aimed primarily for military applications and the telecommunications. As significant progress on the technology is achieved, it later gave the opportunity to extend this milestone for commercial and practical use in various fields especially in the broadcast domain. This opened the possibility of achieving full digital broadcasting from production until transmission.

Plans for the Digital Broadcast started in the 1990's wherein several countries including the United States, Europe and Japan embarked on a research and standardization process to allow the dissemination of the digital technology. Several countries started pilot services and predicted that the rapid deployment is likely to gain widespread adoption in the 21st century.

(1) The United States
In the early part of the 90's, The Advanced Television System Committee (ATSC) was conceptualized[1] and the FCC started accepting proposals to establish a basis for its standardization. After going through a rigorous evaluation of various proposals, eventually, 8-VSB has been chosen for digital modulation and MPEG-2 video coding for the compression. 8-VSB stands for 8 level Vestigial Side Band modulation for a 6MHz channel bandwidth and utilizes a single carrier to broadcast HDTV quality video contents. The commercial service of ATSC started 1997. The U.S. Congress set June 12, 2009 as the deadline for full power television stations to stop broadcasting analog signals, and full-power television stations nationwide have been required to broadcast exclusively in a digital format after the date.

Countries in the North American continent and South Korea adopted ATSC system for their digital television standard.

(2) Europe
In EU, in the early 90's, several digital broadcasting systems were developed. As one of these activities, a

---

[1] A/53, ATSC Digital Television Standard, Part-1 – Part-6

new organization called the "DVB (Digital Video Broadcasting) forum," has been organized to conduct research, standardization and promotions of digital broadcasting. DVB standardized the digital satellite broadcasting (DVB-S), the digital cable broadcasting (DVB-C), etc., until finally they released the digital terrestrial Broadcasting (DVB-T)[2] in 1997.

These digital broadcasting systems adopts a common structure, that is, MPEG-2 for source coding and service multiplexing with different transmission technologies that best fit to each transmission media. For digital terrestrial broadcasting, OFDM technology was adopted to enable Single Frequency Network (SFN) to achieve efficient utilization of the limited frequency resource.

(3) Japan

Japan was also one of the pioneers of Hi-vision or High Definition Television (HDTV), and many key technologies associated for HDTV which has been developed since the early 1990's. Experimental HD broadcast service started through satellite broadcasting, which led to the widespread recognition and acceptance of HDTV in the field of Outside Broadcast/Electronic News Gathering and even Satellite News Gathering.

Within the same period, the concept of making efficient use of the frequency spectrum gained strong support, this allowed new services to be delivered including rich audio/video content toward mobile and portable receivers. Since a new generation of services has been conceived to be possible in digital broadcast, the appropriate selection of the transmission system and video compression needed further studies that led to the consideration and adoption of the Band Segmented Transmission Orthogonal Frequency Division Multiplexing (BST-OFDM) in 1998 right after the standardization of the ATSC standard. ISDB-T (Integrated Services Digital Broadcasting-Terrestrial) finalized its transmission standard[3] and started commercial digital terrestrial broadcast service in Japan in 2003 with the One-Seg service being introduced in 2006.

One-Seg is a very convenient and useful service in the ISDB-T system that targets people always on the move. One-Seg enables them to access their favorite programs through their portable receivers without paying for an extra service fee.

In addition to the terrestrial HDTV transmission and mobile/portable service, ISDB-T also included the data broadcasting service that uses the BML as its middleware.

Later on, Brazil, after adopting ISDB-T, decided to introduce some variations that they see fit for their region. This variation is concentrated on the selection of MPEG-4/H.264 compression for their video and audio system as opposed to the MPEG-2 compression system has been adopted in Japan. Also, Brazil decided to develop their own middleware called "Ginga" as an alternative to the "Broadcast Markup Language (BML)". Commonality and difference of both Japanese and Brazilian systems are described in the Chapter II.

In 2009 - 2010, many South and Central American countries announced the formal adoption of ISDB-T as their sole digital terrestrial standard. The Philippines (2010), Botswana (2013) and Maldives (2014) followed the adoption.

I.2 Comparison between Analog and Digital Broadcasting

Why go for digital broadcast?

Though this is a very simple question, its relevance can never be understated. In order to answer this question it would be easier to focus on components that comprise a digital broadcasting system and discuss its features.

---

[2] ETSI EN 300 744, Digital Video Broadcasting (DVB); Framing structure, channel coding and modulation for digital terrestrial television

[3] ARIB STD-B31, Transmission System for Digital Terrestrial Television Broadcasting

(1) What are the advantages of digital technology?
   a) No degradation in multi-generation recording/reproduction
   b) Ease of signal processing and has less degradation in processing/editing
   c) Harmonization with computer aided contents and ease of computer processing
   d) Flexibility in supporting various types of TV signal (different resolution, different aspect ratio, etc.)

Due to these advantages, many broadcast studio systems has been digitalized or in the process of being converted to full digital. This trend has resulted in the sudden increase in the technical quality of a TV programs. In addition, new services like complex computer graphics allows the incorporation of computer aided digital processing tool to generate more attractive contents. As described above, these simple reasons justified the digitization of studios which began in the 1980's era.

(2) What are the advantages of digital technology in transmission area?
   a) Less degradation against interference be it man-made or natural noise
   b) Detection and correction of errors occurred in the transmission path through the use of Forward Error Correction techniques
   c) Robustness of the signal even in mobile or portable reception environment
   d) Larger channel capacity compared to analog broadcasting, which enables multiple High Definition or Standard Definition programs to be transmitted in the same channel
   e) Lower power requirement for replicating the analog coverage area while providing superior quality as compared to analog
   f) Multiple transmission/reception service that targets fixed receivers such as those found in the home (flat panels or recorders) or portable receivers (mobile phones/handheld devices) in the same frequency
   g) Effective use of frequency resource by using the Single Frequency Network technology which only works with OFDM

Due to the above stated advantages of the digital transmission systems, broadcast system enables not only high quality TV program service but also stable reception service even in severe receiving condition. Digital signal processing technology can detect and recover errors on the data that was lost in the transmission path.

Europe and Japan placed high importance to the effective use of frequency since most has been assigned already for analog TV broadcast, in this way, new types of TV services can be delivered which will benefit even other countries. In this regard, OFDM technology became the logical choice for digital broadcast.

Chapter II and Chapter III will explain the details of these merits.

(3) What are the types of new broadcast services are available in digital technology?
   a) HDTV service is only available in digital broadcast
   b) Multi-program service (in analog TV service, one TV channel in one frequency band)
   c) Multi-channel audio and/or multi-lingual audio service
   d) Data-casting service
   e) Interactive service

In this new era of communication, broadcast services are expected to harmonize with the previously exclusive telecommunication application of mobile communication, the Internet and the PC world. From above, d) and e) are new features of digital broadcast service that are expected to lead the way for new types of services and new business opportunities.

Through (1) and (3), Table I-1 shows the comparison of analog and digital broadcast as a tool for establishing a general understanding.

Table I-1 Comparison of analog/ digital broadcasting

Items	Analog	Digital
Signal quality	Good	Very Good
Signal degradation		
In recording media	Degradates in process	√
In signal transmission	Degradates in process	√
In signal processing	Degradates in process	√
Flexibility of service		
High quality(HDTV)	X	√
Multi-SDTV	X	√
Multi-channel audio	√	√
Additional service		
Data-casting	Teletext Only	√
Interactivity	Limited	√
Transmission		
Required TX power	High	Low
Mobile/portable reception	√	√
Effective use of frequency resource	X	√
Hierarchical transmission	X	√
Receiver commonality	X	√

√:better/available, X:impossible

Chapter II. Structure of Digital Broadcast Systems

This chapter will discuss how digital broadcast systems are generally structured. ISDB-T, as a digital terrestrial broadcast system, also follows this structure, in which audio visual source are digitally formatted, mixed with other source, made to conform to broadcast signals and transmitted to the receivers. This chapter will further discuss about the major digital terrestrial broadcast standards, followed by the general overview of the ISDB-T standard and the commonalities and differences of Japanese terrestrial broadcast standard and Brazilian terrestrial broadcast standard.

II.1 Structure of Digital Broadcast Systems

Generally the signal flow of digital broadcast consists of "source coding," "multiplexing" and "transmission" as shown in Figure II-1. "Source Coding" block converts analog audio, visual and data source into digital format and feeds to "Multiplexing" block that mixes multiple sources into one stream. Then "Transmission" convers digital data stream into radio wave format. "Source coding" and "multiplexing" are, roughly speaking, independent from transmission coding as "Source coding" and "multiplexing" are about how they handle digital data and in general same technology are used in one standard. However, transmission technology may vary among the transmission media, as the nature and characteristics of transmission path has significant differences.

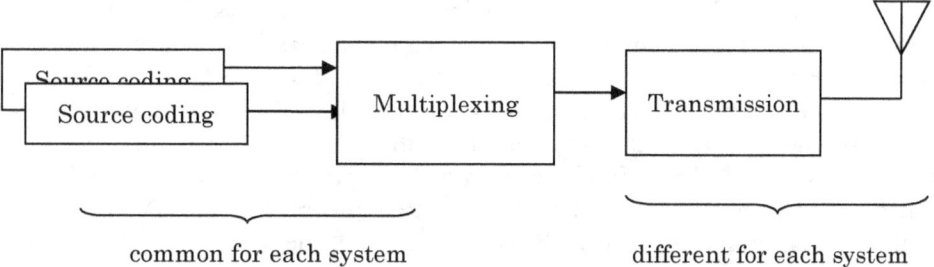

Figure II-1 Structure of digital broadcast systems

Figure II-2 shows an example of the structure of digital broadcast in Japan. Source coding and multiplexing technology use MPEG standard regardless of the media - in satellite, cable and terrestrial broadcast. Transmission technology may vary and has separate standard in satellite, cable and terrestrial broadcast.

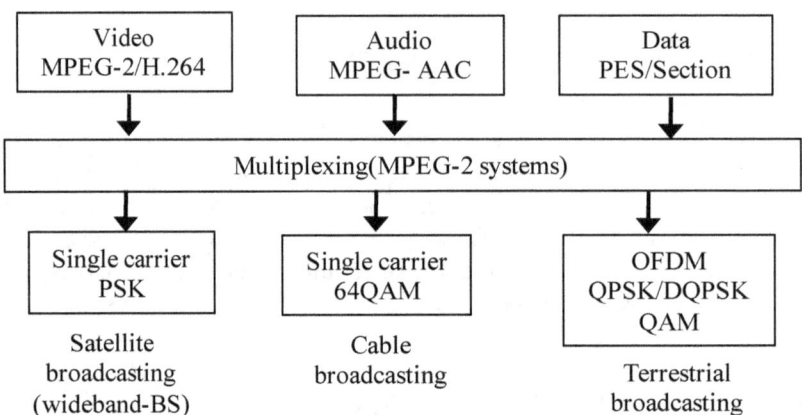

Figure II-2   Fundamental technologies for the Japanese digital broadcast system

II.1.1   Source coding technology

Source coding technology is a key element for the digital broadcast, converting analog audio and visual source into digital format, consists of two technologies – sampling and compression. "Sampling" technology converts analog waveform input into digital discrete data, whereas "compression" technology compresses its digital input into smaller volume as the raw digital input tend to be very large and very hard to handle.

Video/audio compression technology is a key element that enabled digital broadcast to transmit TV signal within limits of the bandwidth available for broadcast. For an example, the required bit rate of an uncompressed source signal of SDTV (Standard definition TV) is as much as 270Mbps (13.5MHz sampling frequency, Y:Pr:Pb = 4:2:2, 10bit/sample). The transmission bandwidth necessary to convey this uncompressed signal is far wider than available analog TV bandwidth. Assuming that a transmission system of 6MHz digital terrestrial broadcast, such as ATSC, DVB-T and ISDB-T with a transmission bit rate of 19-23Mbps is in the maximum range of transmission rate with the maximum parameter sets. In case of an 8MHz bandwidth, the maximum transmission bit rate of such systems increases 8/6 of 6MHz system. This transmission capacity of one channel is too small to transmit an uncompressed YUV signal. Therefore, the introduction of the digital broadcast service had to wait until the development of the compression technology.

Around the early 1990's, video compression technology achieved the reasonable compression ratio with sufficient bit rate and sufficient signal quality to transmit TV services in a limited TV broadcast bandwidth. In the 1990's, MPEG-2 was standardized by the Moving Pictures Expert Group (MPEG) and was adopted as an international standard by ISO/IEC and ITU. MPEG-2 Systems, along with video and audio compression and coding, enabled multi SDTV and/or HDTV transmission in one broadcast channel.

In the around 2000's, more efficient compression technology were proposed including H.264/AVC-10 which has become more practical to use thanks to the progress in semiconductor manufacturing technology that enabled high speed and large scale Integration (LSI). In the meantime, for audio coding technology, the newer coding technology, named MPEG-AAC has been adopted in the late 1990's. Technical details of the source coding technology and related specifications will be discussed in chapter V.

II.1.2   Multiplexing technology

One of the most important features of digital broadcast is "Multiplexing". Adopting digital format, broadcast signals become capable of carrying not only video and audio stream such as those in analog television, but also multiple TV programs and Datacasting in one channel.

When ISDB-T chose what multiplexing technology to adopt, there were following requirements:
(i) flexibility: services should be multiplexed, such as HDTV, SDTV and datacasting, etc.
(ii) expandability: new service and new technologies needs to be supported according to the progress of technology and service.
(iii) commonality: for the cost reduction of the receiver, common technology and hardware should be adopted.
(iv) ease of program/content selection: in digital broadcast system, the number of services increases rapidly, a simple and easy control process should be introduced to avoid such burden for the viewers.
(v) international harmonization: international common technology and hardware should be used to reduce a receiver cost.

In order to answer to these requirements, ISDB-T chose MPEG-2 systems for its multiplexing technology. MPEG-2 systems features the following design concept: harmonization with communication, multiplexing and PSI/SI.

To harmonize video and audio transmission technology with digital communication system, MPEG-2 systems adopted "Transport Stream Packet" format, in which digital data to be carried are divided and placed into fixed size packets so that the packets can fit ATM (Asynchronous Transfer Mode: technology of high speed digital communication, mainly used for online broadband high-speed transmission) structure. The relations of ATM format and MPEG-2 TSP format are shown in the Figure II-3. The packet size of MPEG-2 TSP are fixed to be 188 byte in length so that the data of MPEG-2 TSP are divided into 4 data blocks, and each data block can be precisely stuffed into each ATM cell data area.

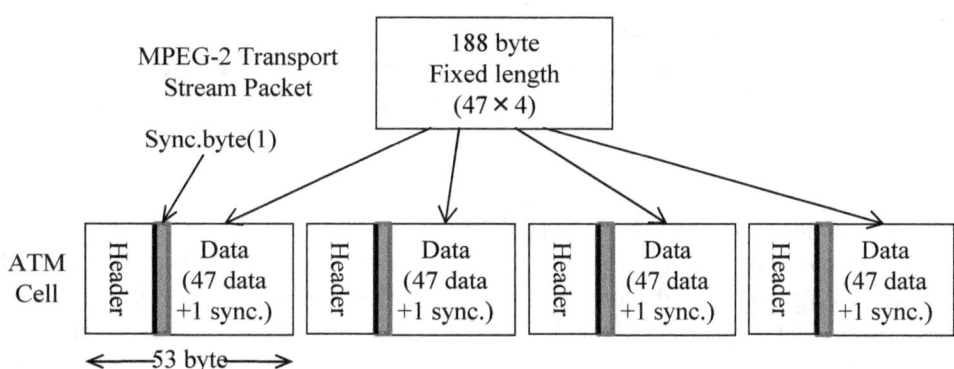

Figure II-3 Relationship between ATM format and MPEG-2 TS format

In digital broadcast systems, various types of contents can be multiplexed, such as, video, audio, stream type data, file type data and PSI/SI (control information) into a stream sent from single transmission. Figure II-4 illustrates the multiplexing structure adopted in ISDB-T. As shown in the figure, there are two types of data format to be carried in the transport stream packet, Packetized Elementary Stream (PES) and Section format. PES is mainly for large volume of continuous data such as audio and video. Section is for small volume data (PSI/SI) whose transmission occurs less frequent than PES. Data casting choses both the PES and Section format depending on its use. These packets are mixed in one continuous data stream called a transport stream. The details of the relationship between each data format and process of data format conversion are described in Chapter IV.

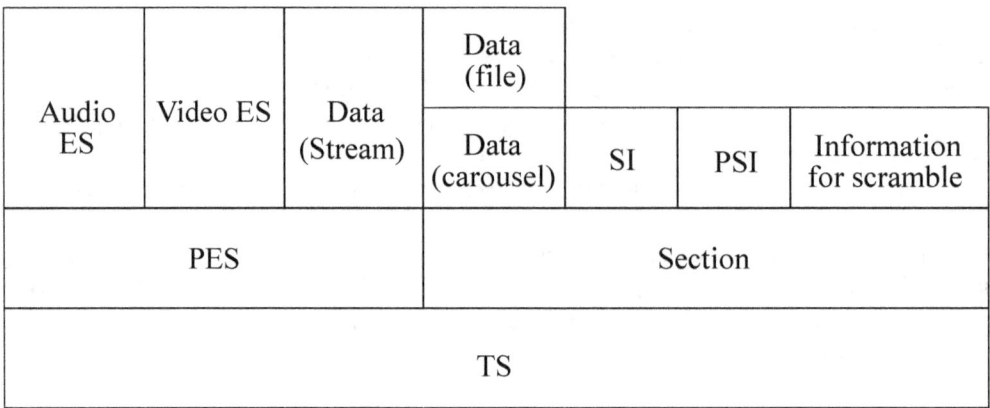

Figure II-4 Multiplexing structure

PSI (Program Specific Information) and SI (Service Information) are introduced along with the multiplexing technology. Digital broadcast enables us to transmit various contents in one channel, therefore the receiver has to filter out unnecessary data based on the viewer's channel selection and reconstruct the stream for the viewer with the desired information inside the transport stream. To support the selection of contents, some control information called PSI/SI is introduced to identify the desired data. The details of PSI/SI will be explained in Chapter IV.

II.2   Transmission Technology

The first half of this chapter explains the source coding and multiplex technology. The latter half will explain on the transmission technology and how the stream is generated through source coding and multiplex technology.

II.2.1   Signal modulation in digital broadcasting

One modulation technique that is commonly used in digital broadcasting is called "I-Q modulation", where "I" is the "in-phase" component of the waveform, and "Q" represents the quadrature component. In its various forms, I-Q modulation works well with digital formats. An I-Q modulator can also create analog signal such as AM, FM and PM.

The process of imposing an input signal onto a carrier wave is called modulation. Transmitters vary the carrier wave, a wave of constant frequency such as a sine wave, in the way of amplitude or phase (or frequency) so that the receivers can detect the difference of the received signal from the original carrier wave and extract information from them. An image of varying signal in phase and amplitude direction is shown in Figure II-5.

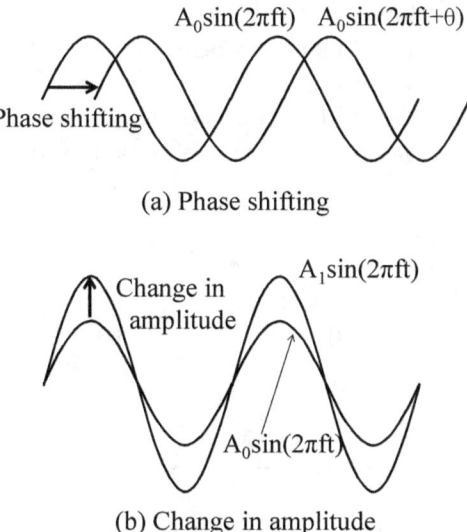

(a) Phase shifting

(b) Change in amplitude

Figure II-5   An image of phase shifting and amplitude changing

Digital modulation in ISDB-T use phase shift and amplitude change. It can be illustrated by placing the phase shift (θ in the figure) and amplitude change (A_1 in the figure) onto a polar coordinates system as illustrated on Figure II-6. Here, on communication systems, a point is placed in the complex plane in Cartesian coordinates for ease of circuit designs, having horizontal axis real part, and vertical axis imaginary part. Please note that on communications design, real part is called In-phase and imaginary part is called Quadrature. The reason is that by preparing circuits to generate carrier signal and quadrature signal, whose phase is 90 degrees different from carrier signal, and adding the two signals together in specific proportions, it can generate digital and analog modulation easily thanks to the Euler's formula.

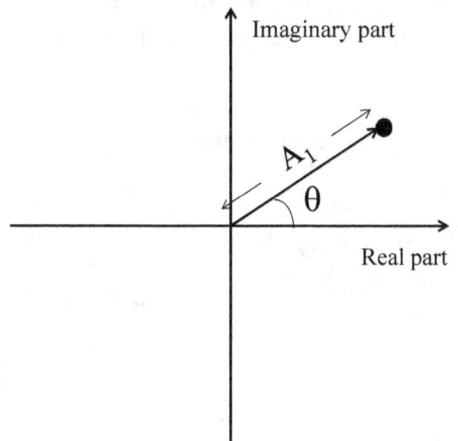

Figure II-6   Placement on a polar coordinates

The difference between the received signal on receivers and carrier wave can represent what information is placed onto the signal. The diagram representing how digital information is placed on IQ coordinates is called constellation. For example, if the received signal varies in four phases, $+\pi/4$, $+3/4\pi$, $-3\pi/4$ and $-\pi/4$ illustrated in Figure II-7, it can carry 2 bits of information. That is, if the signal received falls into the first (upper right) quadrant, it means transmitter intended to send bit 00. The same applies to other quadrants. It is called Quadrature Phase Shift Keying (QPSK). If a signal has 16 states, it is called 16QAM (16 Quadrature Amplitude Modulation). See Figure II-18(a) for 16 QAM constellation.

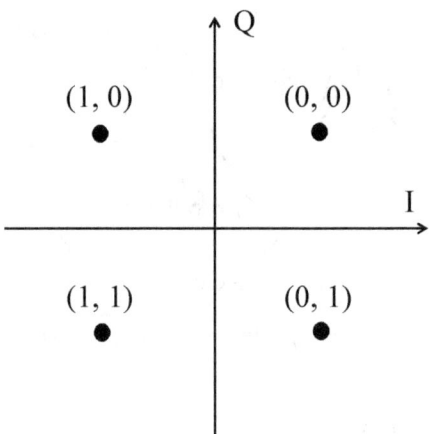

Figure II-7   QPSK constellation on IQ coordinates

II.2.2   Various transmission technology in different media

As briefly described in the first half of this chapter, various transmission technology were adopted for different transmission media. Before going into details of the technology, let's take a look at the difference between the various media of broadcast to understand the basis of the selection of specific technologies that constitute the terrestrial digital broadcast.

Because the transmission characteristics of each transmission media - satellite or terrestrial - are very different, it is natural to adopt different transmission technologies. As an example of radio frequency media, the difference of "satellite broadcast" and "terrestrial broadcast are shown in Table II-1.

Table II-1   Differences between "satellite" and "terrestrial" broadcast

Item	Satellite	Terrestrial
Distance	Very long (>36,000km)	Not long (as far as 100km)
Transmitter power	Critical (limited around several hundred Watts)	Not critical (several thousands of Watts are available)
Usable bandwidth	Not critical (27MHz in Japanese BS)	Critical (6/7/8MHz in VHF/UHF band)
Multi-path disturbance	Negligible	Critical

In case of satellite broadcast system, the distance between transmitter and receiver is significantly large (>36,000km); whereas in case of the terrestrial broadcast, the distance is as long as 100km which is very short compared to satellite broadcast. Regarding the transmitter power, satellite broadcast has upper limit because of the power limitation as satellites draws power from solar panels. For the terrestrial broadcast, transmitter power is not as critical as satellite because commercial power source that can supply enough electricity are generally made available to the transmission sites. Regarding frequency use, satellite broadcast uses a wider bandwidth than that of terrestrial because satellite broadcast uses microwave frequency band whose use is newer and less congested compared to that of VHF and UHF. In the case of Japan, 500MHz (11.7-12.2 GHz) bandwidth is allocated for satellite broadcast. For terrestrial broadcast, 240MHz (470-710MHz) is allocated. In other words, bandwidth limitation is the critical issue for a terrestrial broadcast service.

As described above, "satellite" and "terrestrial" broadcast have different transmission characteristics that is why different technical solutions must be implemented. In choosing the modulation for satellite broadcast, the major constraints of the "transmission link with limited power" aside from bandwidth limitation is the received signal level which tends to be very low compared to the noise level (thermal noise) - in technical terms, the reception is under low C/N ratio (Carrier to Noise ratio) condition. Because of these characteristics, the modulation system called QPSK or 8PSK was chosen for current satellite broadcast in Japan. These modulation systems have robust reception characteristics under low C/N condition although the transmission capacity is not better than Multi-level quadrature amplitude modulation (Multi-level QAM) system that is used for a terrestrial broadcast.

On the other hand, in terrestrial broadcast, the most important consideration of these "transmission link with limited bandwidth" is the transmission of large volume of data in a limited bandwidth which reinforces the selection of a more efficient transmission utilizing a multi-level quadrature amplitude modulation (Multi-level QAM) called 64QAM or 8VSB are used for a terrestrial digital broadcast. In general, in this transmission system, the most important consideration is the limitation of transmission bandwidth.

II.3 Multi Carrier or Single Carrier in Terrestrial Broadcast

The single carrier transmission means single radio frequency carrier is used to carry the information in the form of bits. OFDM (Orthogonal Frequency Division Multiplex), also known as a multicarrier transmission system uses multiple carrier signals, each of which represents bits, at different frequencies in each channel bandwidth. In the field of digital terrestrial broadcast system, ISDB-T, DVB-T and a part of DTMB adopts "multi-carrier" transmission system, whereas ATSC and a part of DTMB adopts "single carrier" system.

The OFDM is one of the multi-carrier transmission systems widely used in the digital communications domain. Figure II-8 illustrates the basic concept of the OFDM. (a) shows a waveform of a pulsed signal with the width of T in the time axis. This time domain signal is converted to a frequency domain waveform illustrated in (b) by Fourier transform process.

As shown in (b), the resulting waveform of the frequency domain crosses the frequency axis for the first time at (1/T). This implies that the energy of this pulse signal is zero at the (1/T) on the frequency axis.

In OFDM transmission system, as shown in (c), each carrier is positioned with the distance of (1/T) on the frequency axis to each other. At every 1/T, the combined carrier energy of the signals is zero. Therefore, it is said that there is no "Inter Carrier Interference (ICI)" in this frequency arrangement. This relationship is named "Orthogonal relation" which is the fundamental concept of "OFDM."

This concept also means that OFDM transmission system does not need any guard band between adjacent carriers. This is currently the most efficient technique for frequency utilization.

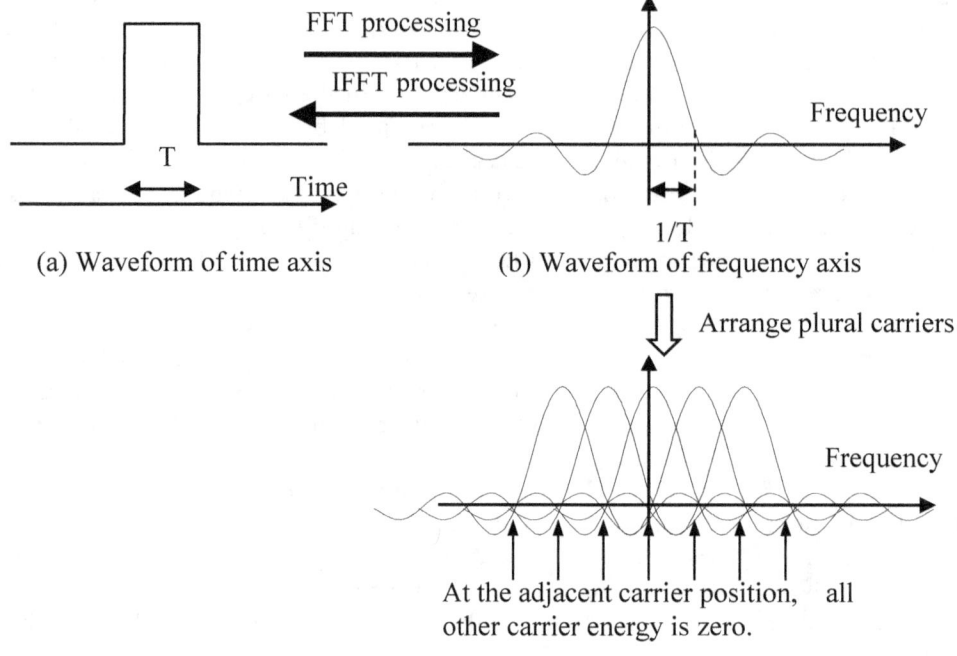

(a) Waveform of time axis

(b) Waveform of frequency axis

(c) Orthogonal arrangement on frequency axis

Figure II-8　Relationship between the waveform of the frequency and time domain for OFDM

II.3.1　Transmission capacity of OFDM and single carrier
In general, there are two types of transmission system which are used in digital communication systems, one is "single carrier" transmission system, the other is "multi-carrier" transmission system. Also, as described briefly in the section II.6, in the field of digital terrestrial broadcast system, ISDB-T and DVB-T adopted OFDM transmission system which is a multi-carrier transmission system, whereas ATSC and a part of DTMB adopts single carrier system.

Firstly, we will look into the transmission capacity of a single-carrier system and a multi-carrier transmission system. Figure II-9 shows an image of a "single carrier" transmission system (left side) and "multi-carrier" transmission system (right side). In this example, a system with four (4) carriers is illustrated in (b) as a multi-carrier system.

In communication theory, the relationship between transmission bandwidth and transmission symbol duration is inverse proportional. In case of single carrier system, the symbol duration is short, as a result, a wider bandwidth is necessary. On the other hand, in case of multi-carrier system, symbol duration of each carrier is long, therefore, the transmission bandwidth of each carrier is narrow. As shown in Figure II-9 (b), multiple carriers (in case of Figure II-9, four carriers) can be allocated into same bandwidth of single carrier system

In the case illustrated in Figure II-9 (b), the bandwidth of each carrier of a "multi-carrier" system (Bw(M)) is 1/4 of the bandwidth of "single carrier" system (Bw(S)), whereas the symbol duration of "multi-carrier" system(Ts(M)) is four times of the symbol duration of "single carrier" system (Ts(S))

The total transmission capacity is calculated as the multiplication of (number of symbols) in a certain period of time and (number of carriers) within the available frequency. In the case of Figure II-9, the transmission capacity of single carrier system is 4, that is equal to the transmission capacity of a "multi-carrier" system.

This statement simply means that the transmission capacity of single carrier system and multi-carrier system is the same with the bandwidth given, therefore from the view point of transmission system efficiency, the two have no difference.

In general, from the view point of hardware size, a single carrier system is simpler than a multi-carrier system, because a multi-carrier system needs frequency-time domain conversion processes. In case of OFDM system as a multi-carrier system, Inverse Fast Fourier Transform (IFFT) and Fast Fourier Transform (FFT) technique needs to be utilized. Development of affordable LSI technology with a high speed processing and a high-density integration capacity enabled IFFT and FFT LSI are also necessary. Therefore, this complexity of multi-carrier system is at first a significant hindrance in its application for adoption as a digital broadcast system.

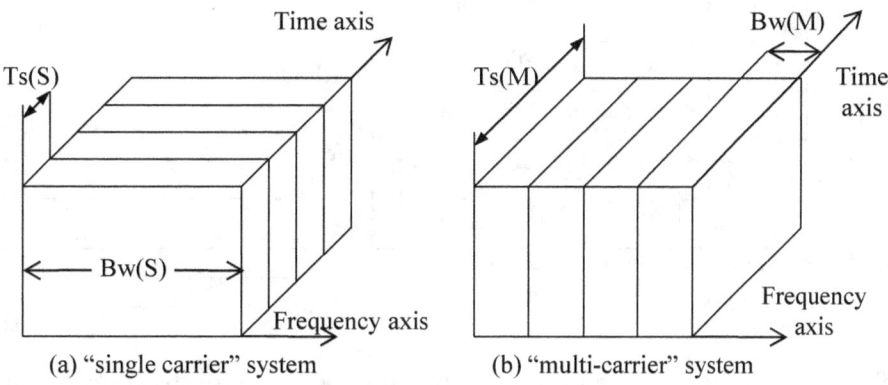

Bw(S): bandwidth of "single carrier" system,
Bw(M): bandwidth of each carriers of "multi- carrier" system
Ts(S): symbol duration of "single carrier" system
Ts(M)): symbol duration of "single carrier" system
Bw(S)=4 x Bw(M),　and　Ts(S)= Ts(M)/4

Figure II-9　Image of "single carrier" and "OFDM"

Then, from which aspect does a transmission system is selected?

Again, the difference of "single carrier" system and "multi-carrier" system is in the symbol duration. For a terrestrial broadcast, multi-path interference is generally observed. Multipath interference is a phenomena in which the desired signal is interfered by the same signal which took different path from the transmitter, for example, signals that are reflected from mountains or buildings and then arrives at the receiver with a certain time delay. This multipath component degrades a received signal quality because of Inter Symbol Interference (ISI), in which the desired signal is interfered by time-delayed signal carrying different information. In general, degradation by ISI is proportional to a ratio of (a length of overlap period)/(a length of symbol). Because of this reason, a longer symbol duration system is better fitted for a terrestrial transmission system.

II.3.2　Robustness against multi-path interference

In this section, we will look into the relationship between a symbol duration and a multipath interference.

Figure II-10 shows an image of a single carrier transmission system and a multiple carrier transmission system. In this Figure, the four carriers system is shown as an example. The symbol length of multiple carrier system is extended to four times of the symbol length of single carrier system.

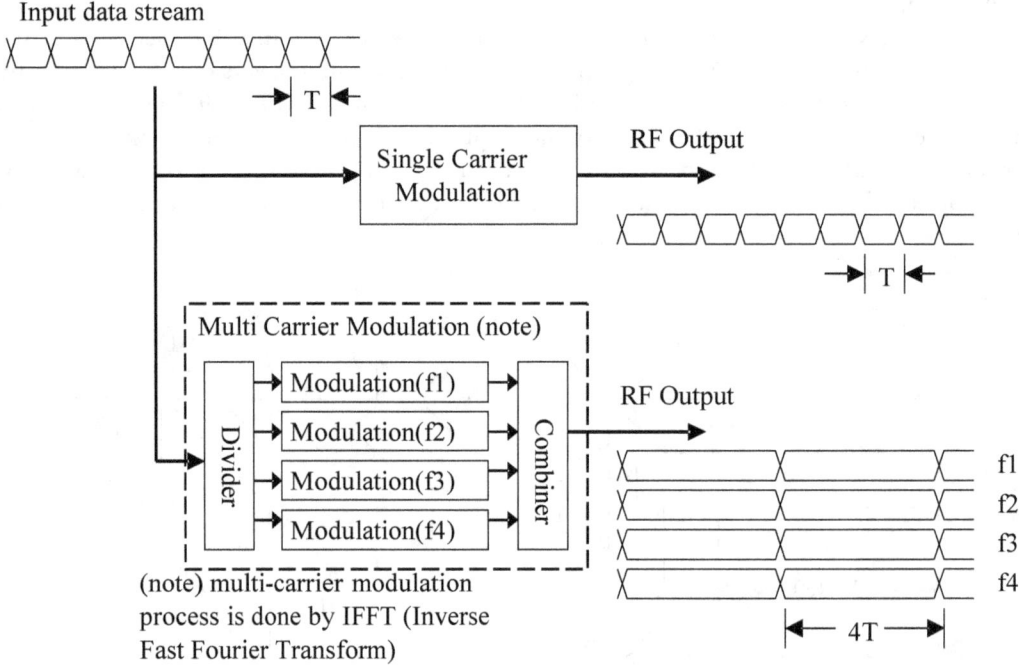

Figure II-10 Conceptual diagram for the relationship between modulation and symbol length

Figure II-11 illustrates the relationship between multipath interference and the signal degradation. There are two cases in the Figure II-11: (a) single carrier system, and (b) multi carrier system. The figure shows the difference in the timing of the desired signal (direct signal) and an interfering multipath signal. During the period, indicated as "Δt" in the figure, a symbol of the desired signal and a different symbol of multipath interference signal are received by the receiver at the same time and degrades the reception performance. This is called "Inter Symbol Interference (ISI)." The degree of ISI is proportional to the ratio of symbol length and the multipath delay time Δt,

In this example, the symbol length of multiple-carrier system is 4 times of that of single carrier system. Therefore, ISI is reduced to 1/4 in 4 carrier transmission system.

This example illustrates how multi carrier system absorbs impact of the multi-path interference. And this is why OFDM was chosen for ISDB-T and DVB-T system.

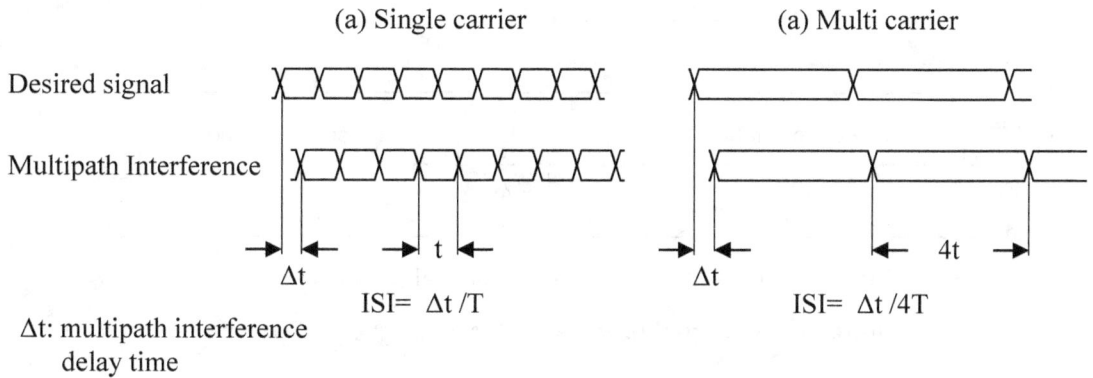

Figure II-11   Relationship of multi-path delay and ISI

Figure II-12 shows an example of the relationship between Δt and the permissible interference signal ratio in case of single multi-path interference. As shown in the figure, a single carrier sample degrades rapidly as time delay between the two signals of the multi-path interference widens.

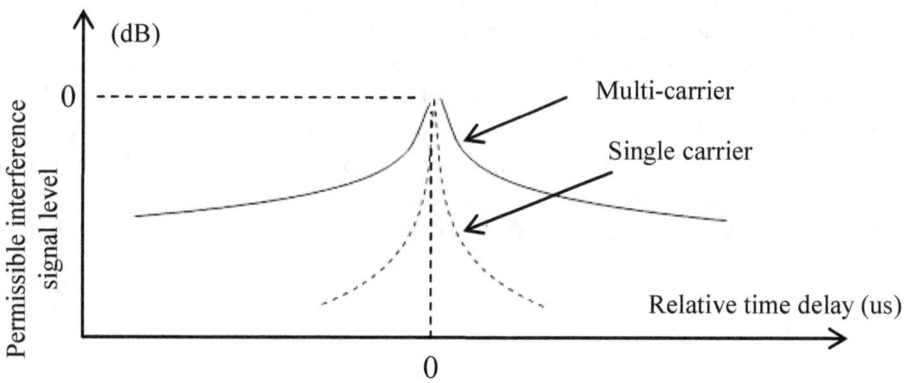

Figure II-12   Relationship of the degradation and relative delay time (no guard interval)

II.3.3   Introduction of the "Guard Interval"

As described above, it is widely recognized that the multi-carrier system reduces the degradation caused by multipath. In addition to that, in the field of terrestrial broadcast, a more robust transmission system is necessary for longer multipath. The technique of introducing "Guard Interval" is adopted in both ISDB-T and DVB-T system.

Guard intervals are inserted between the effective symbols (See Figure II-13 below). As long as the echoes fall within the guard interval, they will not affect the receiver's decoding ability. Further, the guard interval in ISDB-T is formed by copying an end portion of the OFDM symbol to the guard interval portion, so that the receivers can easily detect and synchronize with the effective symbols with correlation method (to be discussed in the section VII.3).

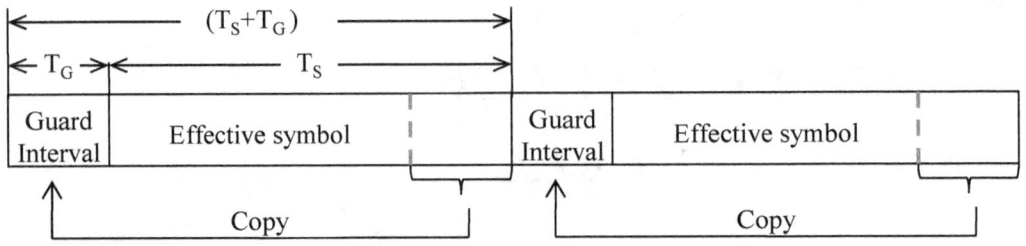

(note) T_S: Effective symbol length, T_G: Guard interval length

Figure II-13 Signal processing for adding guard interval

As shown in the Figure II-13, the end part of the effective symbol is copied and inserted at the front of effective symbol. By this process, the "symbol length" is extended to (T_S + T_G). The extended symbol length is represented as "apparent symbol length" in Figure II-11. In order to explain the effect of the guard interval, two cases are shown in Figure II-14.

In case (a), the delay time Δt_1 is shorter than T_G. On the other hand, in case (b), the delay time Δt_1 is longer than T_G. Please note that a FFT window is the time period to pick up a received signal for FFT signal processing in the receiver. Its length is designed as long as Effective symbol length (=T_S).

If FFT signal processor picks up signal with interference during FFT window, the demodulated signal contains some interference component. Therefore, to reduce the degradation caused by ISI, clear signal without ISI component needs to be put into the FFT window.

In case (a), a period without ISI is longer than Ts, therefore, it is possible to position an FFT window with length of Ts during the period without ISI. As a result, the delay time between desired signal and multi-path signal is not longer than T_G, In this case, the degradation caused by a multipath interference is not critical.

On the other hand, in case (b), period without ISI is shorter than Ts, therefore, a FFT window cannot be located without having ISI. As a result, the received signal is degraded by a multipath interference to a certain extent.

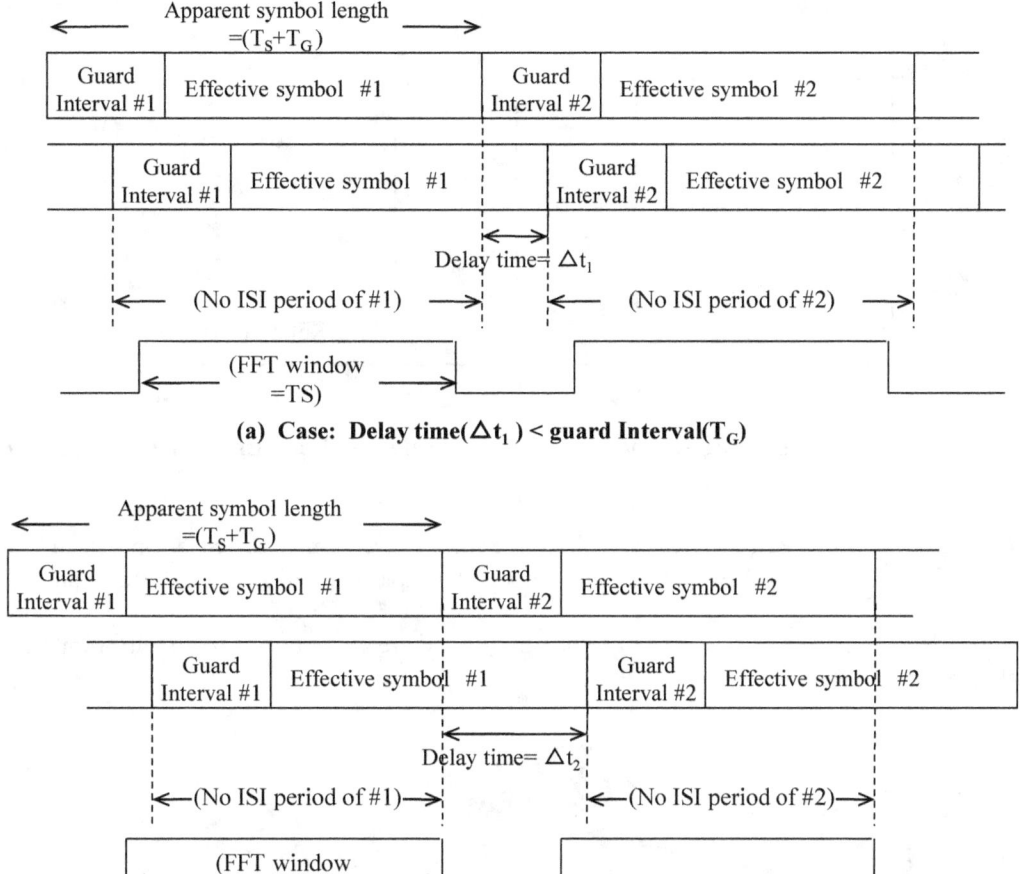

Figure II-14 Relationship between the received signal and the FFT window

In ISDB-T system, a Guard Interval is added to each symbol. As a result, robustness against multi-path interference is achieved to almost 0dB D/U ratio (Desired to Undesired ratio) during the period of Guard Interval length $(+/-T_G)$ in practice.

In Figure II-15, it has been shown that, by adding guard interval, OFDM system achieves robust transmission performance under serious multipath conditions. ISDB-T and DVB-T both achieves robustness during +/- Guard interval length because both utilize OFDM transmission system.

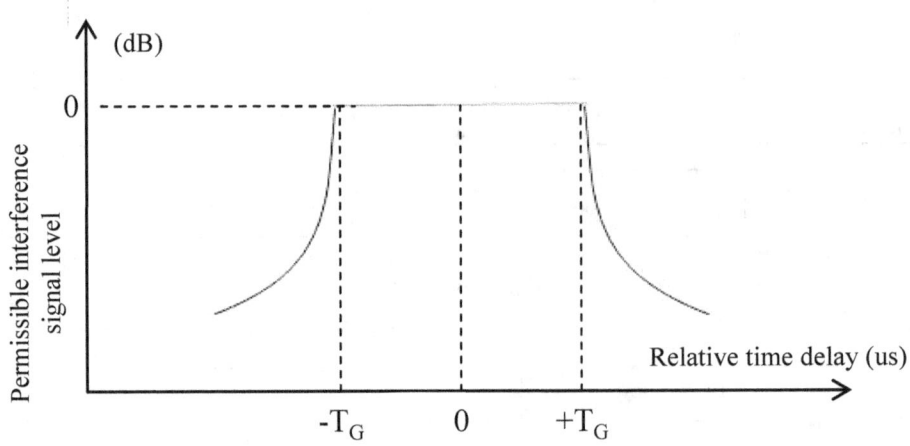

Figure II-15 Relationship of the degradation and relative delay time (with guard interval)

Please note that the multi-path interference phenomenon is very similar to "SFN operation," as illustrated on Figure II-16 (a) and (b).　(a) shows an image of single multipath situation; (b) illustrates an image of SFN operation,

In case of SFN operation, same signal of same frequency are transmitted from different transmitter #1 and #2 within permissible time difference.

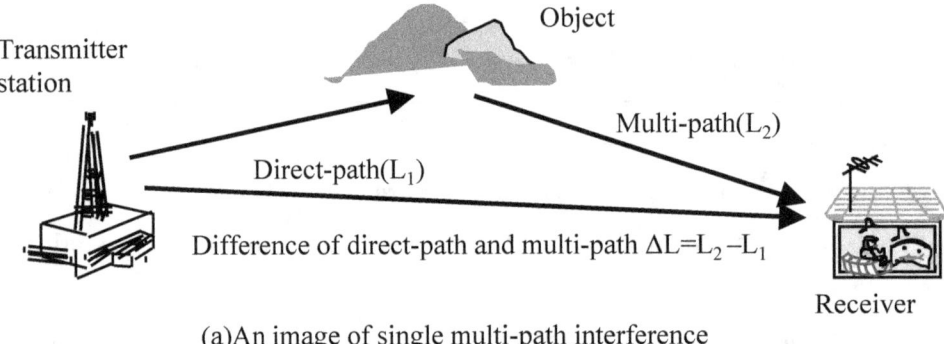

(a) An image of single multi-path interference

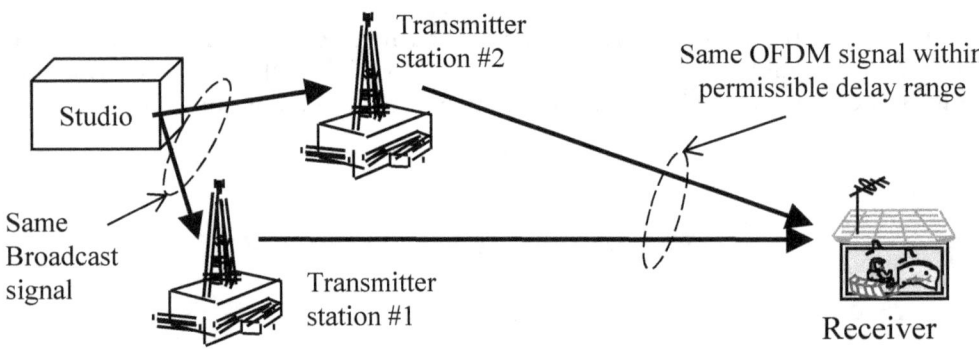

(b) An image of SFN operation

Figure II-16 Similarity of single-multi-path interference and SFN operation

Robustness against multi-path is a very important factor for digital terrestrial broadcast, due to the following reasons:

(1) In VHF/UHF band, multi-path interference always exists. In case of analog TV, multi-path interference results in ghosting, the best description of which is the slight phase difference in the image inside the TV screen where viewers see more than one image with overlapping displays, this ghosting effect creates a serious discomfort to the viewers. Multipath interference occurs due to existence of mountains, buildings and other structures that poses as a reflection point for the signal. This is more pronounced in urban areas. ISDB-T shows excellent reception performance even under Multi-path receiving condition.
(2) By adopting the same concept to counter multipath interference, SFN (Single Frequency Network) can easily be constructed. This leads to the following advantages;
    - efficient frequency use,
    - no change in channel for mobile/portable service,
    - ease of expanding the coverage area, such as those in the shadow of mountain and building, etc, by installing small power repeaters.

II.4 Hierarchical Transmission

This section explains on the hierarchical transmission, one of the major progresses of the digital broadcast technology from the analog system.

II.4.1 What is "Hierarchical transmission"?

Hierarchical transmission service is a concept of transmitting broadcast service for multiple modes of reception service in one frequency and one transmitter. Figure II-17 illustrates the difference between hierarchical transmission service and conventional transmission service. The left side of above figure shows conventional transmission service. As illustrated, a different frequency and a different transmitter is necessary for different reception services adopting the conventional technology. The right side of this figure shows "hierarchical transmission service". In this case, only one frequency and transmitter is required for different reception service such as fixed and portable reception.

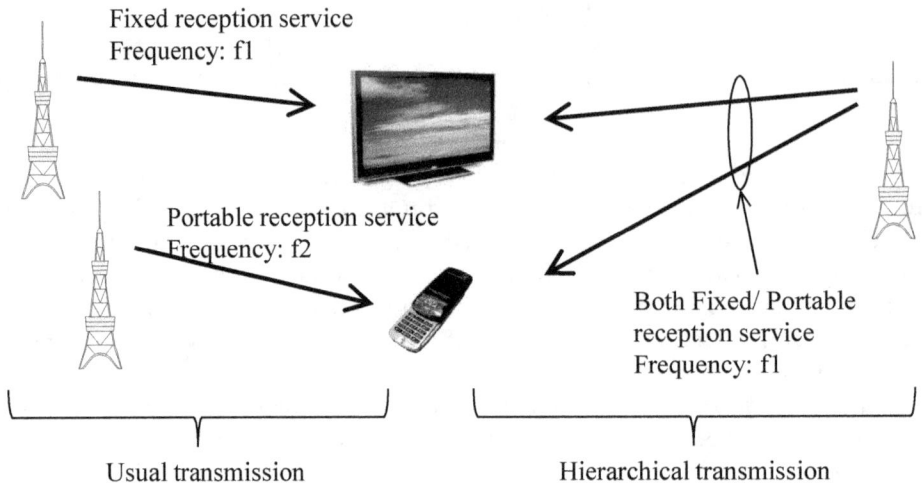

Figure II-17　The concept of transmission system

The next question would be what is required to realize "hierarchical transmission service". Considering the case of two reception service for fixed and portable reception, available antenna gain for the receivers makes a big difference. For fixed reception, a large size directional antenna (such as Yagi antenna) is usually installed. These antennas are generally placed on the roofs for better reception. On the other hand, for portable reception, antenna size is usually limited to the physical size of a receiver unit with the low antenna elevation (about 1.5m high). In a typical case, the received signal level of portable reception is 10-15 dB lower or less compared to fixed reception. (The difference of received signal level depends on the receiving antenna performance and the receiving antenna height. 10-15dB is an example.)

In order to compensate for this difference in the received signal level when transmitting within one frequency band, different modulation parameter sets should be assigned. There are two systems proposed for this purpose. One is "non uniform mapping" proposed as one transmission system for DVB-T. Another is "Segmented OFDM transmission" proposed as one transmission system for ISDB-T

The following sections will discuss on the two transmission systems.

II.4.2　Non uniform mapping (adopted in DVB-T)
In digital communication system, multi-level QAM (Quadrature Amplitude Modulation) modulation technique is very popular, and used not only for single carrier system but also multiple carrier system.. In general, RF signal can be presented by two parameters, amplitude and phase. These parameters can be replaced to another parameter sets, these are in-phase component and quadrature component. Then RF signal is represented a point on In-phase - Quadrature coordinates (usually named I-Q coordinate) by using its in-phase component and quadrature component.

Multi-level QAM means one of digital modulation method in which a RF signal can take M states (I, Q components). For example, 16 QAM is a digital modulation system in the case of M = 16. Figure II-18 (a) illustrates the points that RF signals can take in this case on the I-Q coordinates. This modulation system is named 16QAM, because M=16. Both DVB-T and ISDB-T adopt multi-level QAM (including QPSK: Quadrature Phase Shift Keying) modulation. Another mapping system is also defined in both DVB-T and ISDB-T.

In general, the state of the symbol position on I-Q coordinate plane is named "constellation" such as illustrated in Figure II-18 (a).

DVB-T proposes an unique mapping system named "non-uniform mapping" for hierarchical transmission

system[4]. (b), it shows a constellation of "Non-uniform 16QAM."
In Figure II-18, both constellation of "uniform mapping" and "non-uniform mapping" are illustrated.

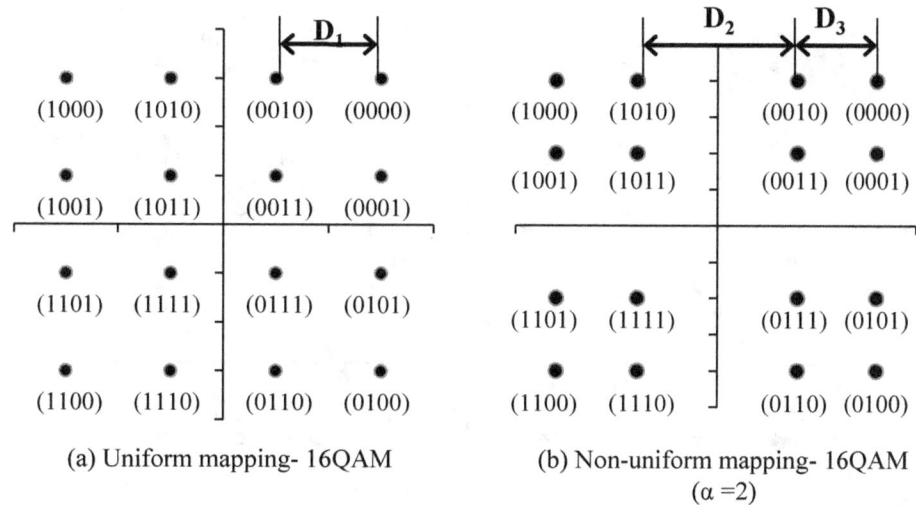

(a) Uniform mapping- 16QAM (b) Non-uniform mapping- 16QAM
(α =2)

Figure II-18   Constellation of uniform and non-uniform mapping (16QAM)

The distance between symbol positions on I-Q coordinate plane represented by the unit vector on the I and Q axis is called the Euclid distance, shown as D_1, D_2 and D_3 in the figure.   Euclid distance represents the noise margin of the system.   Symbols with longer Euclid distance can be distinguished with larger noise level in the communication path.

In case of uniform mapping system, such as ISDB-T, Euclid distance between constellation points are uniform ($=D_1$).   On the other hand, for non-uniform mapping, Euclid distance is different.   In case of DVB-T, the distance between (1010) and (0010), illustrated as D_2 is twice of the distance between (0010) and (0000), illustrated as D_3.

The relation among Euclid distance shown in Figure II-18 is as follows: (suppose that the power of uniform mapping signal is equal to the non-uniform mapping signal)

$$D_2 > D_1 > D_3 \qquad\qquad\qquad [\text{II-1}]$$

D_2 is used to discriminate the upper 2bits of symbol, on the other hand, D_3 is also used to discriminate the lower 2 bits.   In Figure II-18, D2 discriminate (<u>10</u>10) and (<u>00</u>10), D_3 is used to discriminate (00<u>10</u>) and (00<u>00</u>).

Generally, the length of Euclid distance is inverse proportional to the required C/N ($=(C/N)_{REQ}$).   Therefore, following relation is formed.

$(C/N)_{REQ}$ for upper 2 bit of non-uniform 16QAM< $(C/N)_{REQ}$ for uniform 16QAM
< $(C/N)_{REQ}$ for lower 2 bit of non-uniform 16QAM

$$[\text{II-2}]$$

---

[4] ETSI EN 300 744 V.1.5.1(2004-11)   Digital Video Broadcasting (DVB); Framing structure, channel coding and modulation scheme for digital terrestrial television

Non-uniform 64QAM is illustrated in Figure II-19.

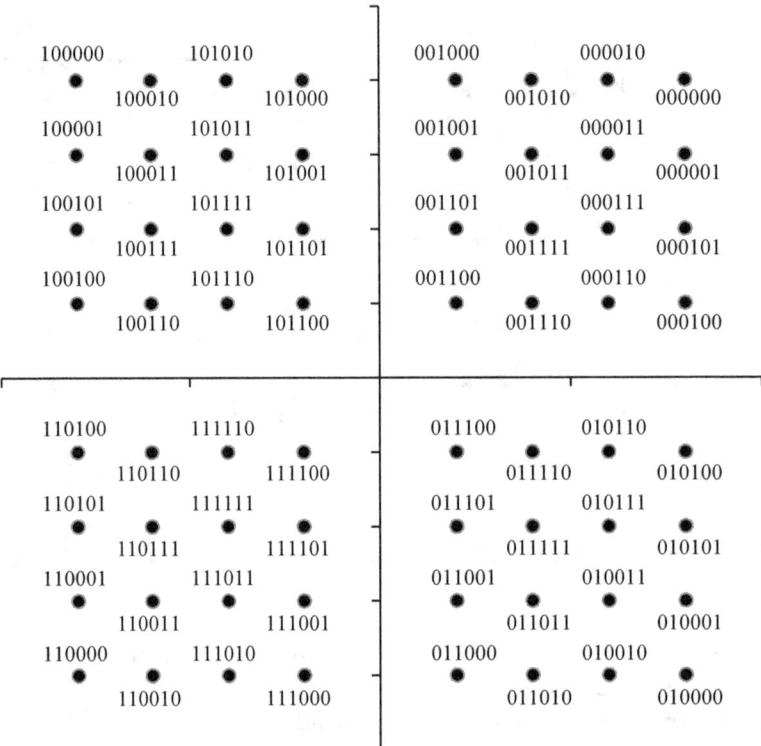

Figure II-19 Constellation chart for Non-uniform 64QAM

As illustrated in the Figure, each state are viewed as the combination of upper 2 bits and lower 4 bits, that is, it is viewed as the combination of 16-QAM (corresponding to lower 4 bits) and 4-PSK (corresponding to upper 2 bits) modulation

This is referred to as 4-PSK in 64-QAM. The bit-rates of the two partial streams together yield the bit-rate of a 64-QAM stream. The image of hierarchical transmission system using Non-uniform mapping is shown in Figure II-20 (from which the dotted-line blocks have been added), the input comprises of two completely separate MPEG-2 transport streams. After combining at the inner interleaver, the composite stream is sent to the transmitting antenna.

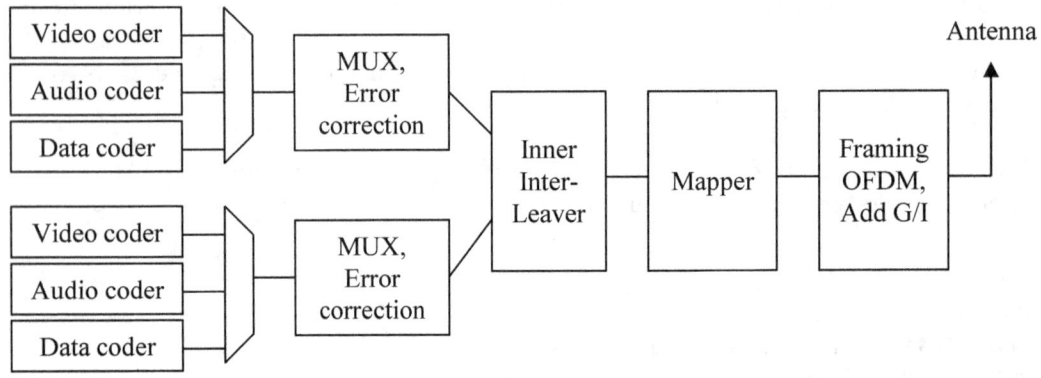

Figure II-20 Image of hierarchical transmission using non-uniform mapping

Please note that the equation in [II-1] and [II-2] were described in the previous page. In case of the Non-uniform 64QAM, the same relationship is also established.

Table II-2 shows the required C/N ratio for uniform/ non-uniform transmission system.

Table II-2 Required C/N for non-hierarchical transmission to achieve a BER = 2 × 10-4 after the Viterbi decoder (in case of the Gaussian channel) [5]

Constellation	Uniform mapping		Non-uniform mapping (α =2)	
QPSK in uniform mapping, or QPSK in Non-uniform 64QAM	r= 1/2	3.5 dB	r= 1/2	6.5 dB
	r= 2/3	5.3 dB	r= 2/3	9.3 dB
	r= 3/4	6.3 dB	r= 3/4	11.1 dB
64QAM in uniform mapping, or 16QAM in Non-uniform 64QAM	r= 1/2	13.8 dB	r= 1/2	17.1 dB
	r= 2/3	16.7 dB	r= 2/3	19.2 dB
	r= 3/4	18.2 dB	r= 3/4	20.4 dB
	r= 5/6	19.4 dB	r= 5/6	21.9 dB
	r= 7/8	20.2 dB	r= 7/8	22.2 dB

As shown in the Table, it is necessary to consider the difference of the C/N penalty when designing a channel plan and estimate the coverage area.

For example, some digital transmission system channel plan is originally designed based on "Non-hierarchical modulation (Uniform mapping), the coverage area may be reduced if it is changed to "Hierarchical modulation (Non-uniform mapping)

In addition, please take note that the maximum data rate of "Low Priority" stream (16QAM in Non-uniform 64QAM) is limited to 2/3 of total data rate of "Non hierarchical transmission" (uniform mapping).

II.4.3   Segmented OFDM transmission in ISDB-T

"Segmented OFDM transmission"[6] is one of hierarchical transmission technology, which is adopted in ISDB-T system. Figure II-21 illustrates the image of the hierarchical transmission service based on "Segmented OFDM transmission"

This Figure shows the case of "2 hierarchy service." Carriers of OFDM signal is grouped into segments in the frequency axis. The number of segments is 13 in case of ISDB-T. While a segment which is intended to be used for handheld reception service is modulated in QPSK, the other for fixed reception service is modulated by 64QAM. In other words, carriers grouped for mobile reception takes only 4 values in constellation, whereas other carriers take 64 values. In case of the Figure, about 12 dB difference exists for required C/N between 2 hierarchies.

ISDB-T can have single hierarchy, 2 hierarchies and 3 hierarchies. In case of 2 hierarchy service, ISDB-T use the center segment for mobile, the rest 12 for fixed.

---

[5] Quoted from "ETSI 300 744 V1.6.1 (2009-01), Appendix A"
[6] ARIB STD-B31, Chapter 2

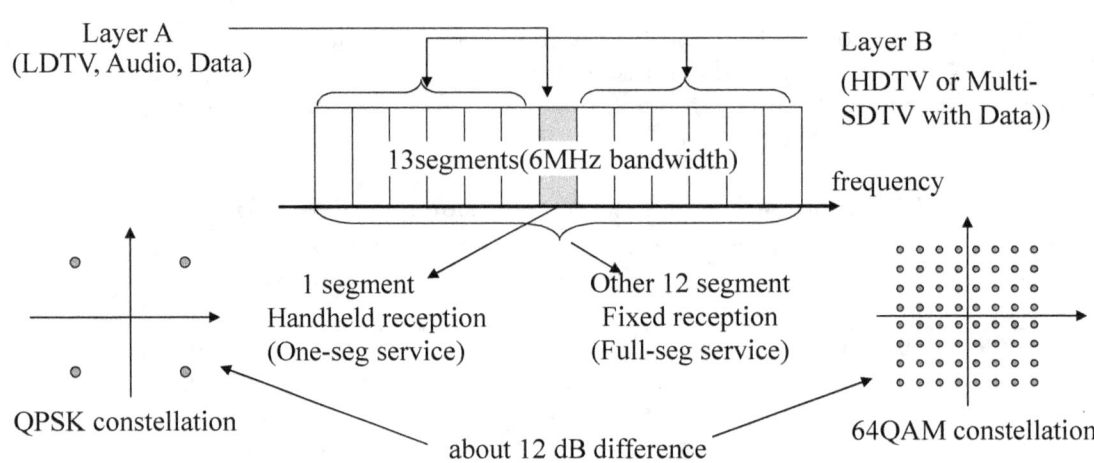

Figure II-21 Image of the hierarchical transmission based on "segmented OFDM transmission"

The key points of "segmented OFDM transmission" adopted in ISDB-T system is as follows:
  (b) transmission band is divided into 13 segment
  (c) maximum number of hierarchy is 3
  (d) the number of segment in each hierarchies are independently selected, but, total number of segment should be 13
  (e) various hierarchy can select different transmission parameter set (mapping ,coding rate, time interleave length)

The example parameters of each hierarchy is illustrated in Figure II-21 above which are further described in Table II-3.

Table II-3   An example of transmission parameter set (in case of Figure II-21)

	Hierarchy A	Hierarchy B
Reception style	Handheld	Outdoor fixed
No. of segments	1	12
Mapping (modulation)	QPSK	64QAM
Inner coding rate	r=2/3	r=3/4
Time interleave	I=4	I=2

As described in Figure II-21 and Table II-3, mapping and coding rates are not different between non-hierarchical and hierarchical transmissions (non-hierarchical transmission in segmented OFDM transmission is the case that all 13 segment belong to single hierarchy.)
The details of hierarchical transmission of ISDB-T will be explained in the section III.2.

II.4.4   Mobile reception service in terrestrial broadcast
As the radio communication system do not need a wire for means of transmission, it can provide services for

the outdoor or moving audiences. Even in the analog broadcast era, FM and AM radio service are available in mobile/portable reception mode. In the case of analog TV reception, a stable reception service is rather difficult because of the deterioration in the signal caused by dynamic multipath. Dynamic multipath effect occurs when the relationship between the phase and the amplitude of the direct wave and the reflected wave changes as the receiver moves around. There are several technologies to improve a signal degradation caused by dynamic multipath.

By adopting segmented OFDM technology, ISDB-T enabled fixed, mobile and portable reception in one frequency slot as shown in the Figure II-22. ISDB-T further incorporated an anti-fading technique for mobile and portable reception service, such as, "Time Interleaving" and "Band segmented OFDM transmission." The details of these techniques will be explained later in Chapter III.

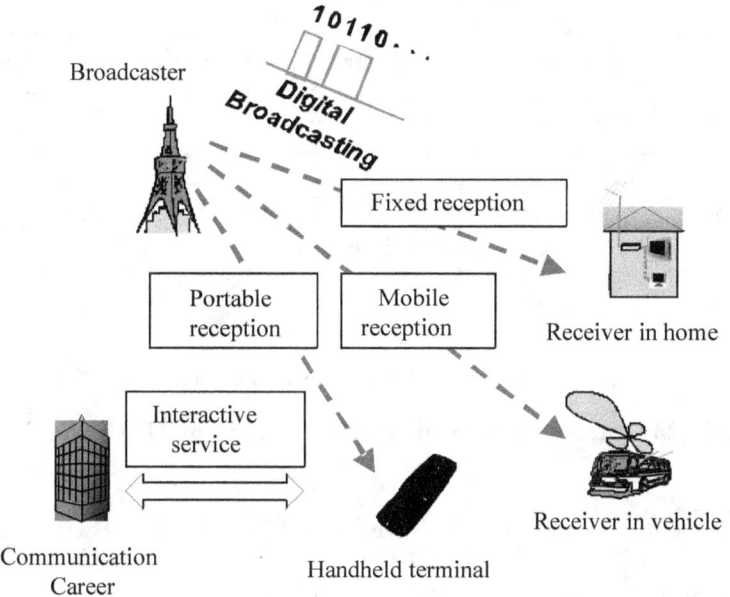

Figure II-22 Service image of digital terrestrial broadcast

II.5 Error Correction System (Channel Coding)

Error correction system is important in digital broadcast system. The error correction system is designed to fit to the characteristics of transmission condition. Error correction is one function of the channel coding system which is composed of various kinds of communication techniques, such as, error correction system, interleaver, mapping, etc. The composition of ISDB-T channel coding system will be explained in the section III.3 later. Especially in terrestrial digital broadcast, a lot of factors that degrade receiving quality should be considered such as interference (both co-channel and adjacent channel, static multipath, urban noise, dynamic multi-path (fading), etc.).

Error correction coding systems are designed to work best for random error. Therefore, these error correction coding systems are used with the interleaver, which randomizes the burst error caused by some of the degrading factors.

Figure II-23 shows the transmission coding portion of DVB-T and ISDB-T.

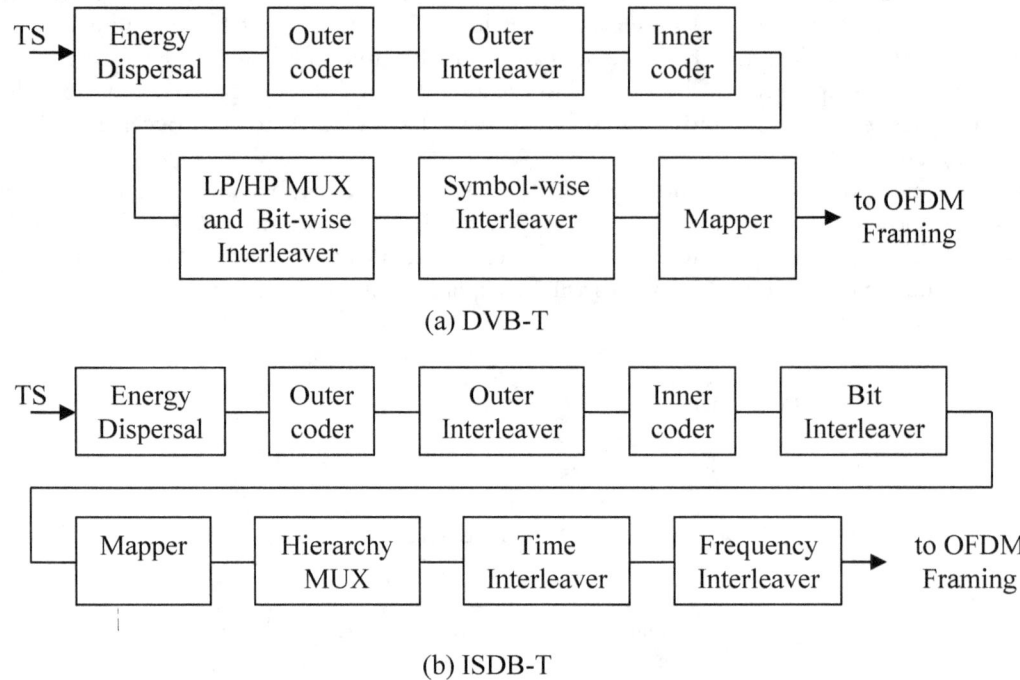

(a) DVB-T

(b) ISDB-T

(note) non-hierarchical transmission case both for DVB-T and ISDB-T

Figure II-23　Structure of transmission coding of DVB-T and ISDB-T

Both system adopts the three types of interleaver, outer interleaver, bit-wise interleaver and frequency interleaver.

The effect of "symbol interleaver" of DVB-T is similar to "frequency interleaver" of ISDB-T. This interleaver randomizes the burst error caused by the degradation in the frequency domain that is generated by static multipath. As a result of introducing these interleavers, both ISDB-T and DVB-T exhibits strong robustness against static multipath interference, enabling SFN operation.

The other type of interleaving technique called time interleaver is the unique feature of ISDB-T. This type of interleaver improves the reception performance under dynamic multipath condition, such as fading. Therefore, the time interleaving is very effective for time fluctuated signal such as mobile/portable reception and impulse noise. Section III.3.7 will discuss on this technique in detail.

II.6　Digital TV Standards

This section will briefly introduce four representative digital terrestrial TV standards[7], DVB-T, ATSC, ISDB-T and DTMB, which have been already in commercial service. For the details of each systems, please refer to the references indicated and the standard documents of each system. As the video and audio coding technology and multiplexing technology adopts similar technologies, this section focuses on transmission system.

---

[7] ITU-R BT.1306-4 Error-correction, data framing, modulation and emission methods for digital terrestrial television broadcasting

II.6.1 ISDB-T

ISDB-T (Integrated Services Digital Broadcasting-Terrestrial)[8] is one of the representative digital terrestrial TV standard, developed and standardized in Japan and improved in Brazil[9]. Now, many Central, South America countries, Philippines, Maldives and Botswana have adopted.

In 1990s, Japan started to develop and standardize of digital broadcasting in satellite, cable TV and terrestrial broadcasting. ISDB-T was developed as an advanced digital terrestrial broadcasting system that satisfy many requirement, such as, high quality/flexible service, efficient frequency utilization, fixed/mobile/portable reception service in one frequency. The key technologies of ISDB-T in Japan are as follows:

(a) Video compression system: MPEG-2 for fixed reception; H.264 for One-Seg service
(b) Multiplexing system: MPEG-2 systems
(c) Transmission system: Segmented OFDM (Orthogonal Frequency Division Multiplex) with time interleave

Brazil improved ISDB-T system by adopting new technologies, such as H.264 for video coding system and GINGA for data casting on top of the same transmission system.

Figure II-24 shows the conceptual block diagram of ISDB-T system. The broadcast contents, video/audio/data, are digitized and encoded in the "Source coding block", then multiplexed in the "Service multiplexing block". The output of the multiplexer is then fed into the TS Re-Multiplexing circuit that is located in "transmission coding block". The output of the TS Re-multiplexing circuit, called "Broadcast TS," is fed to the ISDB-T modulator, which includes hierarchical transmission processing circuit, error correction/interleaving circuit, modulation, TMCC insertion and framing circuit.

While the function of the "TS Re-multiplexer" is defined in ARIB STD-B31 as a function of "ISDB-T transmission system, this function has already been incorporated into the Multiplexer in the products available in the market, which is located in the final stages of Studio production.

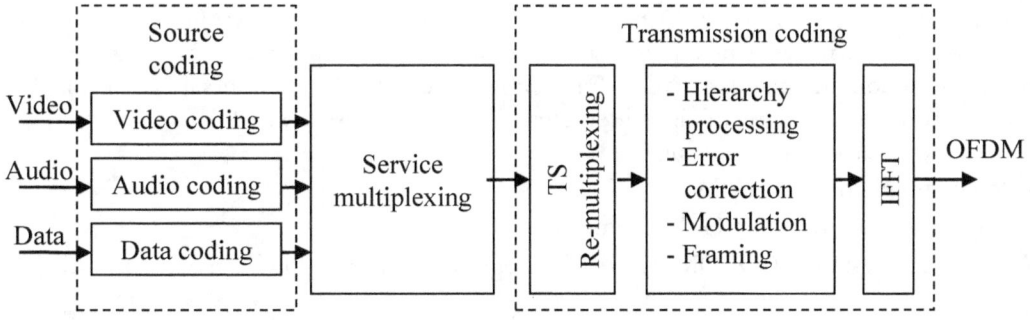

Figure II-24   Block diagram of ISDB-T system

Table II-4 shows the details of the features of the ISDB-T transmission system.

---

[8] Refer Figure II-30 of this book for ISDB-T related standards
[9] Refer Figure II-31 of this book for ISDB-Tb related standards

Table II-4　Features of the ISDB-T transmission system[10]

Item	Effects of the technology	Description in this book
Multi carrier system	- reduces the degradation caused by multi-path - makes SFN possible in order for efficient frequency spectrum use	III.3.1
Guard interval	Same effect as above	Same as above
Three modes of operation	2k, 4k and 8k mode are prepared.　These modes are selectable depending on the service type and requirement.	III.1.-III.4
Hierarchical transmission by "Segmented OFDM"	-enables both fixed and mobile reception service in a single channel -saves both frequency resource and broadcaster's infrastructural cost -utilizes "Segmented OFDM" for hierarchical transmission	III.2
Robust error correction system	- maintains signal integrity in severe transmission and reception condition.　Concatenated error correction with multiple interleaving was introduced.	III.3
Time Interleaving	- Improves the receiving quality under fading condition. Time interleaving is one of the error correction system employed in digital communication.	III.3.7

II.6.2　DVB-T

DVB (Digital Video Broadcasting) is a consortium established for the standardization and commercialization of digital broadcasting in early 1990s in Europe. A number of equipment manufacturers, network operators, software developers, regulatory agencies, etc. participate in DVB consortium. The members of DVB were working for a promotion and standardization for digital broadcasting system in satellite, cable TV, terrestrial broadcasting. DVB-T is the name of digital terrestrial TV standard issued by DVB.

DVB-T standard bases on the following core technologies.

(a) Video compression system : MPEG-2
(b) Multiplexing system : MPEG-2 systems
(c) Transmission system : COFDM (Coded Orthogonal Frequency Division Multiplex)

Figure II-25 shows the block diagram of DVB-T transmission system[11]. As described above, DVB-T adopts concatenated error correcting coding, which is also adopted in the digital satellite broadcast system (DVB-S) and the digital cable broadcasting system (DVB-C) to maximize a commonality between these systems.

To enable a flexible network configuration, two mode, named "2k" and "8k" are specified. In addition, a flexible guard interval length is specified. Regarding carrier modulation, different levels of QAM modulation and different inner code rates are adopted in DVB-T system. For hierarchical transmission

---

[10] ARIB STD-B31 Version 1.6(Nov.30 2005); Transmission System for Digital Terrestrial Television Broadcasting
[11] ETSI EN 300 744 V.1.5.1(2004-11)　Digital Video Broadcasting (DVB); Framing structure, channel coding and modulation scheme for digital terrestrial television

system, non-uniform mapping system is prepared.

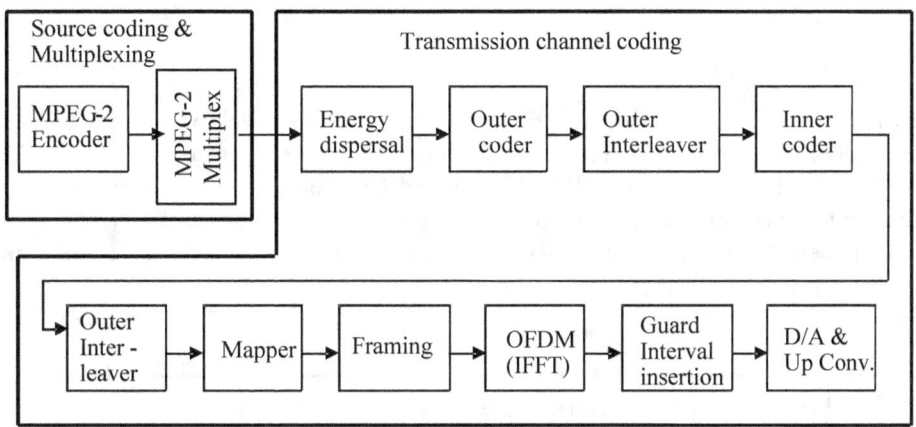

(note) this block diagram shows the non-hierarchical transmission case.

Figure II-25   Functional block diagram of DVB-T system

Table II-5 explains main transmission parameters of 8MHz DVB-T system. For the transmission parameters of 6MHz and 7MHz system, please refer ITU. BT.1306.

Table II-5   Main transmission parameters of 8MHz DVB-T system

Parameter	2k mode	8k mode
Outer coding	Reed-Solomon RS (204,188, t = 8) shortened code, derived from the original systematic RS (255,239, t = 8)	
Outer Interleaving	convolutional byte-wise interleaving with depth I = 12	
Inner coding	punctured convolutional codes, based on a mother convolutional code of rate 1/2. punctured rates of 2/3, 3/4, 5/6 and 7/8.	
Inner interleaving (1) Bit-wise interleaving	separately defined for "non-hierarchical" mode and "hierarchical" mode	
(2) symbol interleaving	Mapping words onto the 1512 active carriers	Mapping words onto the 6048 active carriers
Mapping	Uniform mapping: QPSK, 16QAM, 64QAM Non uniform mapping: α =2 : 16QAM, 64QAM, α =4 : 16QAM, 64QAM	
OFDM framing		
(1) number of carriers	1705	6817
(2) carrier spacing	1.116 kHz	4.464 kHz
(3) active symbol length	224us	896us
(3) guard interval ratio	1/4, 1/8, 1/16, 1/32 of active symbol length	

II.6.3 ATSC

ATSC[12] is one of the representative digital terrestrial TV standard, developed and standardized in United States. The Advanced Television Systems Committee (ATSC) was established in 1980s as the Joint Committee on Inter-Society Coordination (JCIC). Federal Communication Committee (FCC) requested to develop for advanced television service. After many proposals and evaluation tests, a digital terrestrial Television system named ATSC system has been standardized in 1995. Figure II-26 illustrates the block diagram of ATSC system.

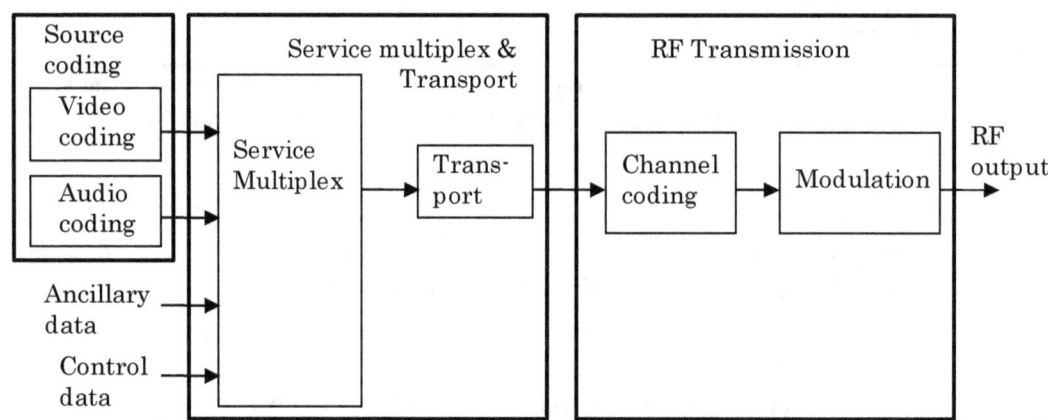

Figure II-26   Block diagram of ATSC system

Figure II-27 shows the block diagram of ATSC transmission system. As described in the figure, ATSC adopts concatenated error correcting coding, composed of Reed Solomon (RS) encoding and Trellis encoding. DVB-T and ISDB-T adopt Convolutional encoding for inner coding, while ATSC adopt Trellis coding.

---

[12] A/53, ATSC Digital Television Standard, Part-1 – Part-6

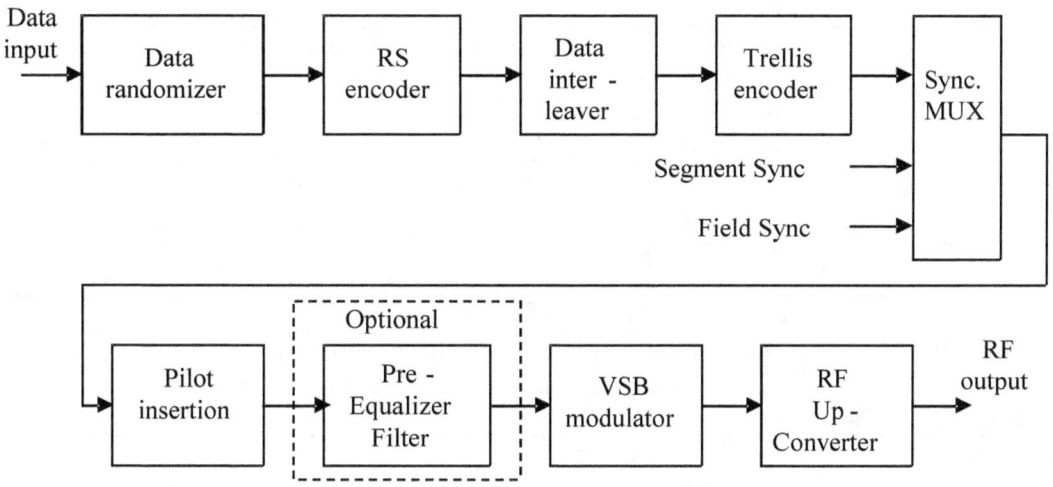

Figure II-27 Block diagram of ATSC transmission system

In addition, ATSC adopts unique single carrier modulation system, while DVB-T and ISDV-T adopt multi-carrier transmission system. These are the main difference between ATSC and DVB-T/ISDB-T.
The single carrier modulation system of ATSC is named 8-VSB (8-level Vestigial Sideband Modulation). 16-VSB system is also specified as one of ATSC standard, however 8-VSB system is used as for the terrestrial broadcasting. ATSC adopts this modulation system due to its efficiency.

Figure II-28 illustrates a RF frequency spectrum of 6MHz ATSC system. A linear phase raised cosine Nyquist filter response is used for the spectrum shaping As illustrated in the figure, Vestigial Sideband modulation technique is utilized to transmit over 19Mbps data stream within 6MHz bandwidth.

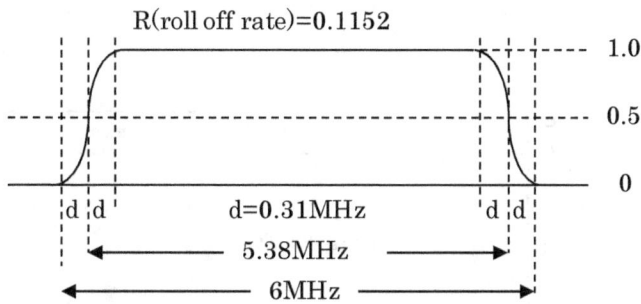

Figure II-28 RF spectrum of 6MHz ATSC system

The main transmission parameters for 6MHz ATSC system (8-VSB) are shown in Table II-6. For the transmission parameters of 7MHz and 8MHz system, refer ITU. BT.1306.

Table II-6 Main transmission parameters of 6MHz ATSC system

Parameter	contents	remarks
Data randomizer	XORs all the incoming data bytes with a 16-bit maximum length pseudo random binary sequence	

	(PRBS)	
RS coding	The Reed Solomon code used in the VSB is a t = 10 (207,187) code. RS data block size = 187 bytes, RS parity block=20 bytes. Total block size =207 byte(Data segment)	
Data interleaving	52 data segment (intersegment) convolutional byte interleaving	
Trellis encoding	two-thirds rate (R=2/3) trellis code for 8-VSB modulation	
Sync. MUX	Add Data Segment Sync and Data Field Sync.	
Pilot insertion	A small in-phase pilot of which frequency is same as the suppressed-carrier shall be added to the data signal.	
VSB modulation	8level VSB modulation. Symbol rate =10.76 M Symbol/s	

In ATSC system, data for transmission are organized in data frame as shown in Figure II-29. Each data frame consists of two data fields, each containing 313 data segments. Each data field starts with synchronization signal called data field sync. Also, each segment starts with synchronization signal called segment sync. The remaining carry data plus its associated RS-FEC.

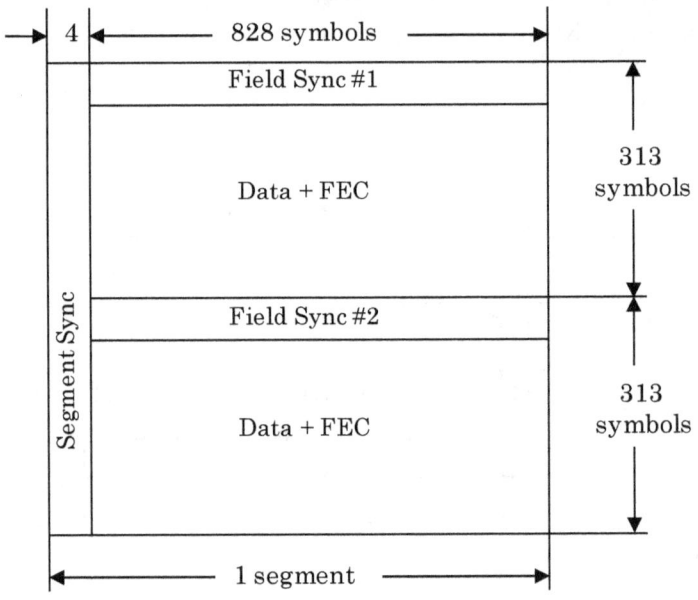

Figure II-29   A structure of 8-VSB data frame

II.6.4   DTMB

Digital Terrestrial Multimedia Broadcast system[13] was developed in China and now is the digital terrestrial

---

[13] GB 2060 – 2006, Framing structure, channel coding, and modulation for digital television terrestrial broadcasting system

TV broadcasting standard.   Around 2000, China started to accept proposals for a digital terrestrial television standard.   Three proposal below were chosen:
 - ADTB-T (Advanced Digital Terrestrial Broadcasting-Terrestrial) based on single carrier system,
 - DMB-T (Digital Multimedia Broadcasting-Terrestrial) based on multicarrier system,
 - TiMi (Terrestrial Interactive Multiservice Infrastructure) based on multicarrier system.

Chinese Academy of Engineering led a working group to merge these leading proposals, and finally the working group agrees a merged standard (Chinese National Standard Organization) in 2006. The official name of this Chinese standard is "Framing structure, channel coding, and modulation for digital television terrestrial broadcasting system" (GB 20600-2006), however the name of DTMB which is the abbreviation of "Digital Terrestrial Television Multimedia Broadcasting" is more popular. The Chinese digital terrestrial television standard supports both single-carrier system and multi-carrier system.
   DTMB transmission system adopts following key technologies:
   (1) TDS-OFDM (Time Domain Synchronous –OFDM)
   (2) Hierarchical frame structure
   (3) BCH + LDPC

TDS-OFDM (Time Domain Synchronous –OFDM) is a one deviation of the OFDM technology, which is characterized by the use of time signal. As shown at the bottom of in Figure II-30, a signal frame is composed of a "frame body" and a "frame header" The frame header includes PN sequence which is used for both symbol synchronization and channel estimation. While, the frame body is composed of OFDM signal.
   DTMB is also characterized by its unique hierarchical frame structure. As illustrated in Figure II-30 ,at the top level of hierarchy is the calendar day(24 hours), next hierarchy is the minute frame (1 minutes), then third hierarchy is super frame, its length is exactly 125ms. The bottom hierarchy is the signal frame. Three kinds of frame length are prepared.

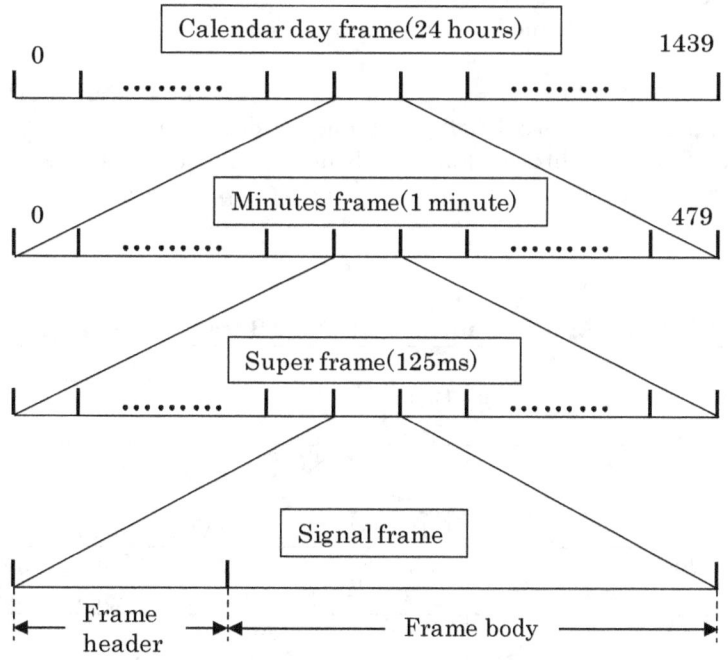

Figure II-30   Frame structure of DTMB system

Another feature is use of BCH and LDPC. DTMB system adopts concatenated error correction system, however, a different error correction technique from DVB-T/ISDB-T transmission system, BCH + LDPC. Figure II-31 shows a block diagram of DTMB transmission system.

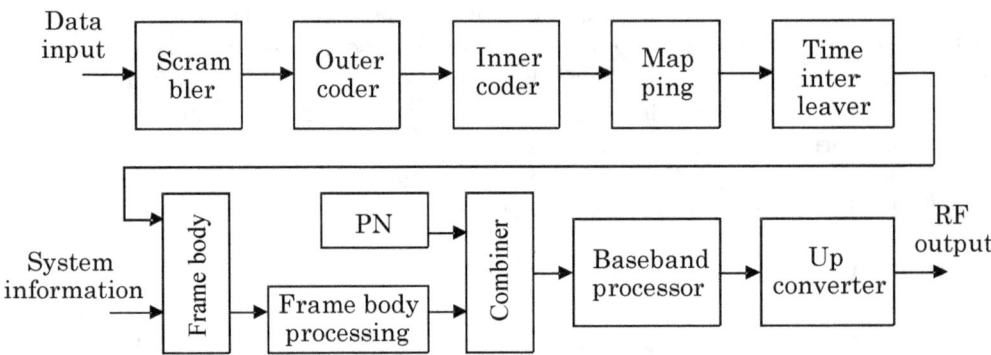

Figure II-31 Block diagram of DTMB transmission system

Table II-7 explains the main transmission parameters of DTMB system. As described in Table II-7, DTMB system has many parameters. For the, three FEC rate, 0.4(7488,3008), 0.6(7488,4512) and 0.8(7488,6016) are prepared. As an example, FEC rate of 0.4 (7488,3008) is made by 4 BCH(762,752) as outer codes and LDPC (7493,3048) as inner code, then first 5 parity check bits of LDPC(7493,3048) code are punctured. For the mapping (carrier modulation), five parameters are prepared.

According to the system requirement, any parameter sets for the FEC rate and the mapping parameter can be chosen. Two parameters are prepared for time interleaving, any of two parameters is chosen according to a system requirement. In the frame body-processing block, any of single carrier or multi-carrier is selected.

For the multi-carrier system, 3780 carrier system is provided. The output of frame body processor is combined with the frame header. Two length of frame header are prepared. As described above, the frame header is used for frame synchronization and channel estimation in a receiver side. The combined signal (signal frame) is fed to the baseband processor, and via up converter RF output of DTMB system comes out.

Table II-7 Main parameters of DTMB transmission system

Parameters	Contents
Outer coding	BCH(762,752)
Inner coding	LDPC(7488,6096), (7488,4572), (7488,3048)
Mapping	QPSK,16QAM,32QAM,64QAM
Time interleaving	-52/240 with delay of 170 Signal Frames -52/720 with delay of 510 Signal Frames
Frame body processing	-Hierarchical frame structure -selection of single carrier(1) or

	multi-carrier(3780)
Multi-carrier system parameters	-carrier spacing: 2kHz -symbol duration: 500us -number of carriers: 3780 -symbol rate: 7.56 MSPS
Frame header	PN420, PN595, PN920

II.7 Structure of the ISDB-T Standards

In this section, firstly, the standard structure of an ISDB-T system (Japan) is explained, followed by the explanation of the ISDB-T system (Brazil).

II.7.1 Requirement for ISDB-T and solution

Before going into the composition of the technical standard, it would be appropriate to understand the background and structure of the technology elements in the ISDB-T systems.

Figure II-32 shows the relationship between the requirement and technical solutions.

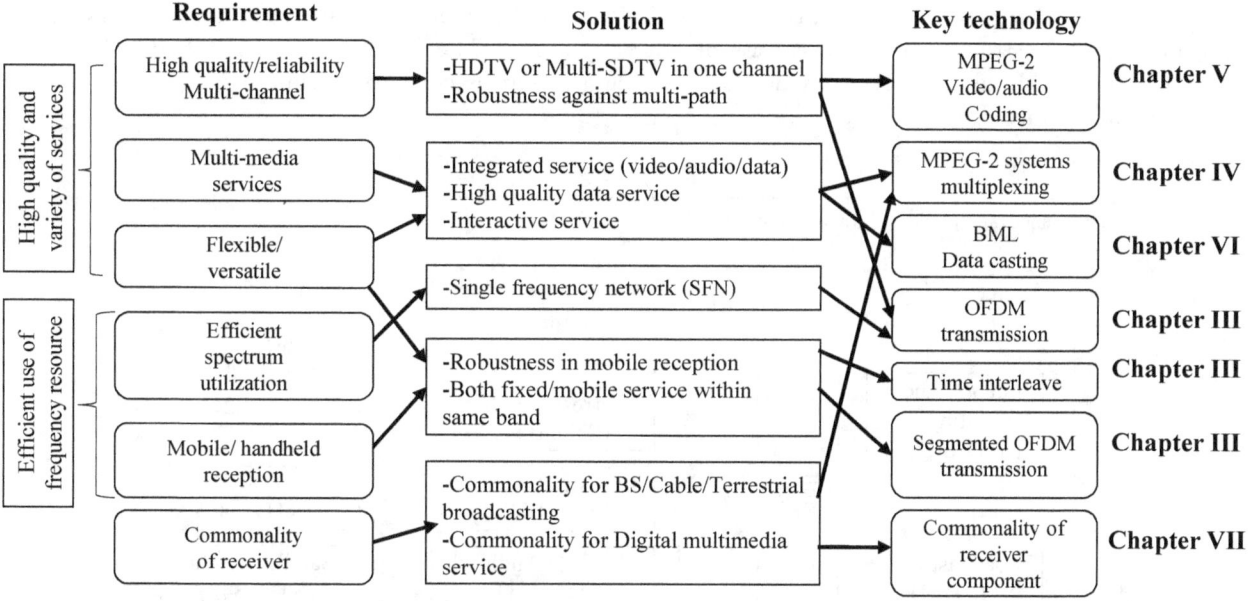

Figure II-32   Relations of "Requirement" and "Solution" for ISDB-T (Japan)

The first step of the study and development of ISDB-T and the requirement written on left side were identified, followed by the solution requirement shown on right side.

Requirement can be categorized in three groups: "high quality/variety of service", "efficient use of frequency resource" and "commonality of receiver." In the digital era, high quality video/ audio service, such as HDTV are expected, and at the same time, more service in a channel is also desired. In addition, a variety of service, such as interactive services, high quality data service, etc., are also expected in the digital era. Flexibility of service is also an advantage of digital broadcast as realized by the multiplexing technology described in section II.1.2 the key technology of digital broadcast.

Since frequency spectrum is a precious and finite resource, its efficient utilization is a key issue of digital broadcast. A digital broadcast system which enables fixed and mobile/ portable reception service in a single frequency (Channel) in one transmitter is an attractive proposition for the effective use of frequency resource. This technology is called "hierarchical transmission system." ISDB-T adopts one of the hierarchical transmission technology named "segmented OFDM transmission." The details of hierarchical transmission system in ISDB-T will be explained in the section III.2.

Finally, commonality with digital audio broadcast is also an important issue to save on receiver cost.

Next step, solutions for these requirement described above were investigated and developed. The solutions are shown in the middle column of Figure II-32.

II.7.2 Structure of ISDB-T technical standards (Japan)

In Japan, the Radio Law and Broadcast Law set forth the technical requirements for broadcasters in opening and operating the broadcasting stations. Especially, Broadcast Law article 93 and 102 requires broadcasting stations in compliance with the technical standards set in the ordinance "standard transmission method on digital television."[14] This ordinance covers most of the topics to be discussed in chapters III through VI including RF, multiplexing, video/audio coding and datacasting.

Association of Radio Industries and Business (ARIB) issues standards for digital broadcast that supplements further details of the ordinance. These standards for broadcast are named "ARIB STD-Bxx" according to the ARIB naming convention.

Figure II-33 shows the composition of the technical standard for digital broadcast in Japan. As shown in the figure, digital broadcast system in Japan consists of several Standards. These standards are standardized for each each individual technical area such as video/audio coding system is common technologies that are used in various broadcast media.

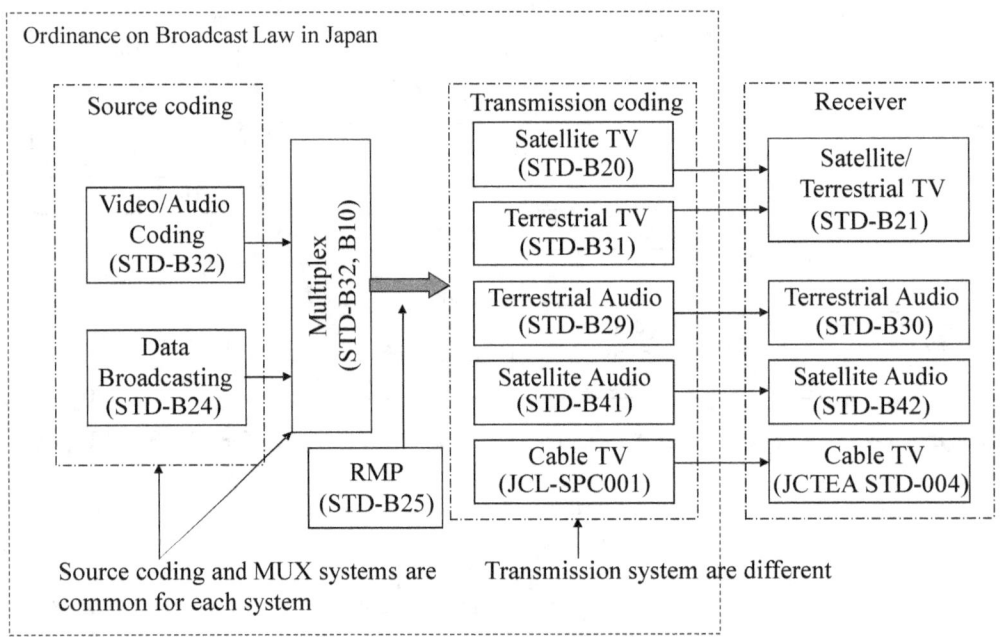

Figure II-33 Composition of ARIB Standard for digital broadcast (including JCL standard)

---

[14] Ministry of International Affairs and Communications, Ministerial ordinance No. 87 (2011), "standard transmission method on digital television"

In addition, "Operational Guideline" is also developed and issued from ARIB as "Technical Report". These technical reports for broadcast are named "ARIB TR-Bxx" While All the core technical elements required are written in ARIB STD, the details for operation of broadcast are defined separately in the "Operational Guideline", which is based on ARIB STD. Some of the ARIB STDs are available on the ARIB website[15]. It should be noted that the standards and guideline introduced in this table are for Japanese system and that some specifications for each countries may vary. It is necessary to check the standards/guidelines for each country.

Table II-8 Standards and guideline of ISDB-T (Japan)

Name	Contents	Chapters in this book
Video/Audio coding (STD-B32)	-video coding system(MPEG-2) -Audio coding systemMPEG-AAC) -Multiplexing system(MPEG-2 systems)	Chapter V for video and audio coding. Chapter IV for multiplexing
Data Broad-casting (STD-B24)	-Data broadcast description(BML) -Data transmission format -Small size Video coding (MPEG-4,H.264)	Chapter VI
Program line-up information (STD-B10)	-description of PSI/SI for transport multiplexing and Electronic Programming Guide (EPG)	Chapter IV
Transmission coding (STD-B31)	-ISDB-T transmission structure -segmented ODFM transmission system -error correction system & interleave -network management	Chapter II and III
Receiver (STD-B21)	-structure of ISDB-T receiver -specifications of ISDB-T receiver -specifications of video/audio output -definition/specification for other functions	Chapter VII
Operational guideline For terrestrial TV broadcast (TR-B14)	-actual operational parameters and functions	See Appendix

II.7.3 Structure of ISDB-T technical standard (Brazil)
Brazil decided to adopt ISDB-T system for Brazilian digital terrestrial TV broadcast in 2006.

Brazil took effort incorporating new technologies into ISDB-T to build a more advanced and sophisticated system. As a result, ISDB-T system (Brazil) has been established in 2007. The standard of datacasting has been finished in 2010.

Both system are based on common transmission technology and built on the same technology platform. Standard documents of Brazilian ISDB-T system has been officially issued from ABNT (Associação Brasilia de Normas Técnicas), which is the official organization for standardization in Brazil.

---

[15] ARIB, "List of ARIB Standards in the Field of Broadcasting," http://www.arib.or.jp/english/html/overview/sb_ej.html

Figure II-34 shows the composition of the Brazilian digital terrestrial standards, named ABNT. The structure of Brazilian ISDB-T standard look similar to the structure of Japanese ISDB-T standards.

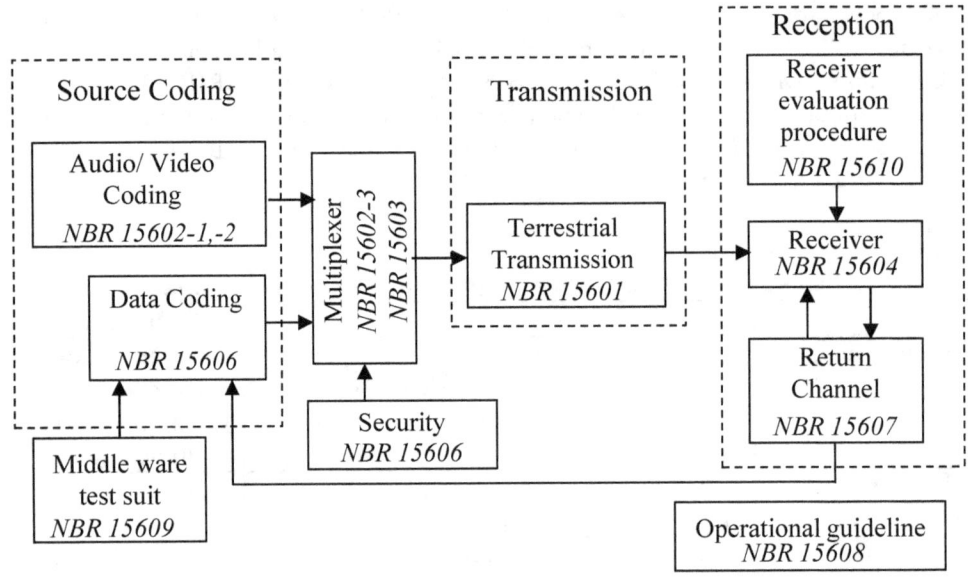

Figure II-34   Composition of ABNT Standard for digital terrestrial broadcast

II.7.4   Commonalities and differences between ARIB/ABNT standards

Brazilian standards bases on ISDB-T (Japan), but several points are different because of the difference in the period of its standardization including the difference of culture, regulation, etc.

Table II-9 shows the commonalities/ differences of ARIB/ABNT standards. For details, please refer "ISDB-T INTERNATIONAL HARMONIZATION DOCUMENTS"[16], which has been issued by the Joint Working group of Japan and Brazil. These documents are freely downloadable from the DiBEG homepage.

Table II-9   Commonalities/ differences of ARIB/ABNT standards

Subsystem	Standards and Commonalities/ differences
Transmission	Standard: ARIB STD-B31(Japan), ABNT-NBR 15601 (Brazil)

---

[16] DiBEG, "Harmonization Documents", http://www.dibeg.org/techp/aribstd/harmonization.html

	Both Brazilin and Japanese transmission system stands on "Segmented OFDM" transmission system, and adopts the same transmission parameters (i.e. sampling frequency, mode, guard interval, error correction, interleaving, mapping, etc.). Therefore, both systems are almost the same except following items: (i) Channel allotment(footnote 1): Japan :UHF Channel 13-62, Brazil: VHF Channel 07-13 and UHF Channel 14-69 (ii) Spectrum mask: 3 types (Brazil), 2 types (Japan) (note) two items (i) and (ii) above are related to the channel plan of each countries (iii)Intermediate frequency of receiver : in Japan, its 57MHz (defined in ARIB STD-B21), in Brazil, its 44MHz (note) this difference is based on the difference of analog TV receivers (for details, see" ISDB-T International Harmonization" volume 1)
Video coding	Standard: ARIB STD-B32 part 1(Japan), ABNT-NBR 15602-1(Brazil)
	Regarding a video signal, both Brazilian and Japanese system are basically common. In addition, the One-Seg video coding system uses the same coding system, H.264│MPEG-4 AVC. The difference is in the full-seg video coding system, Japan adopts H.262│MPEG-2, while Brazil adopts H.264│MPEG-4 AVC. (for details, see" ISDB-T International Harmonization" volume 2)
Audio coding	Standard: ARIB STD-B32 part 2(Japan), ABNT-NBR 15602-2(Brazil)
	Both Brazilian and Japanese audio coding system utilize a common technology named AAC. However, because of the difference of video coding system, the audio coding in the Japanese system is based on MPEG-2 AAC while the Brazilian system is based on MPEG-4 AAC It should be noted that the basic coding algorithm of MPEG-4 AAC and MPEG-2 AAC is almost the same, these are only the difference in the standards. (for details, see" ISDB-T International Harmonization" volume 3)
Multiplexing	Standard: ARIB STD-B32 part 3, STD-B10 (Japan), ABNT-NBR 15602-3, NBR-15603 (Brazil)
	a)multiplexing system: Both Brazilian and Japanese Multiplexing scheme/signal format uses MPEG-2 systems. Both systems are basically the same, however, some parts are different: b)Descriptor(PSI/SI): The technical bases of "Descriptor" are MPEG-2 systems standard (ISO/IEC 13818-1), DVB standard and ISDB-T original standard. Both Brazilian and Japanese system are basically the same only with some minor differences : (for details, see" ISDB-T International Harmonization" volume 4, volume 5, volume 6 and volume 7))
Receiver	Standard: ARIB STD-B21 (Japan), ABNT-NBR 15604 (Brazil)

	The receiver's specification depends on the standards that are applied to each system. As explained above, Brazilian and Japanese ISDB-T system is different in some points. The big differences are in the tuner portion (difference of frequency band), video decoder portion (difference of coding system) and conditional access (see below in this table). Brazilian and Japanese receivers have a lot of common technology, such as hierarchical transmission service, HDTV/SDTV reception, datacasting service, etc. (for details, see" ISDB-T International Harmonization" volume 8)
Datacasting	Standard: ARIB STD-B24 (Japan), ABNT-NBR 15606-1,-2,-3,-5 (Brazil)
	a) Mono-media: Basically common but several difference exist in video coding, video clipping etc. b) Multi-media coding: Brazilian multimedia coding system is composed of two categories, a presentation engine (XML) and an execution engine (JAVA); in Japan, XML is used. c) Data transmission: For data transmission system, it is basically common except for the object carousel.

(note) it is noted that the frequency of Japanese UHF channel is different from Brazil's. The center frequency of Channel 13 in Japan is 473MHz, this frequency is the center frequency of Channel 14 in Brazil

Chapter III. ISDB-T Transmission System

In the previous sections, the key technologies of the ISDB-T transmission system were explained. This chapter will explain the structure and technical details of ISDB-T transmission system based on the ARIB STD-B31[17]. The main purpose of this chapter is to establish a basic understanding of the technologies and systems defined in the standard rather than going immediately to the intricate details of the related technical parameters that will be discussed further in the succeeding chapters. Please refer to ARIB STD-B31, while reading this chapter in order to gain further understanding.

III.1 Technical Overview

In terms of the transmission spectrum of a digital television broadcast, ISDB-T basically is made up of 13 contiguous OFDM blocks (hereinafter referred to as "OFDM segments"). Each segment will have a bandwidth that is equal to one fourteenth of a television-broadcast channel bandwidth. The structure of an OFDM segment will be explained in the section III.4.

In ISDB-T, it utilizes an MPEG-2 System which can have a single or multiple transport stream (TS) inputs that is re-multiplexed to create a single TS. This TS is then processed by several channel-coding steps to form a single OFDM signal. A key channel-coding step among other is the time interleaving which intends to increase the channel robustness especially for the mobile receiving condition to which variations in field strength are unpredictable and inevitable.

III.1.1 Transmission parameters

Before looking more into the details of the transmission technology, the transmission parameters of ISDB-T will be discussed first in this section. ISDB-T transmission system was designed to enable flexible transmission depending on the requirements of each broadcaster, which may range from mobile to fixed reception or even both. In the Rec. ITU-R BT.1306-1 ANNEX 1 Table 1 case c, the transmission parameter set for 6MHz, 7MHz and 8MHz ISDB-T transmission system are described. In this text, as examples, the transmission parameters applicable for a 6MHz and 8MHz ISDB-T transmission system are shown below[18].

1) Segment structure and segment bandwidth
The RF signal of ISDB-T consists of 13 segments. A segment bandwidth is exactly defined as 6/14 MHz =428.57..kHz in 6MHz system and 8/14 MHz = 571.428…kHz in 8MHz system respectively. This segment bandwidth is common for each available mode which will be explained in further detail below.

2) Number of carriers in one segment
The number of carriers varies for the available modes 1, 2 and 3. For mode 1, the number of carrier is 108; for mode 2, the number of carrier is 216, twice that of mode 1; and for mode 3, the number of carrier is 432, 4 times that of mode 1.

---

[17] ARIB STD-B31 Version 2.2(Mar..18 2014); Transmission System for Digital Terrestrial Television Broadcasting
[18] ITU-R BT.1306-4 Error-correction, data framing, modulation and emission methods for digital terrestrial television broadcasting

3) Mode:
In consideration of the distance of each transmitter in an SFN and the robustness to Doppler shift during mobile-reception, ISDB-T offers three different frequency spacing among OFDM carriers. This carrier spacing are identified as system modes. The available carrier spacing among each OFDM carrier in 6MHz system are approximately 4.0 kHz, 2.0 kHz, and 1.0 kHz in modes 1, 2, and 3, respectively. In 8MHz system, they are 5.3 kHz, 2.6 kHz and 1.3 kHz.

4) Guard interval ratio:
As described in the section II.3.3, the Guard Interval (GI) is added into the front of each symbols. The length of the guard interval (the guard time) is selected depending on to the extent of the desired tolerance to counter the effects of multipath delay. Guard interval is defined as the ratio of the Guard interval length and Symbol length.

The guard interval length is calculated as the product of the symbol length and the guard interval ratio. Therefore, the guard interval length of mode 3 is longer than that of either mode 1 or mode 2 due to its closer carrier spacing resulting into a longer symbol length. The range, in which a multipath interference is countered, is proportional to the guard interval length. Therefore, in considering SFN operation, mode 3 is more suitable than mode 1 and mode 2 because of its longer guard interval length.

5) Frame length:
One OFDM frame consists of 204 symbols for mode 1, 2 and 3. The total symbol length is given by following equation:

$$\text{Total symbol length} = (\text{effective symbol length}) + (\text{guard interval})$$

Please see Table III-1 for the corresponding values. Please try to calculate the frame length for any of the available Modes and guard interval. The answers are written in Table III-1 below.

Following 6) and 7) Explains hierarchical transmission.

6) Number of hierarchies:
3 as the maximum[19]

7) Number of segments for each hierarchy:
Any number can be assigned until 13, but take note that the total number of segments should be equal and not to exceed 13.

8), 9) and 10) are to be selected independently for each hierarchy.

8) Inner coding rate:
Inner error correction coding systems are made up of convolutional coding with puncturing. For the puncture sequence, more details can be found on the section III.3.

The punctured ratio of the inner coder is selectable to any of the available code rates, i.e. 1/2, 2/3, 3/4, 5/6 and 7/8

9) Carrier modulation (mapping):
Selectable from any of QPSK, DQPSK, 16QAM and 64QAM.

---

[19] See section II.4.3 of this book

As for DQPSK, it is not included in the Operational Guidelines since this modulation is not used in practice. For the carrier modulation, more details can be found on the section III.3.

10) Depth of time interleaving:
Depth of time interleaving represents time length in which OFDM carriers are interleaved so that errors caused mainly by dynamic multipath[20] in transition path can be scattered and error corrected. Longer time interleaving generally shows more robustness, but longer time delay. For the depth of time interleaving, a more thorough discussion can be found on section III.3.7.

Segment parameters and transmission signal parameters of 6MHz system are shown in Table III-1 and Table III-2 below, and 8MHz system in Table III-3 and
Table III-4 respectively.

Table III-1 OFDM-Segment Parameters for 6MHz system[21]

Mode		Mode 1		Mode 2		Mode 3	
Bandwidth		3000/7 = 428.57… kHz					
Spacing between carrier frequencies		250/63 = 3.968… kHz		125/63 = 1.9841… kHz		125/126 = 0.99206… kHz	
Number of carriers	Total count	108	108	216	216	432	432
	Data	96	96	192	192	384	384
	SP*1	9	0	18	0	36	0
	CP*1	0	1	0	1	0	1
	TMCC*2	1	5	2	10	4	20
	AC1*3	2	2	4	4	8	8
	AC2*3	0	4	0	9	0	19
Carrier modulation scheme		QPSK 16QAM 64QAM	DQPSK	QPSK 16QAM 64QAM	DQPSK	QPSK 16QAM 64QAM	DQPSK
Symbols per frame		204					
Effective symbol length		252 μs		504 μs		1008 μs	
Guard interval		63 μs (1/4), 31.5 μs (1/8), 15.75 μs (1/16), 7.875 μs (1/32)		126 μs (1/4), 63 μs (1/8), 31.5 μs (1/16), 15.75 μs (1/32)		252 μs (1/4), 126 μs (1/8), 63 μs (1/16), 31.5 μs (1/32)	
Frame length		64.26 ms (1/4), 57.834 ms (1/8), 54.621 ms (1/16), 53.0145 ms (1/32)		128.52 ms (1/4), 115.668 ms (1/8), 109.242 ms (1/16), 106.029 ms (1/32)		257.04 ms (1/4), 231.336 ms (1/8), 218.484 ms (1/16), 212.058 ms (1/32)	
IFFT sampling frequency		512/63 = 8.12698… MHz					
Inner code		Convolutional code (1/2, 2/3, 3/4, 5/6, 7/8)					
Outer code		RS (204,188)					

---

[20] See section X.3.3 in this book
[21] Quoted from ARIB STD-B31, Chapter 3, table 3-1

*1: SP (Scattered Pilot) and CP (Continual Pilot) are used by the receiver to establish synchronization and demodulation purposes.
*2: TMCC (Transmission and Multiplexing Configuration Control) is the control information that shows the current state of the ISDB-T signal. This information is used to initialize the ISDB-T receiver's functions to receive services.
*3: AC (Auxiliary Channel) is used to transmit additional information aside from the default services of ISDB-T. AC1 is available in equal number in all segments, while AC2 is available only for differentially modulated segments. Discussions pertaining to Differential modulation can be found on ARIB STD-B31[22].

Table III-2 Transmission signal parameters for 6MHz system[23]

Mode		Mode 1	Mode 2	Mode 3
Number of OFDM segments N_s		13		
Bandwidth		$3000/7$ (kHz) $\times N_s$ + $250/63$ (kHz) = $5.575…$MHz	$3000/7$ (kHz) $\times N_s$ + $125/63$ (kHz) = $5.573…$MHz	$3000/7$ (kHz) $\times N_s$ + $125/126$ (kHz) = $5.572…$MHz
Number of segments of differential modulations		n_d		
Number of segments of synchronous modulations		n_s ($n_s + n_d = N_s$)		
Spacings between carrier frequencies		$250/63 = 3.968…$kHz	$125/63 = 1.984…$kHz	$125/126 = 0.992…$kHz
Number of carriers	Total count	$108 \times N_s + 1 = 1405$	$216 \times N_s + 1 = 2809$	$432 \times N_s + 1 = 5617$
	Data	$96 \times N_s = 1248$	$192 \times N_s = 2496$	$384 \times N_s = 4992$
	SP	$9 \times n_s$	$18 \times n_s$	$36 \times n_s$
	CP*1	$n_d + 1$	$n_d + 1$	$n_d + 1$
	TMCC	$n_s + 5 \times n_d$	$2 \times n_s + 10 \times n_d$	$4 \times n_s + 20 \times n_d$
	AC1	$2 \times N_s = 26$	$4 \times N_s = 52$	$8 \times N_s = 104$
	AC2	$4 \times n_d$	$9 \times n_d$	$19 \times n_d$
Carrier modulation scheme		QPSK, 16QAM, 64QAM, DQPSK		
Symbols per frame		204		
Effective symbol length		252 μs	504 μs	1.008 ms
Guard interval		63 μs (1/4), 31.5 μs (1/8), 15.75 μs (1/16), 7.875 μs (1/32)	126 μs (1/4), 63 μs (1/8), 31.5 μs (1/16), 15.75 μs (1/32)	252 μs (1/4), 126 μs (1/8), 63 μs (1/16), 31.5 μs (1/32)

---

[22] ARIB STD-B31 Version 2.2;. Transmission System for Digital Terrestrial Television Broadcasting, Chapter 3.9.3.1
[23] Quoted from ARIB STD-B31, Chapter 3, table 3-2

Frame length	64.26 ms (1/4), 57.834 ms (1/8), 54.621 ms (1/16), 53.0145 ms (1/32)	128.52 ms (1/4), 115.668 ms (1/8), 109.242 ms (1/16), 106.029 ms (1/32)	257.04 ms (1/4), 231.336 ms (1/8), 218.484 ms (1/16), 212.058 ms (1/32)
Inner code	Convolutional code (1/2, 2/3, 3/4, 5/6, 7/8)		
Outer code	RS (204,188)		

*1: SP (Scattered Pilot) and CP (Continual Pilot) are used by the receiver for synchronization and demodulation purposes

---

Calculation process for carrier spacing and effective symbol length of 6MHz system

As an example, the calculation process for carrier spacing and effective symbol length for mode 1 of a 6MHz ISDB-T system is given below.

In case of mode 1, 2k FFT (Fast Fourier Transform) and IFFT (Inverse Fast Fourier Transform), circuits are used for the conversion from/to frequency domain and to/from time domain. The exact number of sampling point of FFT and IFFT is defined as 2048(=2^{11}).
For mode 2, 4096(=2^{12}). For mode 3, 8192(=2^{13}).

a) Calculation of carrier spacing:
As described in 1) and 2) above, the total carrier in a 6MHz bandwidth is calculated as follows:

{6/14 MHz(bandwidth of one segment)}/{108(number of carrier in one segment)}= 6000/(14×108)=3.968…kHz

b) Sampling frequency of FFT and IFFT
The carrier spacing as calculated above is equal to the spacing of the FFT sampling point. Therefore, the sampling frequency of an FFT and IFFT are calculated as follows:

3.968..kHz(carrier spacing) × 2048(number of FFT point)=8.12698.. MHz

The FFT sampling frequency is identical for all modes (1,2, and 3). Try to calculate the rest for your verification.

c) Effective symbol length:
As explained in the section II.3, in OFDM transmission system, the effective symbol length is given as reciprocal of the carrier spacing.

Effective symbol length =1/(carrier spacing)=1/3.968..kHz =252 us

For mode 2 and mode 3, please use the above example and try to calculate the resulting Effective Symbol Length. The answers can be found on table III-1.

The major difference between 6MHz, 7MHz and 8MHz ISDB-T system is in the difference of a bandwidth. While maintaining the same OFDM segment structure and number of carriers, ISDB-T fits into these differences of bandwidth by changing the carrier spacing. Comparing 8MHz system to 6MHz system, carrier spacing widens by 8/6 and symbol duration shortens by 6/8.

In case of 6MHz system, the segment bandwidth is calculated below:
Segment bandwidth of 6MHz =6MHz/14=428.57…kHz
On the other hand, the segment bandwidth of 8MHz system is calculated by a same calculation process:
Segment bandwidth of 8MHz =8MHz/14=571.428…kHz

Based on this, transmission parameters of 8MHz system are calculated shown in Table III-14 below. Note that parameters different from 6MHz system are shown in bold.

Table III-3   OFDM segment parameters of 8MHz system[24]

Mode		Mode 1		Mode 2		Mode 3	
Bandwidth		4000/7 = 571.428…kHz					
Spacing between carrier frequencies		8000/(14×108)= 5.291… kHz		8000/(14×216)= 2.645… kHz		8000/(14×432)= 1.322… kHz	
Number of carriers	Total count	108	108	216	216	432	432
	Data	96	96	192	192	384	384
	SP[*1]	9	0	18	0	36	0
	CP[*1]	0	1	0	1	0	1
	TMCC[*2]	1	5	2	10	4	20
	AC1[*3]	2	2	4	4	8	8
	AC2[*3]	0	4	0	9	0	19
Carrier modulation scheme		QPSK 16QAM 64QAM	DQPSK	QPSK 16QAM 64QAM	DQPSK	QPSK 16QAM 64QAM	DQPSK
Symbols per frame		204					
Effective symbol length		189 μs		378 μs		756 μs	
Guard interval		47.25 μs (1/4), 23.625 μs (1/8), 11.8125 μs (1/16), 5.90625 μs (1/32)		94.5 μs (1/4), 47.25 μs (1/8), 23.625 μs (1/16), 11.8125 μs (1/32)		189 μs (1/4), 94.5 μs (1/8), 47.25 μs (1/16), 23.625 μs (1/32)	
Frame length		48.195 ms (1/4), 43.3755 ms (1/8), 40.96575 ms (1/16), 39.760875 ms (1/32)		96.39 ms (1/4), 86.751 ms (1/8), 81.9315 ms (1/16), 79.52175 ms (1/32)		192.78 ms (1/4), 173.502 ms (1/8), 163.863 ms (1/16), 159.0435 ms (1/32)	
IFFT sampling frequency		0.005291 × 2048= 10.8359… MHz					
Inner code		Convolutional code (1/2, 2/3, 3/4, 5/6, 7/8)					
Outer code		RS (204,188)					

[24] Quoted from ARIB STD-B31, Version 2.2 Annex A Table A-5

Table III-4 Transmission signal parameters for 8MHz system[25]

Mode	Mode 1	Mode 2	Mode 3
Number of OFDM segments N_s	13		
Bandwidth	$4000/(7 \times 108)$ (kHz) × $(108 \times N_s +1)$ = 7.433…MHz	$4000/(7 \times 216)$ (kHz) × $(216 \times N_s +1)$ = 7.431…MHz	$4000/(7 \times 432)$ (kHz) × $(432 \times N_s +1)$ = 7.429…MHz
Number of segments of differential modulations	n_d		
Number of segments of synchronous modulations	n_s ($n_s + n_d = N_s$)		
Spacings between carrier frequencies	$8000/(14 \times 108)$ = 5.291… kHz	$8000/(14 \times 216)$ = 2.645… kHz	$8000/(14 \times 432)$ = 1.322… kHz
Number of carriers — Total count	$108 \times N_s + 1 = 1405$	$216 \times N_s + 1 = 2809$	$432 \times N_s + 1 = 5617$
Number of carriers — Data	$96 \times N_s = 1248$	$192 \times N_s = 2496$	$384 \times N_s = 4992$
Number of carriers — SP	$9 \times n_s$	$18 \times n_s$	$36 \times n_s$
Number of carriers — CP[*1]	$n_d + 1$	$n_d + 1$	$n_d + 1$
Number of carriers — TMCC	$n_s + 5 \times n_d$	$2 \times n_s + 10 \times n_d$	$4 \times n_s + 20 \times n_d$
Number of carriers — AC1	$2 \times N_s = 26$	$4 \times N_s = 52$	$8 \times N_s = 104$
Number of carriers — AC2	$4 \times n_d$	$9 \times n_d$	$19 \times n_d$
Carrier modulation scheme	QPSK, 16QAM, 64QAM, DQPSK		
Symbols per frame	204		
Effective symbol length	189 μs	378 μs	756 μs
Guard interval	47.25 μs (1/4), 23.625 μs (1/8), 11.8125 μs (1/16), 5.90625 μs (1/32)	94.5 μs (1/4), 47.25 μs (1/8), 23.625 μs (1/16), 11.8125 μs (1/32)	189 μs (1/4), 94.5 μs (1/8), 47.25 μs (1/16), 23.625 μs (1/32)
Frame length	48.195 ms (1/4), 43.3755 ms (1/8), 40.96575 ms (1/16), 39.760875 ms (1/32)	96.39 ms (1/4), 86.751 ms (1/8), 81.9315 ms (1/16), 79.52175 ms (1/32)	192.78 ms (1/4), 173.502 ms (1/8), 163.863 ms (1/16), 159.0435 ms (1/32)
Inner code	Convolutional code (1/2, 2/3, 3/4, 5/6, 7/8)		
Outer code	RS (204,188)		

---

[25] Quoted from ARIB STD-B31, Version 2.2 Annex A Table A-6

> **Calculation process for carrier spacing and effective symbol length for 8MHz system**
>
> As an example, the calculation process for carrier spacing and effective symbol length for mode 1 of a <u>8MHz ISDB-T system</u> is given below:
>
> In case of mode 1, 2k FFT(Fast Fourier Transform) and IFFT(Inverse Fast Fourier Transform), circuits are used for the conversion from/to frequency domain and to/from time domain.(the exact number of sampling point of FFT and IFFT is defined as 2048(=2^{11}).
> For mode 2, 4096(=2^{12}).   For mode 3, 8192(=2^{13})
>
> a) Calculation of carrier spacing(for mode 1):
> as described in 1) and 2) above, the total carrier in a <u>8MHz bandwidth</u> is calculated as follows:
>   *{6/14 MHz(bandwidth of one segment)} / {108(number of carrier in one segment)}=*
>   <u>*8000/(14x108)=5.291…kHz*</u>
>
> b) Sampling frequency of FFT and IFFT
> The carrier spacing as calculated above is equal to the spacing of the FFT sampling point. Therefore, the sampling frequency of an FFT and IFFT are calculated as follows:
>
>   <u>*5.291..kHz*</u>*(carrier spacing) x 2048(number of FFT point)=*<u>*10.8359.. MHz*</u>
>
> The FFT sampling frequency is identical for all modes (1,2, and 3).   Feel free to calculate the rest for your verification.
>
> c)Effective symbol length:
> As explained in the section II.3, in OFDM transmission system, the effective symbol length is given as reciprocal of the carrier spacing, therefore:
>
>   *Effective symbol length =1/(carrier spacing)=1/*<u>*5.291..kHz*</u> *=*<u>*189 us*</u>
>
> For mode 2 and mode 3, please use the above example and try to calculate the resulting Effective Symbol Length.   The answers can be found on Table III-4.

As shown in the table, the "Code rate of the Outer coder" and the "Number of symbols per frame" are common for all modes and in any of the available case of the hierarchical transmission.

III.1.2   Functional block diagram of an ISDB-T transmission system
In this section, the coding process based on the illustrated ISDB-T Transmission block diagram in Figure III-1.   ISDB-T transmission system is made up of three basic functional blocks, "TS re-multiplexing", "Channel coding" and "OFDM framing & modulation"[26],[27]

---

[26] Takada, M. Saito, M. Proceedings of the IEEE, Vol. 9, Issue.1,Jan. 2006, Transmission System for ISDB-T
[27] Nakahara, S Moriyama, S.   Kuroda, T.   Sasaki, M.   Yamazaki, S   Yamada, O. IEEE transactions on Broadcasting, Vol.42,

Please take note that there are hierarchies in the middle that makes ISDB-T unique from other systems. In order to differentiate coding parameters for different reception modes, i.e. fixed or mobile, hierarchies were introduced. As shown in Figure III-1, three circuits, which are composed of convolutional coding, bit interleaving and mapping, are made available for hierarchical transmission. The details of these functional blocks will be explained further in the section III.2 – III.4.

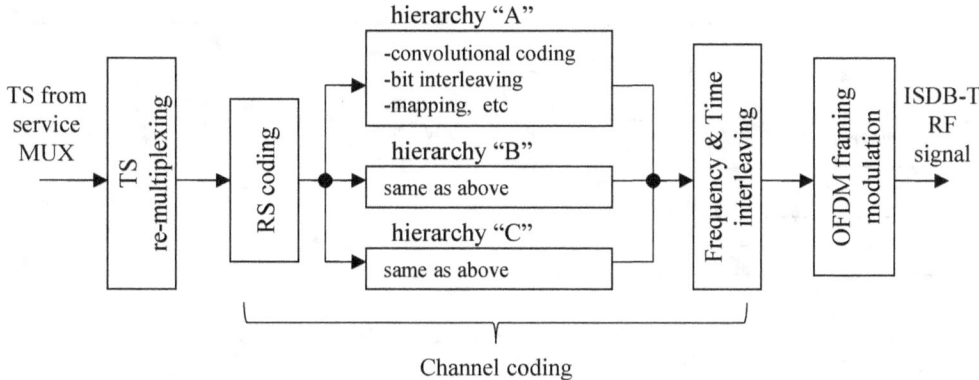

Figure III-1   Conceptual block diagram of ISDB-T system

(a) Hierarchical transmission block indicated as "TS re-multiplexer" in the Figure III-1 which will be explained in the section III.2.2 later
(b) Channel coding block (a dashed line box in Figure III-1): explained in the section III.3
(c) OFDM transmission block (a right side box in Figure III-1): explained in Section III.4

Please note in that in both the actual hardware implementation and published ARIB STD, there is no such division as shown above. This division serves only as a functional illustration to facilitate a better technical appreciation.

In this text, each function will be explained according to the above division and some supplemental explanations which are not provided in the ARIB STD-B31 for better understanding.

Below, Figure III-2 shows a block diagram which is basically the same as that illustrated in Figure III-1 but goes into a more detailed explanation of all the functional blocks that comprises the ISDB-T transmission system.

The section III.2 through III.4 will cover the explanation of this block diagram.

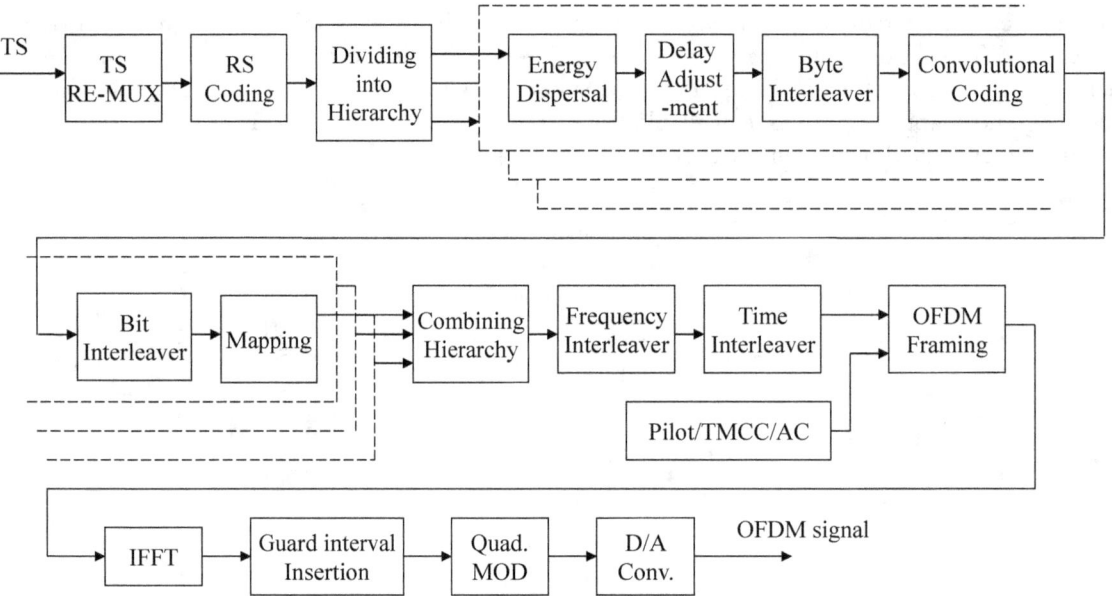

Figure III-2   An overall block diagram of ISDB-T transmission system

III.2   Hierarchical Transmission in ISDB-T

Figure III-3 shows the same figure as Figure III-2 with the functional boxes related to hierarchical transmission highlighted in bold.   This section will discuss the functions of "Hierarchical transmission"[28] [29] whose specifications are defined in ARIB STD-B31, Section 3-1, 3-2, 3-4, 3-6, 3-10.

---

[28] Uehara, M. Takada, M. Kuroda, M. IEEE transactions on Consumer Electronics, Vol.45, Issue 1, Feb. 1999, Transmission scheme for the terrestrial ISDB system
[29] Uehara, M.   Proceedings of the IEEE, Vol.94, Issue 1, Jan.2006, Application of MPEG-2 Systems to Terrestrial ISDB(ISDB-T)

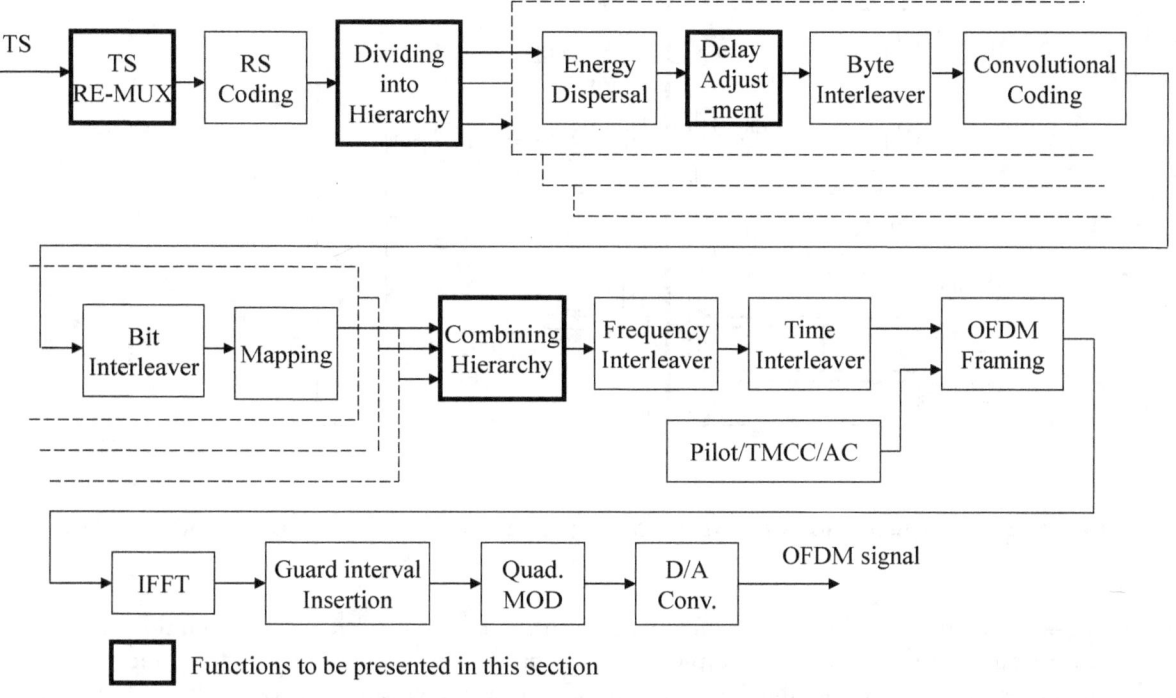

Figure III-3   Related functional block for hierarchical transmission

III.2.1   Concept and rules of hierarchical transmission in ISDB-T
As described in the section II.4.3, ISDB-T adopts "Segmented OFDM transmission system" for hierarchical transmission. To simplify the receiver construction/operation in hierarchical transmission service, the transmission path encodes and modulates the data in a pre-defined process and the receiver traverses the process in the opposite order.   In this process, the receiver's demodulator output should be a single transport stream with constant bit rate of 4fs (four times of the sampling frequency) in any combination of hierarchy and parameters applied for each hierarchy.   To achieve a simple operation of the receiver, several rules ((1) to (8)) are written below[30].

(1) Broadcast TS
In order to simplify the interface and operation of a receiver, all transport stream packets that belong to different hierarchical layers should be multiplexed in one transport stream. This transport stream is called a "Broadcast TS."

---

[30] Uehara, M.   Proceedings of the IEEE, Vol.94, Issue 1, Jan.2006, Application of MPEG-2 Systems to Terrestrial ISDB(ISDB-T)

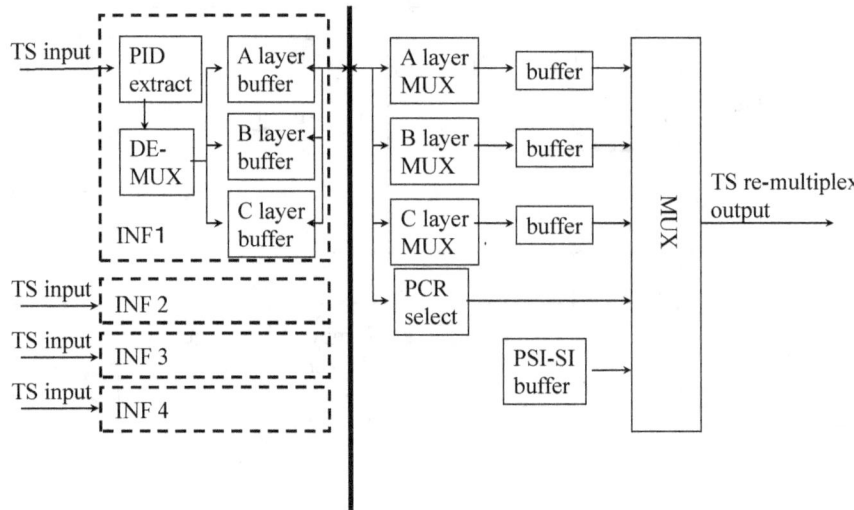

Figure III-4   Illustration of a Multiplexer (includes both service MUX and re-MUX)

Figure III-4 shows the functions of a Multiplexer from the transmitter side. This multiplexer is placed at the input portion of the modulator. Typically, a multiplexer is composed of two functions. One is "service multiplexing", found on the left portion of figure above and the other is re-arrangement of Transport Stream Packets (TSPs) into each layer(s) and "re-multiplexing", shown in right portion of the figure.

Service multiplexer combines various services into one transport stream based on the specification of MPEG-2 systems. TS re-multiplexing on the other hand is defined as follows: "signal processing which re-multiplexes the packets that belong to a different hierarchy including the null packets into one transport stream". Figure III-5 shows an example of a Broadcast TS with Layer A and Layer B. As shown in the figure, the Broadcast TS is composed of A layer TSPs (indicated as TSP_A in the figure), B layer TSPs (indicated as TSP_B in the figure) and Null packets (indicated as TSP_{null} in the figure). The null packet is defined as the packet that contains no data. A null packets is inserted into the Broadcast TS to adjust the data rate (see (7)).

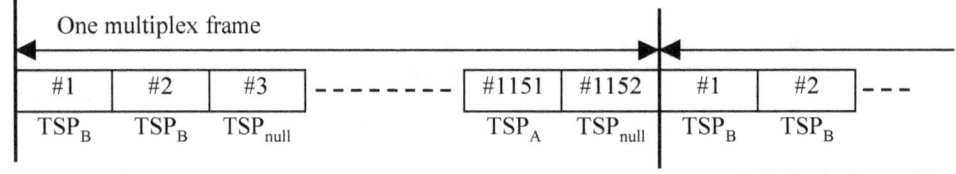

(Mode 1, Guard Interval of 1/8)

Figure III-5   Example of a Re-Multiplexed Transport Stream[31]

(2) TSP arrangement in the Broadcast TS

The arrangement of TSPs of a "Broadcast TS" is defined so that a "Model receiver" can reproduce the TS by simply following a pre-defined process. The concept of a "Model receiver" will be explained in the section

---

[31]Quoted from ARIB STD-B31 fig. 3-3

III.2.2. This arrangement is called "multiplex frame" pattern. The length of "multiplex frame" is same as that of an "OFDM frame."

(3) Strongest hierarchy
As described in the section II.4, in ISDB-T transmission system, the reception performance of each hierarchy is different depending on the transmission parameters selected. As an example illustrated in Figure III-6, a layer of QPSK modulation is more robust in comparison of a layer of 64QAM modulation, the effects are even more pronounced in a real world receiving condition. In this case, a layer of QPSK modulation is defined as "strongest hierarchy". The strongest hierarchy should be able to be received, demodulated and decoded alone. The basis is that there are cases when the strongest hierarchy can be received alone due to its superior performance in reception whereas other hierarchies dismissed because of weaker reception performance. In this case, receivers need to demodulate and decode the PCR and minimum required PSI information. The strongest layer should transmit this information. In the example in Figure III-6, TSPs of layer A should include PCR and minimum required PSI which are necessary to recover TSP. This rule is very important to ensure "partial reception" for One-Seg handheld reception service.

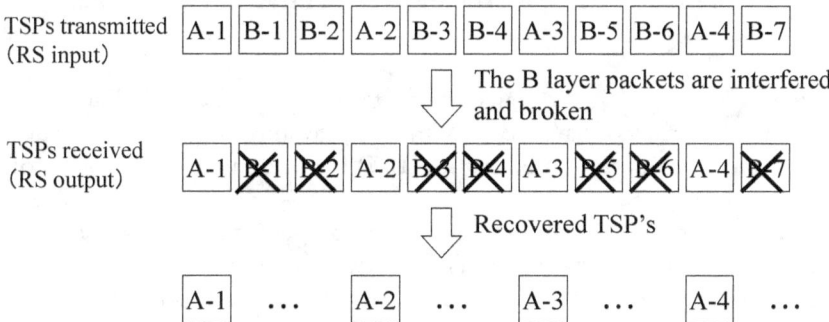

Figure III-6   Concept of a hierarchical transmission (the strongest layer should be able to be demodulated alone)

(4) Delay adjustment
The transmission speed of each hierarchical layer depends on the transmission parameter used (carrier modulation, coding rate, number of segment). This difference in transmission speed may cause confusion in the order of arrival to the receiver, in other words, packets in high-speed layer overtakes packets in slow-speed layer and arrives faster than the expected order. Therefore, it is necessary to adjust the transmission delay of each layer at the transmitter side to ensure that the data arrives at the receiver in the proper order.

For a better understanding of the above statement, two Figures are prepared. Figure III-7 illustrates the case of 2-layer transmission with different transmission speed.

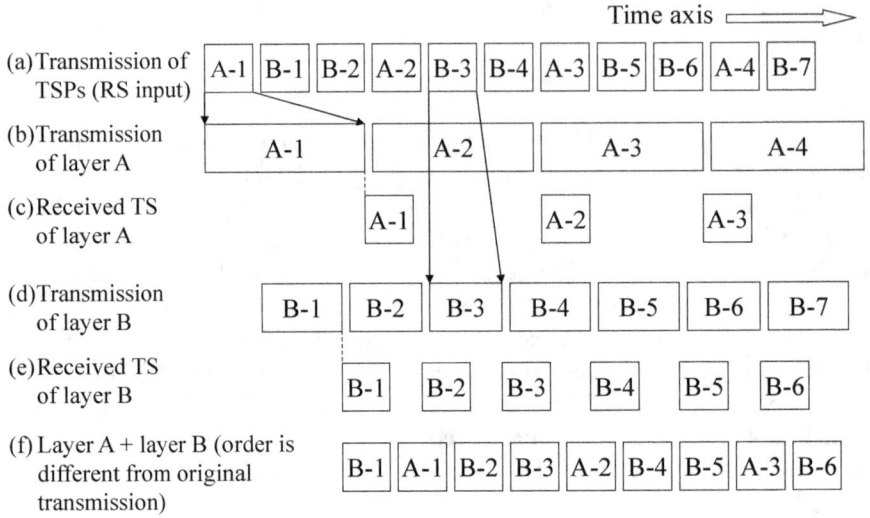

Figure III-7 Image of a hierarchical transmission (1/2)

In the case illustrated in Figure III-7, the transmission time of layer A, including the transmission coding and decoding process, is twice that of layer B. Without delay time adjustment (time compensation), the order of the TSP of the receiver output ((f) of Figure III-7) is different from the order of the TSP from the transmitter input ((a) of Figure III-7).

To get the same arrangement of TSPs, it is necessary to compensate the delay between different layers at the transmitter side. Figure III-8 shows the image of delay compensation at the transmitter side. By compensating the delay, the same order of TSP can be expected at the demodulator output.

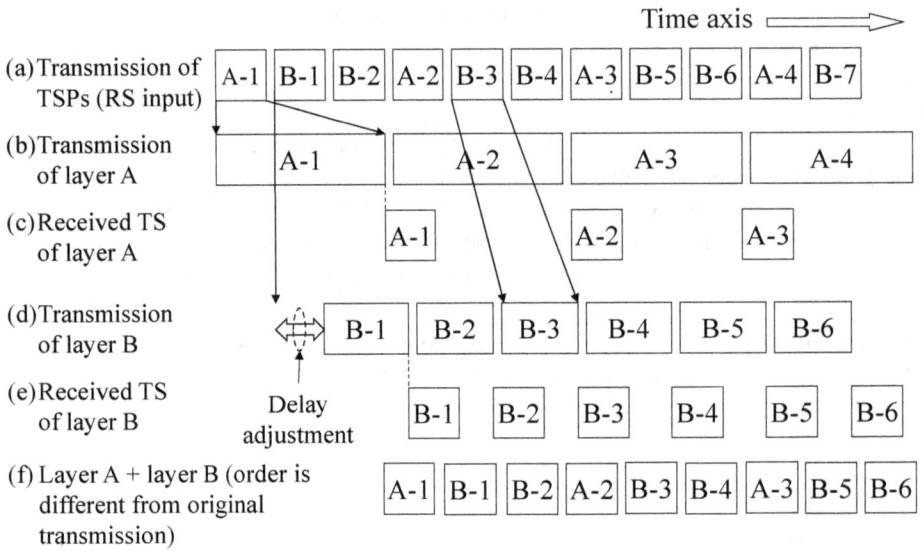

Figure III-8 Image of hierarchical transmission (2/2)

For adjusting the delay, two "delay adjustment" functions are inserted in the transmitter side. One is inserted after "Byte interleave", the other is after "Carrier modulation" (mapping). More details of "delay adjustment" can be found in the section III.2.4 and III.2.5 later.

(5) Multiplex frame
The TSPs in one OFDM frame period are grouped into so called "multiplex- frame". It terms of duration, the multiplex frame length is set to be identical to an OFDM frame length so that receivers can generate multiplex-frame synchronization signal from OFDM frame synchronization signal. For the technical details of "multiplex frame," see the section III.2.2.

(6) The number of packets in 1 segment
The number of packets in a segment in a frame should be an integer in any combination of the transmission parameter and coding rate in order to simplify the arrangement of TSPs according to the Multiplex-frame pattern rule. It is defined that this arrangement should be complete in one OFDM frame (equal length of Multiplex-frame). For this reason, the number of packets for each layer in one OFDM frame should also be an integer. Since the minimum unit of hierarchy is the segment, it is therefore necessary to satisfy the condition that the number of TSPs in one segment is an integer. The number of TSPs in one OFDM frame per One-Segment is shown in Table III-5 below.

Table III-5   Number of TSPs in (one frame)/(One-Segment)

Carrier modulation	Convolution coding rate	Number of TSPs		
		Mode 1	Mode 2	Mode 3
QPSK/DQPSK	1/2	12	24	48
QPSK/DQPSK	2/3	16	32	64
QPSK/DQPSK	3/4	18	36	72
QPSK/DQPSK	5/6	20	40	80
QPSK/DQPSK	7/8	21	42	84
16QAM	1/2	24	48	96
16QAM	2/3	32	64	128
16QAM	3/4	36	72	144
16QAM	5/6	40	80	160
16QAM	7/8	42	84	168
64QAM	1/2	36	72	144
64QAM	2/3	48	96	192
64QAM	3/4	54	108	216
64QAM	5/6	60	120	240
64QAM	7/8	63	126	252

(7) Clock rate
The information data rate of an ISDB-T transmission system varies with its transmission parameters, such as guard interval ratio, modulation index, coding rate, number of hierarchies and number of segments for each hierarchy. If a receiver need to re-generate a clock which corresponds to the information data rate, the resulting requirement for receiver control will be complex. In order to avoid this situation, the ISDB-T transmission system utilizes an identical clock rate for the Broadcast TS (common for a modulator input and a demodulator output) clock rate. Despite of the difference in the transmission data rate as a result of the difference in the transmission parameter settings, this simple implementation guarantees that the receiver

operation can be made simple.

For the clock rate, 4fs[32] is chosen because for any combination of the transmission parameters, the maximum data rate is always lower than 4fs.

Figure III-9 illustrates the clock rate adjustment at both the input of modulator and the output of the demodulator. Any variation in time is compensated by inserting null packets.

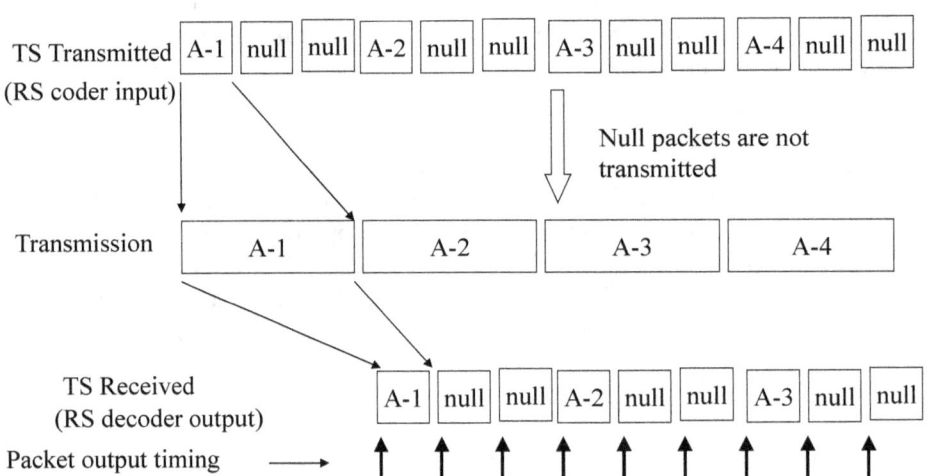

Figure III-9 Transmission rate adjustment in both the transmitter and the receiver

As illustrated in Figure III-9 above, the null packets are not transmitted, rather, it is inserted at the receiver side as a means to adjust the TS transmission rate. At the "packet output timing" (shown in the bottom part of Figure III-9), a packet that has been completely received will generate an output data, on the other hand for an incompletely received packet, it will unfortunately generate a null packet output. Through this operation, the same transport stream coming from the transmitter side can be re-generated at the receiver side.

(8) Number of TSPs in one OFDM frame

The length of an OFDM frame (same length to multiplex frame) depends on mode and guard interval ratio and not on carrier modulation and convolutional coding rate. The clock rate of a broadcast TS at the output of the TS re-multiplexer is constant, therefore, the number of TSPs (including null packets) changes according to the frame length. Table III-6 shows the number of TSPs in one frame depending on the mode and guard interval ratio.

Table III-6   Multiplex-Frame configuration[33]

Mode	Number of transmission TSPs included in one multiplex frame			
	Guard-interval ratio 1/4	Guard-interval ratio 1/8	Guard-interval ratio 1/16	Guard-interval ratio 1/32
Mode 1	1280	1152	1088	1056

---

[32] 32.5…MHz for 6MHz system, 37.92… for 7MHz system and 43.34…MHz for 8MHz system
[33] Quoted from ARIB STD-B31 table 3-5

| Mode 2 | 2560 | 2304 | 2176 | 2112 |
| Mode 3 | 5120 | 4608 | 4352 | 4224 |

For functional blocks relating to hierarchical transmission, "TS re-multiplexer," "Division to hierarchy," "Delay adjustment" and "combine hierarchy" are described below.

III.2.2   TS re-multiplexer
This section will discuss on the TS re-multiplexer whose details are defined in ARIB STD-B31, Section 3.2.

(1) Configuration of multiplex-frame
TS re-miltiplexer (refer to Figure III-3) makes a "Multiplex frame" from packetized broadcast signal ("TS RE-MUX" input signal in the figure or "information TSPs") and null packets which does not transmit information.   The null TSPs are inserted into Multiplex frame to adjust the frame size.
   As described in Subsection III.2.1 above, Multiplex frame is generated based on the following rules:
1) The Multiplex frame length is determined only by the Mode and the Guard Interval length, and is independent from parameters such as the number of layers, the modulation scheme of each layer.   The total number of information TSPs and null TSPs in the Multiplex frame is as shown in Table III-6 above.
2) As shown in the preceding figure III-9, only information TSPs are transmitted from the transmitter, and null TSPs are not transmitted.   Null TSs are reproduced on the receiving side and a Multiplex frame is re-constructed.
3) The placement of each information TSPs and null TSPs in Multiplex frame is specified indirectly on the transmitter side by defining the operation of the model receiver described in (2) below.
   Fig. III - 5 shows an example of arrangement of TSPs in Multiplex frame in the case of two hierarchies.

(2) Model receiver for forming multiplex frame patterns
An arrangement of TSPs is defined as a model receiver behavior, that is, TSPs should be arranged so that model receiver can reproduce TSPs.   The FFT sampling clock is used for signal processing in various functional blocks in practical designs, in which various clocks are generated from FFT sampling clock.   Figure III-10 shows a block diagram of the model receiver.

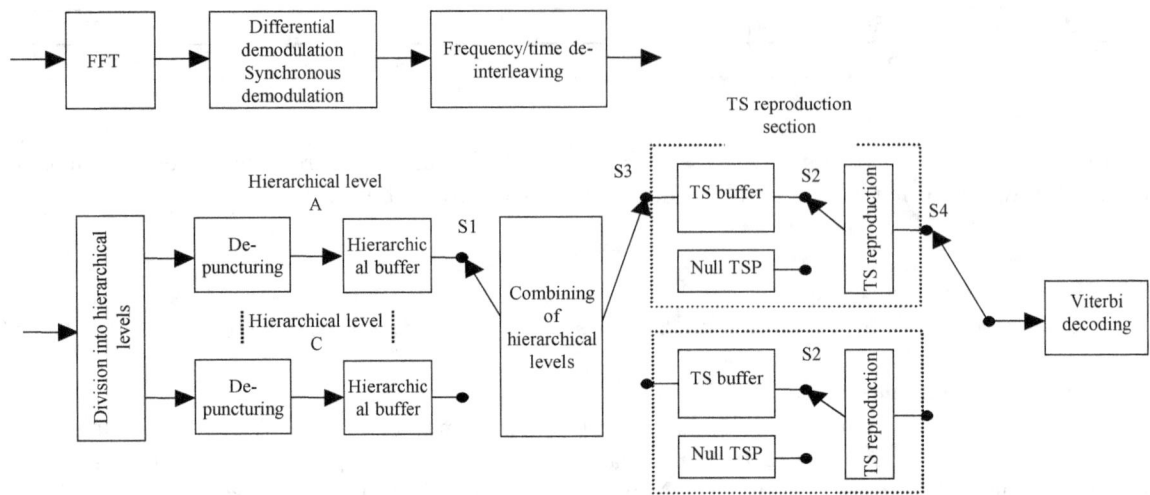

Figure III-10   Model Receiver for Forming Multiplex Frame Patterns[34]

The received signal is input to the functional block in the figure "Division into hierarchical levels" after FFT, demodulation and de-interleaving in order to divide data stream into hierarchical layers.

As shown in the following Figure III-11 as an example, the segment number is assigned in the order of the center segment 0, then down, up, down, up manner on frequency axis, because the carrier demodulation in FFT is done from the center frequency and alternately and sequentially adjacent like lower, upper, lower, upper manner on the frequency axis. The input of the hierarchical divider of receiver receives each segments arranged in ascending order of the segment number. For example in the figure, segment 0, segment 1 to 4 and segment 5 to 12 are grouped in each hierarchy.

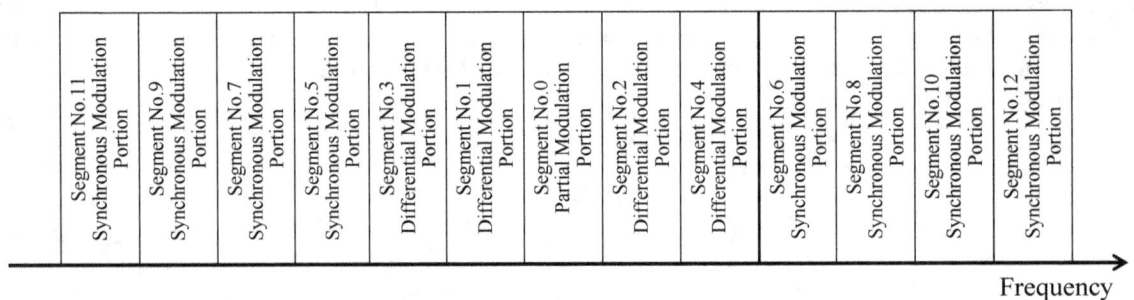

**Figure III-11   OFDM-segment numbers on the transmission spectrum
and an example of usage**

Figure III-12 shows the time arrangement for a case of three hierarchies illustrated in Figure III-11, with A layer being QPSK, r=1/2, 1 segment, B layer being DQPSK, r = 1/2, 4 segments, C layer by 64 QAM, r = 7/8, 8 segments. Mode and guard interval ratio are common for both layers, mode 1 and 1/8. Here, the sample clock in one OFDM symbol is calculated as follows.

---

[34] Quoted from ARIB STD-B31, Chapter 3.2.2, fig. 3-4

2048 (FFT sampling points in a effective symbol)
 + 256 (FFT sampling points in a guard interval)
 = 2304 clock

Therefore, the data size (in this case, same as a number of sample clock) of information TSPs and null TSPs in each hierarchy in one OFDM symbol is calculated as follows

A hierarchy: 96 (= 96 *information carriers*×1 *segments*)
B hierarchy: 384 (= 96 *information carriers*×4 *segments*)
C hierarchy: 768 (= 96 *information carriers* ×8 *segments*)
Null TSPs: 2304-96-384-768 =1056

Each layer's data of which size are calculated by the above formula are sent through de-puncturing and are accumulated in the hierarchical buffer of each hierarchy. A number of sampling clock in one OFDM symbol varies according to mode and guard interval ratio. For other case, shown in Figure III-7 below. Please, try to calculate by yourself.

Table III-7 Number of sampling clock in one OFDM symbol

Mode	Number of sampling clock in one OFDM symbol			
	Guard-interval ratio 1/4	Guard-interval ratio 1/8	Guard-interval ratio 1/16	Guard-interval ratio 1/32
Mode 1	2560	2304	2176	2112
Mode 2	5120	4608	4352	4224
Mode 3	10240	9216	8704	8228

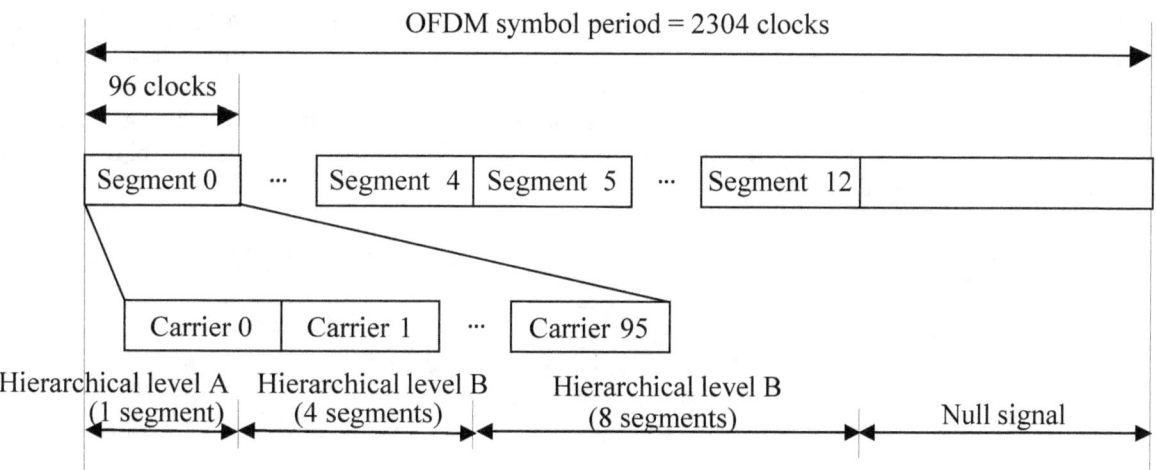

Figure III-12   Time arrangement for the input signals to the hierarchical divider in receiver

Next, the operation of each switch S1 - S4 in the model receiver is explained.

When data of 1 TSP (408 bytes) is input to the hierarchical buffer, S1 forwards the data stored in the hierarchical buffer to the buffer in TS reproduction section. S1 and S3 are linked; when S1 selects Layer A input, then S3 also selects Layer A output. Next, S2 checks the content of the buffer in TS reproduction section in every TSP transmission (408 byte) units timing. If 1 TSP data has been stored, then S2 outputs this data to Viterbi decoder through TS reproduction. The Viterbi decoder is the most widely used decoding method when decoding a convolutional code. If not stored, S2 selects "Null TSP" and forwards null TSP to Viterbi decoder. In this way, TSPs constituting the multiplex frame are output to Viterbi decoder.

III.2.3   Division of TS into hierarchical layers

This circuit, defined in ARIB STD-B31 Section 3.4, is placed between the outer coder (RS coder) described in section III.3.2 and energy dispersal circuit for each layer. The function of this circuit is to divide the TSPs (204 bytes in length).

Figure III-13 illustrates a signal process in a hierarchical divider. The top of this figure shows a Broadcast TS which is composed of TSPs and Null TSPs. Each TSPs is composed of one synchronization byte (illustrated as "S" in the figure), 187 information bytes (illustrated as "I" in the figure) and 16 parity bytes (illustrated as "P" in the figure). An OFDM frame is composed of TSPs that belong to one of layers and Null TSPs. The order of these TSPs in an OFDM frame is pre-determined uniquely according to the hierarchy settings and transmission parameter set. Therefore, the hierarchical divider distributes the TSP for each hierarchy based on this order. At this time, Null TSPs are discarded and is not transmitted.

In the figure, for example, as TSP #1 belongs to layer A, the hierarchical divider diverts the TSP #1 to layer A circuit. Next TSP is null Packet and will not be transferred to any of the following circuits. TSP #2, which belongs to layer B, will be transferred to layer B circuit.

As described above, Null TSPs are only inserted at the output of the Re-multiplexer to adjust the transmission clock rate, but not transmitted. The null packets are reproduced at a receiver side independently. (see the section III.2.1) As a result, the output of the RS decoder from the receiver side is the same Transport Stream to that of the output of the TS re-multiplexer including the null packets.

Figure III-13 shows an example of two hierarchical transmission case. The top of the figure shows the input TS of the hierarchical divider; next is the input of Hierarchical layer A with only TSP #1 is picked up. The bottom of the figure shows the input of Hierarchical layer B with only TSP #2 is picked up. A null TSP is then removed at this stage. At this time, a hierarchical divider shifts the OFDM frame synchronization position in one byte length to change its position to the head of information bytes. As shown in the lower part of figure III-13, a sync byte is placed at the end of the TSP. This sync byte can be used as a termination code of Viterbi decoding on the receiver side. In this case, one Viterbi decoder of the receiver can be used for decoding a data of multi-layers.

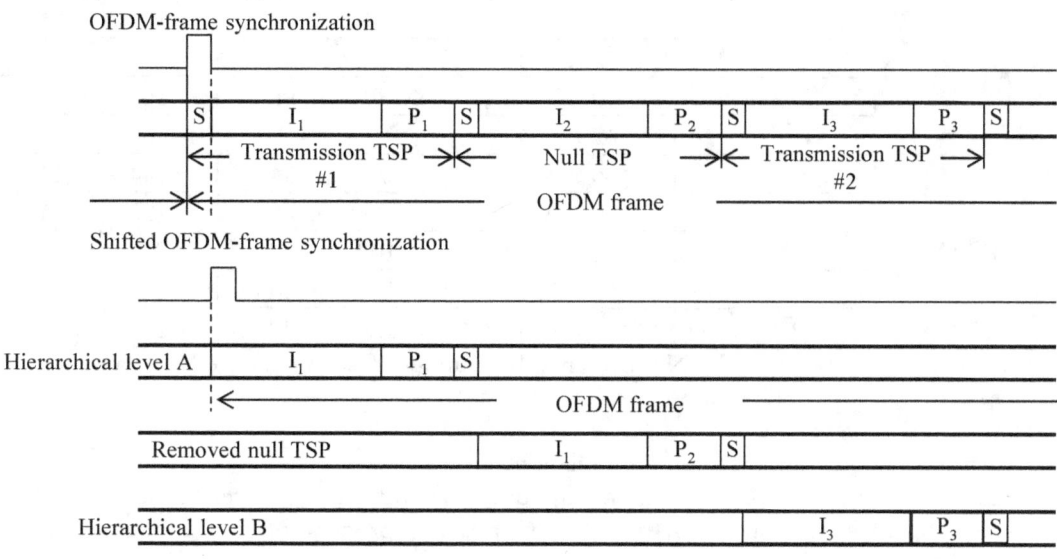

S: Synchronization byte I: Information P: Parity

Figure III-13   Example of a Hierarchical Divider operation[35]

III.2.4   "Delay adjustment" before the byte interleaver

The delay adjustment circuit defined in ARIB STD-B31, Section 3.6 is placed on the transmitter side. The function of the delay adjustment is to compensate for the delay time introduced by the signal processing in both the transmitter and receiver for each hierarchical layer. The total delay including the delay caused by byte interleaving /de-interleaving (11 transmission TSPs) should be adjusted to one OFDM frame length.

As an example, in case of a QPSK coding rate=1/2, the number of TSPs in one OFDM frame is 12 (see Table III-5). Therefore, the number of TSPs in one OFDM frame is equal to 12N (N is number of segment). On the other hand, the delay caused by byte interleaving/de-interleaving is 11 TSPs. Therefore, the delay of this layer as calculated and adjusted by the circuit is (12N-11).

Please try to calculate the other cases. The answers are shown in Table III-8.

Table III-8   Delay-Adjustment values required as a result of Byte Interleaving[36]

Carrier modulation	Convolutional code	Delay-adjustment value (number of transmission TSPs)		
		Mode 1	Mode 2	Mode 3
DQPSK	1/2	12 × N-11	24 × N-11	48 × N-11
	2/3	16 × N-11	32 × N-11	64 × N-11
	3/4	18 × N-11	36 × N-11	72 × N-11
QPSK	5/6	20 × N-11	40 × N-11	80 × N-11
	7/8	21 × N-11	42 × N-11	84 × N-11
16QAM	1/2	24 × N-11	48 × N-11	96 × N-11

---

[35] Quoted from ARIB STD-B31 fig.3-7
[36] Quoted from Ordinance No.87 of the Ministry of Internal Affairs and Communications (2011), Annexed Table 15, Annexed Statement 2, Item 2

	2/3	32 × N-11	64 × N-11	128 × N-11
	3/4	36 × N-11	72 × N-11	144 × N-11
	5/6	40 × N-11	80 × N-11	160 × N-11
	7/8	42 × N-11	84 × N-11	168 × N-11
64QAM	1/2	36 × N-11	72 × N-11	144 × N-11
	2/3	48 × N-11	96 × N-11	192 × N-11
	3/4	54 × N-11	108 × N-11	216 × N-11
	5/6	60 × N-11	120 × N-11	240 × N-11
	7/8	63 × N-11	126 × N-11	252 × N-11

N represents the number of segments used by that hierarchical layer.

III.2.5 "Delay adjustment" before carrier modulation

This functional block[37] is placed between the energy dispersal and the byte interleaver. In order to adjust the delay time difference introduced by the bit interleaving process of the modulator side and bit de-interleaving on the demodulator side.

As described in section III.6, the signal delay introduced by the bit interleaving/de-interleaving process is defined as 120 carrier symbols. The delay time depends on the transmission parameters for each hierarchy.

In order to compensate this delay difference, ARIB STD-B31 defines the total delay due to bit interleaving/ de-interleaving to be equal to 2 OFDM symbols. To adjust the total delay time, the "delay adjustment value", which is shown in Table III-9, is added at the transmitter side, for any carrier modulation system.

> For better understanding, an example is shown below:
>   Mode 1, carrier modulation: 64QAM,
>   Number of data carrier in one segment for mode 1 = 96   (see Table III-1)
>   Bit / symbol = 6   (64QAM, see section III.6 later)
>   2 OFDM symbols = 2 × 96 × 6 × N = 1152N
>   120 carrier symbol = 120 × 6 = 720
> Therefore: delay adjustment value=1152N-720 clocks

Please try to calculate for parameters for your exercise. The answers are shown in Table III-9.

Table III-9 Delay-Adjustment values required as a result of bit interleaving

Carrier modulation	Delay-adjustment value (number of bits)		
	Mode 1	Mode 2	Mode 3
DQPSK QPSK	384 × N-240	768 × N-240	1536 × N-240
16QAM	768 × N-480	1536 × N-480	3072 × N-480
64QAM	1152 × N-720	2304 × N-720	4608 × N-720

---

[37] Defined in ARIB STD-B31, Section 3.9.2

N represents the number of segments used by that hierarchical layer.

III.2.6   Combining hierarchical layers

In this functional block defined in ARIB STD-B31 Section 3.10, the data (carrier modulation form) from the mapping circuit output of each layer are combined and then fed to a time interleaver as a first step in the channel coding process.

III.3   Channel Coding

Channel coding block is composed of many sub-functions which are indicated as boxes with bold line in Figure III-14. This section will focus mainly on the explanation of the error correction system with interleaving and carrier modulation[38] [39] [40]

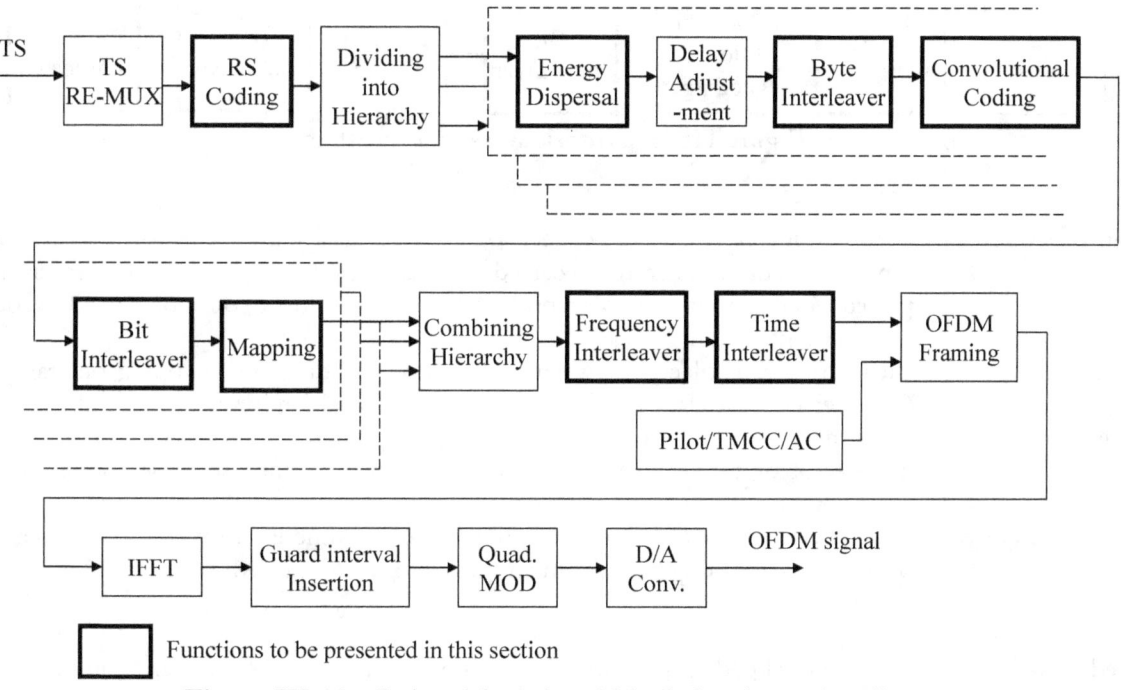

Figure III-14   Related functional block for channel coding

III.3.1   Overview of channel coding

As shown in Figure III-14, the channel coding system of ISDB-T is composed of a concatenated error correction encoding, multiple interleavers and carrier modulation (mapping).

Error correction coding system are composed of 2 encoding system, Reed-Solomon (RS) coding for the

---

[38] ARIB STD-B31, Transmission System for Digital Terrestrial Television Broadcasting, sections 3-3, 3-5, 3-7, 3-8, 3-9, 3-10, 3-11
[39] ETSI EN 300 744 V.1.5.1(2004-11), Digital Video Broadcasting (DVB); Framing structure, channel coding and modulation scheme for digital terrestrial television
[40] Reimers, U. Electronics & Communication Engineering Journal, Vol.9, issue 1,Feb.1997, DVB-T: the COFDM-based system for terrestrial television

outer coder and Convolutional coding for the inner coder. RS Coding is a popular technology especially for packed media such as CD's. This coding system is common for any of the available hierarchy, therefore, it is located before the "hierarchy division".

On the other hand, convolutional coding was developed for digital communication specifically for satellite links. The convolutional coder is placed in each hierarchy in order for the independent selection of the appropriate error correction parameters.

The purpose of interleaving is to randomize the data in order to get the best error correction result. In general, both error correction systems, RS coding and convolutional coding, produce the best error correction performance especially for random errors. Therefore, these interleavers are inserted at the front portion of each error correction decoder of the receiver. Figure III-15 shows the position of each interleaver and gives a brief explanation how it works.

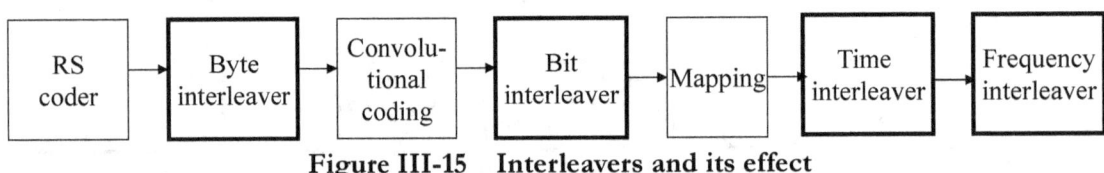

Figure III-15   Interleavers and its effect

Byte interleaver: Byte interleaver is placed between outer coder and inner coder on transmitters. It randomizes the burst error received over the transmission path. On receivers, it is placed to take input from Viterbi decoder output, a most commonly used decoder for convolutional coding.

Bit interleaver: Bit interleaver is placed between convolutional coding and mapping on transmitters. On receivers, it is placed before Viterbi decoder and randomizes the symbol error.

Time interleaver: Time interleaver is placed at the output of mapping (modulation). It randomizes the burst error of time domain that is mainly caused by impulse noise, fading of mobile reception.

Frequency interleaver: Frequency interleaver is located at the output of time interleaver. It randomizes the burst error of frequency domain that is mainly caused by multi-path or carrier interference.

Reed-Solomon and convolutional coding have the ability to detect and correct a certain number of errors in their input data stream. These error correction systems shows its best error correction performances when the errors on the input is in randomly placed. On receivers, even if there are block transmission errors caused by noise and interference, and if the de-interleaver spread these block errors across the input, the coding system have a chance to correct these randomized errors. As Reed-Solomon coding is byte oriented and convolutional coding is bit oriented, byte and bit interleavers are effective respectively.

In addition, OFDM has two directions in its framing – frequency and time direction. Frequency interleaver spreads the symbols in the frequency direction so that the coders can correct the errors caused by the frequency distortion. Time interleaver spreads symbols in the time direction. It is effective on errors caused by interference, such as dynamic fading, whose characteristics changes over time. Details of these interleavers will be discussed in the following sections.

III.3.2   Outer coder

Reed-Solomon coding is one of well-known and widely used error correction system. This coding system provides more robustness against random and burst error. Because of this, RS coding system is widely used and is common for communication system and packed media.

In the outer coder, as shown in Figure III-16, a 16 byte parity code is added after the data (1 sync byte + 187 data bytes), and a TSP of 204 byte is constructed. For an error correction code, a shortened Reed-Solomon (204, 188) code is adopted. This shortened Reed-Solomon code can be obtained by performing Reed-Solomon (255, 239) encoding by adding 51 bytes of 00 HEX before input data and removing the first 51 bytes after encoding.

As an element of this Reed-Solomon code, the element of GF (2^8) is used[41], and the following formula p (x) is defined as a primitive polynomial[42] of GF (2^8):

$$p(x) = x^8 + x^4 + x^3 + x^2 + 1$$

Note also that the following polynomial g (x) is used to generate (204,188) the shortened Reed-Solomon code:

$$g(x) = (x - \lambda^0)(x - \lambda^1)(x - \lambda^2) \cdots (x - \lambda^{15}) \text{ provided that } \lambda = 02 \text{ HEX}$$

Sync. 1 byte	Data (187 bytes)

(a) MPEG2 TSP

Sync. 1 byte	Data (187 bytes)	Parity 16 byte

(b) TSP with RS code (transmission TSP)

Figure III-16   MPEG2 TSP and Transmission TSP

III.3.3   Energy dispersal

If a same data sequence occurs frequently, some spectrum component may have high energy concentration. The energy concentration of some portions of the spectrum might create interference and possibly inter modulation. To avoid the concentration of energy in a certain part of the spectrum, input data should be dispersed.

Energy dispersal[43] of ISDB-T is conducted at each hierarchical layer by applying exclusive-OR with a PRBS (Pseudo Random Bit Sequence) stream and a data stream excluding the synchronization byte. The PRBS is generated by the circuit shown in Figure III-17.

In the PRBS generation circuit, firstly, an initial value of "100101010000000" is set on the shift register. The first output is xor result of bit 15 and bit 14 of the shift register. Next, the data is shifted using the data shifting clock. Output changes sequentially along with the change of the register value at bit 14 and 15. The shift register resets in every OFDM frames. Please note the shift register for generating the PRBS signal also operates during the synchronization byte.

The PRBS generating formula is written below.

---

[41] GF: Galois Fields, A field with a finite number of elements.   First invented by E. Galois.
[42] A primitive polynomial is a polynomial that generates all elements of an extension field from a base field
[43] Defined in ARIB STD-B31 Section 3.5

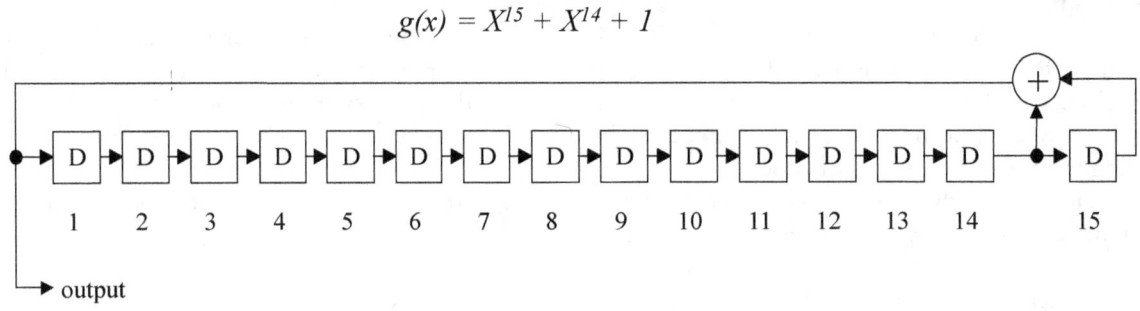

Figure III-17 PRBS-Generating polynomial and circuit[44]

III.3.4 Byte interleaving

As illustrated in Figure III-18, the byte interleaver defined in ARIB STD-B31 Section 3.7 is inserted between outer coder (RS coder) and inner coder (Convolution coder). As explained in section III-2, the outer coder described above improves the performance of error correction against random and burst error. For this purpose, a byte interleaver is inserted between the RS coder and inner coder to randomize the burst error that is caused over the transmission path in order to get the best error correction performance out of an RS decoder.

The input of byte interleaving is TSP with RS error correction code (204 bytes) and after energy-dispersal process. The depth of interleaving is fixed to be 12 bytes in length. Figure III-18 shows the byte interleaving circuit.

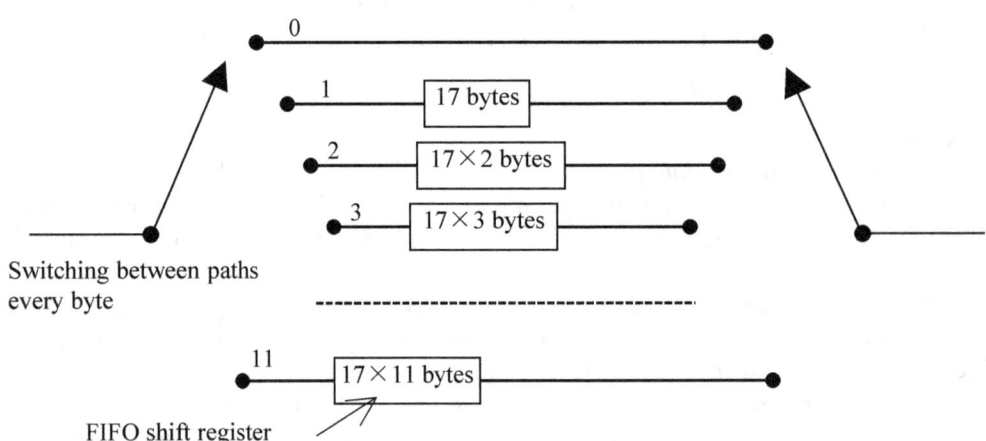

Figure III-18 Byte Interleaving circuit[45]

---

[44] Quoted from Ordinance No.87 of the Ministry of Internal Affairs and Communications (2011), Annexed Table 15, Annexed Statement 1
[45] Quoted from Ordinance No.87 of the Ministry of Internal Affairs and Communications (2011), Annexed Table 15, Annexed Statement 2, Item 1

The byte interleaving circuit starts with the next byte to sync byte of the transmission TSP which is set to go through path 0 with no delay. Then the switch changes in every byte and channels the next byte to the next path that has buffer size of 17 bytes. The switches change sequentially byte by byte and come back to path 0 after path 11. By going through this process, the byte interleaver rearranges the byte stream so that once contiguous data is now spaced further apart into a non continuous stream as illustrated in Figure III-19.

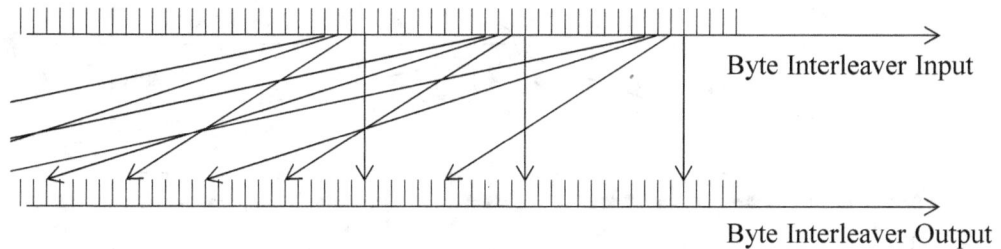

Figure III-19   Rearrangement of stream by Byte Interleaver

In the de-interleaving circuit found in the receiver, the position of the paths are symmetrical i.e. the position of path 0 is the lowest and path 11 is the highest. The total delay of each path are equal: $17 \times 12 \times 11$ byte (11 TSPs).

III.3.5   Inner coder

As for the inner error correction coding, a convolutional coding system with a punctured pattern is adopted[46] for the following reasons:
  (i) Effective error correction performance against random error,
  (ii) Flexibility of adopting a "soft decision Viterbi decoding", which uses an analog demodulating value as the index of data reliability. A signal decoding process in which a demodulated value is treated as analog value is called "soft decision decoding", on the other hand, a signal decoding process in which a demodulated signal is treated as 2 digit value, such as "0" or "1" is called "hard decision decoding". A "soft decision decoding" improves error correction performance compared to a "hard decision decoding",
  (iii) By making use of a "punctured pattern" explained in Table III-10, code rate can be made variable and selectable.

Convolutional coding is a method that takes input and stores them into a k bit shift register (whose width is called constraint length) and outputs the sequence generated by a polynomial (called mother code) based on the stored values in the shift register. Figure III-20 shows the mother code generation circuit of ISDB-T with a constraint length k of 7, and code rate of 1/2.

The polynomial generator of the mother code must be $G_1 = 171_{OCT}$ for X output and $G_2 = 133_{OCT}$ for Y output. 171 Oct means "1111001" in binary expression, that means the XOR result with input of 1st, 2nd, 3rd, 4th and 7th bit of shift register value goes to X output.

In ISDB-T system, in order to improve the transmission efficiency, some bits of the X and Y outputs generated by the convolutional coding circuit shown in Figure III-20 are "punctured", meaning not all but some bits shown in Table III-10 are transmitted. This transmission pattern is called a puncture pattern.

---

[46] See ARIB STD-B31 Section 3.8

For example, convolutional coding generates 2 bit output per 1 bit input. When you choose code rate of 2/3, 2 bit input generates 4 bits: X1, X2, Y1 and Y2. However the punctured circuit outputs only X1, Y1, Y2 in order for reducing bitrates based on the patters defined in the standards.

Note that the puncturing pattern is reset so as to start with frame synchronization. This is a measure to facilitate synchronization acquisition of the puncture pattern in the receiving side.

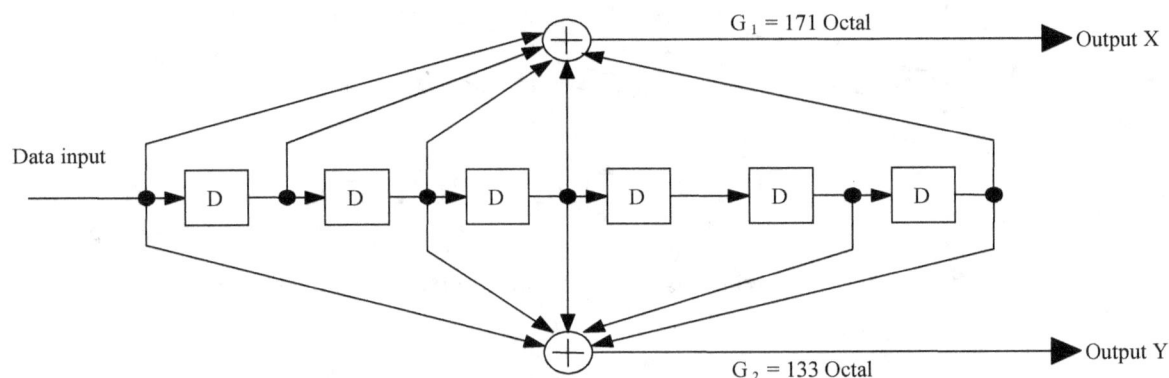

Figure III-20 Convolutional Coding circuit with Constraint Length k of 7 and a Code Rate of 1/2[47]

Table III-10 Inner-coder code rates and transmission-signal sequence[48]

Coding rate	Puncturing pattern	Transmission-signal sequence
1/2	X : 1 Y : 1	X1, Y1
2/3	X : 1 0 Y : 1 1	X1, Y1, Y2
3/4	X : 1 0 1 Y : 1 1 0	X1, Y1, Y2, X3
5/6	X : 1 0 1 0 1 Y : 1 1 0 1 0	X1, Y1, Y2, X3 Y4, X5
7/8	X : 1 0 0 0 1 0 1 Y : 1 1 1 1 0 1 0	X1, Y1, Y2, Y3, Y4, X5, Y6, X7

Figure III-21 shows an example of Carrier to Noise ratio (C/N) vs Bit Error Rate (BER) relations for various code rates. Here, C/N represents a ratio of signal power (represented as carrier power) and noise power. Bit Error Ratio represents how many error bits are included against the total volume of the output data stream. As shown in the figure, lower code rates exhibits better performance (requires less C/N – tolerates lower quality signal) but yields lower data transmission capacity. Meanwhile, higher code rates tend to require higher C/N in order to be able to deliver a higher data transmission capacity. With the above explanation, the code rate should be chosen carefully with consideration to the trade off of the required C/N and data rate.

---

[47] Quoted from Ordinance No.87 of the Ministry of Internal Affairs and Communications (2011), Annexed Table 12, Item 3
[48] Quoted from Ordinance No.87 of the Ministry of Internal Affairs and Communications (2011), Annexed Table 12, Item 3 and 4

Please note that C/N vs. BER data provided above was a result of the test conducted during the initial development stages of ISDB-T. Doing the same measurement now will likely provide an improved result due to the advancement and better performance of the equipment available today. Kindly treat this figure as an illustrative example on the difference of C/N vs. BER based on the punctured coding rate.

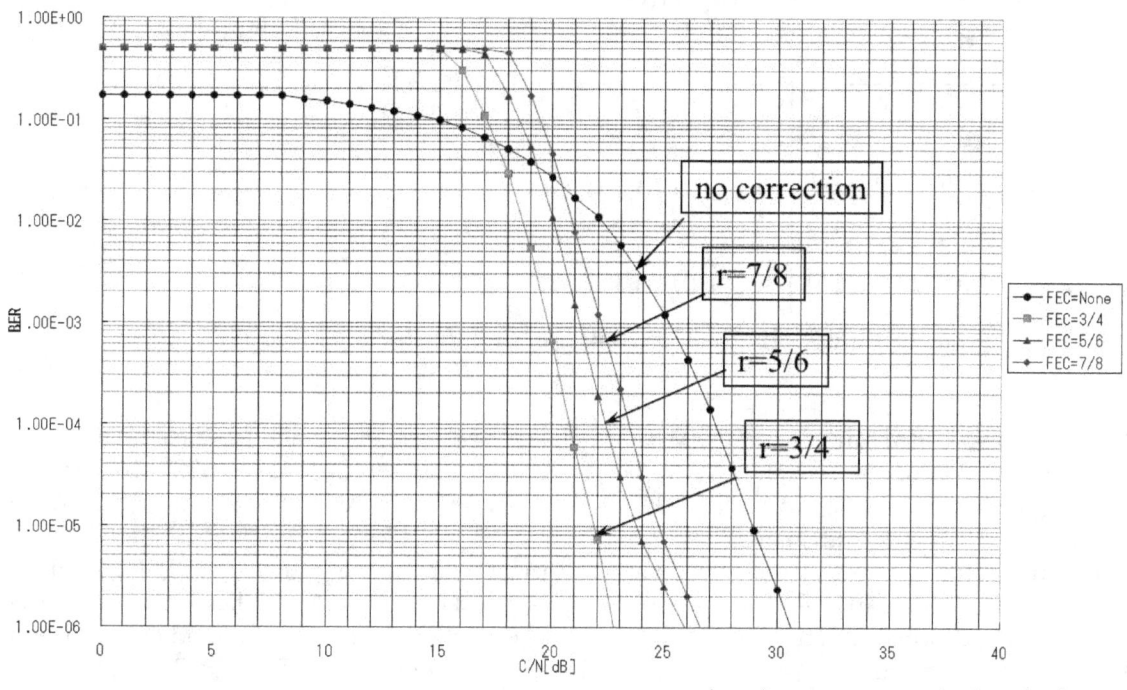

(note) measurement is done before RS decoding

Figure III-21   An example of data BER for convolutional coding/ viterbi decoding

III.3.6   Mapping and bit interleaving

After convolutional coding, bit unit data is arranged to symbol data. Symbol data corresponds to carrier modulation of each OFDM carrier. This signal process is called "mapping" or carrier modulation.

The purpose of mapping is to assign the position of the symbol data on the I-Q coordinate (complex coordinate). If symbols are composed of 4 bits, 16 positions should be assigned on an I-Q coordinate. This mapping (carrier modulation) is called "16QAM", 16 means number of position on I-Q coordinate, and QAM means Quadrature Amplitude Modulation.

Figure III-22 shows an example of the position that corresponds to a 4 bits data on an I-Q coordinate of 16QAM modulation.

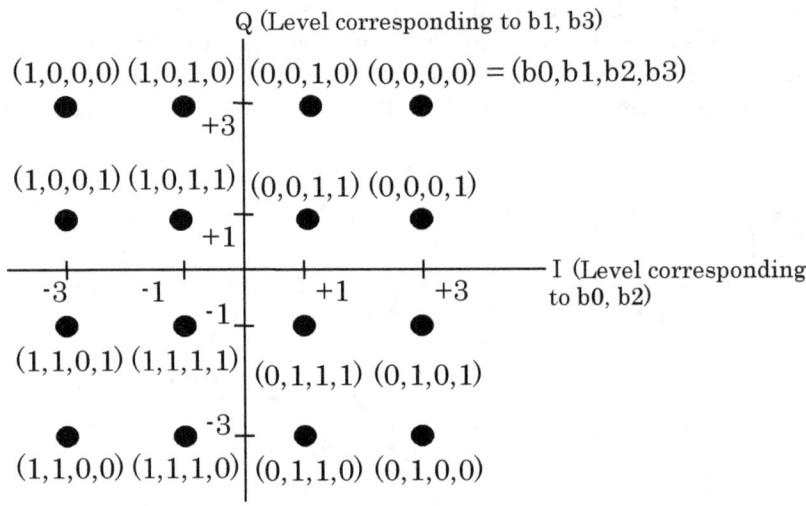

Figure III-22　Constellation of 16QAM[49]

This figure is called a "constellation of 16QAM." As shown in figure, the symbol(s) (data set of 4 bits) are mapped on the position of the I-Q graph according to its 4 bit data set (b0, b1, b2, b3). 16 states are defined on the I-Q graph. The adjacent position differs in one bit in binary representation from each other. This mapping in named "Gray code mapping"

In ISDB-T transmission systems, 4 kinds of mapping (carrier modulation), such as QPSK, DQPSK, 16QAM and 64QAM are prepared for the various services of digital broadcast, these types are fixed reception and/or mobile/portable reception service. As an exception to the carrier modulation proved above, DQPSK is not adopted in the Japanese Operational Guidelines (defined as ARIB TR-B14)[50], therefore, there's no actual service of DQPSK in Japan.

For QPSK and 64QAM constellation and its equivalent circuit, please see ARIB STD-B31[51].

Before mapping, the bit unit data stream is interleaved by the bit interleaver shown in Figure III-23. The idea of bit interleaving is to disperse a consecutive bit sequence randomly so that a burst error occurred at a transmission path can be scattered to get a best performance of an error correction of Viterbi decoder in a receiver. Figure III-23 (a) shows the bit interleaving circuit for 16QAM, and (b) shows the image of a bit interleaver randomized data.

---

[49] Quoted from Ordinance No.87 of the Ministry of Internal Affairs and Communications, Annexed Table 10, Statement 1
[50] ARIB TR-B14, Version 2.8 (May 29,2006),　Operational Guidelines for Digital Terrestrial Television Broadcasting
[51] ARIB STD-B31, Transmission System for Digital Terrestrial Television Broadcasting, Section 3.9

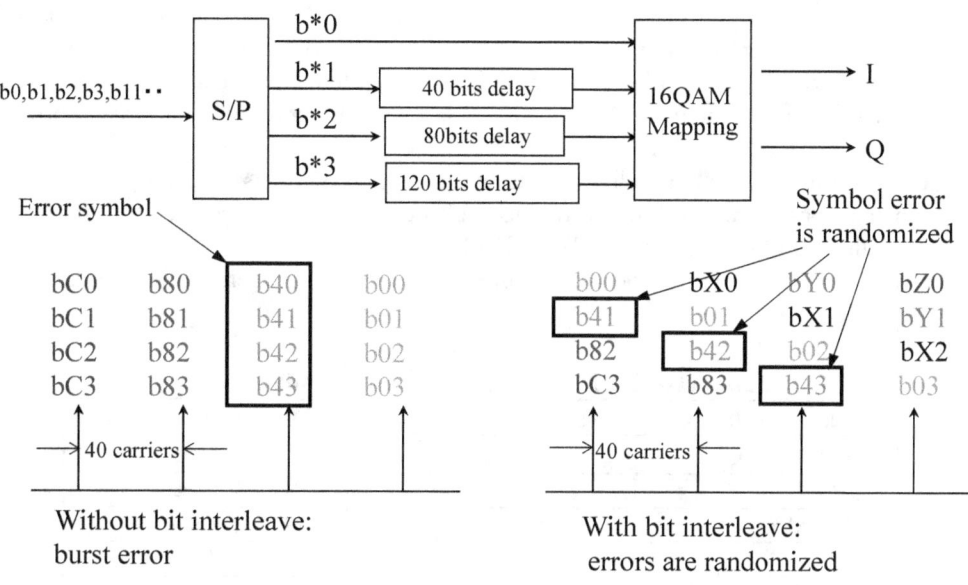

Figure III-23 Mapping and bit interleaving (16 QAM case)

Figure III-24 shows an example of C/N vs BER data for each carrier modulation (mapping).   As shown in the figure, QPSK shows the best reception performance, while 64QAM shows the worst in terms of the required C/N.

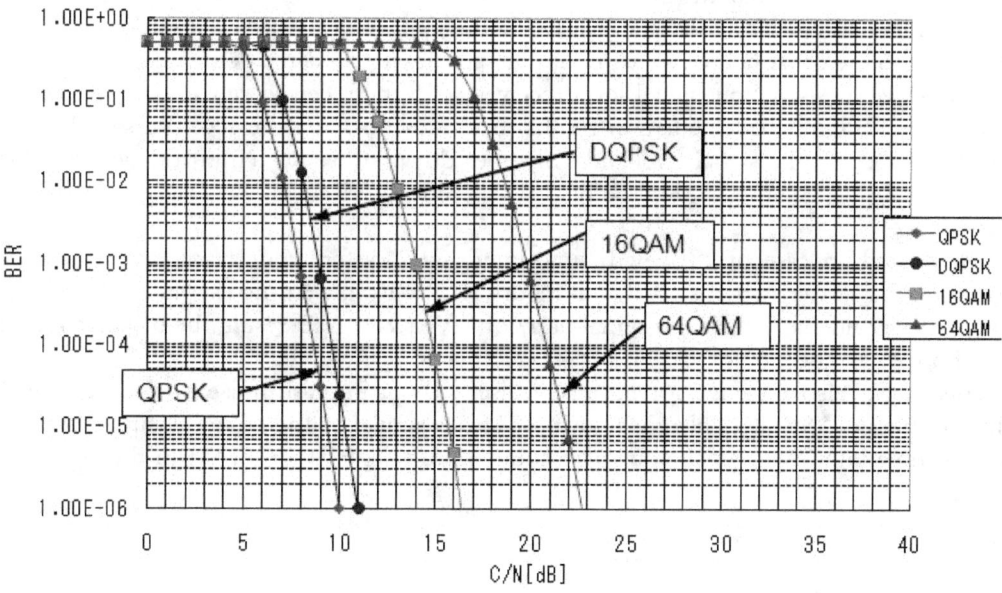

Figure III-24 C/N vs BER for each carrier modulation

From the transmission capacity perspective, 64 QAM can send 6 bits per symbol; and QPSK can send 2 bits per symbol.   In other words, the more transmission capacity requires higher C/N.   In case of mobile and portable reception, it is difficult to provide services with higher data rate services such as high definition video because of a lower reception sensitivity. On the other hand, in a fixed reception system that can increase reception sensitivity by using a larger antenna, it is possible to realize a high definition service which needs a large transmission capacity. As discussed above, transmission capacity and the reception environment are in trade off relationships.

When we place the points in the constellations shown in Figure III-22 as an example be expressed as Z (= I + jQ), the transmission-signal level must be normalized by multiplying each of these points by the corresponding normalization factor shown in Table III-11.

As a result, the average OFDM symbol power becomes 1 regardless of what modulation scheme is used.

Table III-11 Modulation level normalization[52]

Carrier modulation scheme	Normalization factor	After normalization
π/4-shift DQPSK	$1/\sqrt{2}$	$Z/\sqrt{2}$
QPSK	$1/\sqrt{2}$	$Z/\sqrt{2}$
16QAM	$1/\sqrt{10}$	$Z/\sqrt{10}$
64QAM	$1/\sqrt{42}$	$Z/\sqrt{42}$

III.3.7 Time interleaving

The signal after combining each layers is placed into the "time interleaving" circuit. As described in section II.4.4, "Time interleaving" is very effective to reduce the degradation effects from time-varying interference, such as, signal fading, impulse noise. Figure III-25 below illustrates the effect of time interleaving.

The left side (A) shows an example without time interleave, while the right side (B) shows one "with time interleave".

The figures on the top, (A-1) and (B-1), show the structure of an OFDM signal without and with time interleave. With time interleave, illustrated in (B-1) in the figure, certain delays are added carrier by carrier in OFDM signals. On the other hand, A-1 without time interleave, no signal delay are added.

The figure (A-2) and (B-2) show signal level received. If the received signal level fluctuates in the time domain, known as "fading", this fluctuation results to serious error on the received signal. An example is illustrated in Figure III-25, the signal level suddenly goes down at "t_5" which introduces errors in the carrier symbols present at "t_5". As a result, in case of "(A) no time interleave", burst error is introduced as shown in Figure III-25 (A-3).

On the other hand, in case of "(B) with time interleave", different delays added for each carriers will disperse errors on the carrier symbols, shown in Figure III-25 (B-3). In the examples above, the signal delay of "f_4" at the receiver side is 1, while the signal delay of "f_3" is 8. As a result, the distance between errors in the carrier symbol of frequency slot "f_4" and "f_3" is calculated as 7 (=8-1).

Time interleaving is also effective for impulse noise. If an impulse noise occurs at time slot "t_5", the signal quality at time slot "t_5" suddenly goes down. This is similar to the fading disturbance described above. In other words, time interleaving technique also counters the degradation effect in the signal quality due to impulse noise.

---

[52] Quoted from Ordinance No.87 of the Ministry of Internal Affairs and Communications (2011), Annexed Table 10 statement 1 note 4

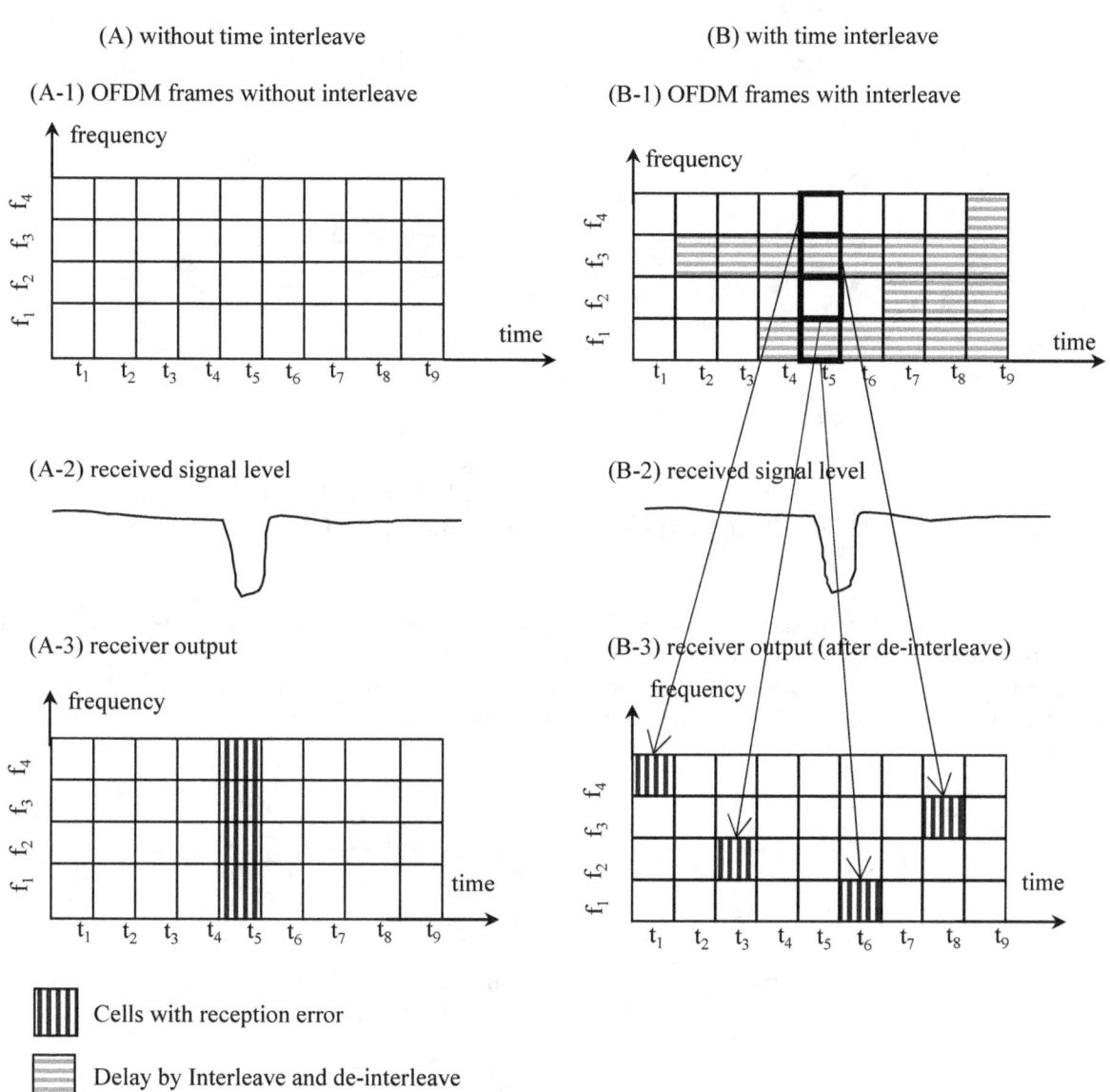

Figure III-25   Effect of time interleaving

Figure III-26 shows the conceptual diagram of "Time interleaving". The signal with each layer combined is fed to each intra-data segment time interleaving circuit and interleaved in units of the modulation symbols (for each pair of the I and Q symbols). In order for time interleaving depth set for each segment, "intra data-segment time interleaving section", a circuit to apply time interleaving in unit of each segment, needs to be prepared.

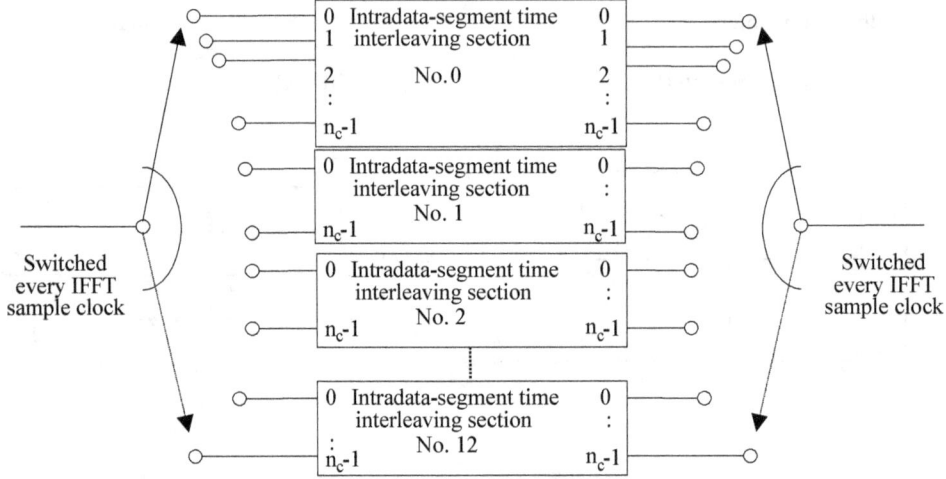

nc is 96, 192, and 384 in modes 1, 2, and 3, respectively.

Figure III-26   Configuration of "Time interleaving" functional block[53]

Figure III-27 shows the detailed structure of each intra data segment time interleave section shown in Figure III-26. This circuit is a set of buffers whose length can be set based on the parameter "I". "I" shown in the Figure III-27 is a parameter to define the time interleave length and can be set independently for each hierarchy. The values of "I" that can be selected for each mode are shown in Table III-12.

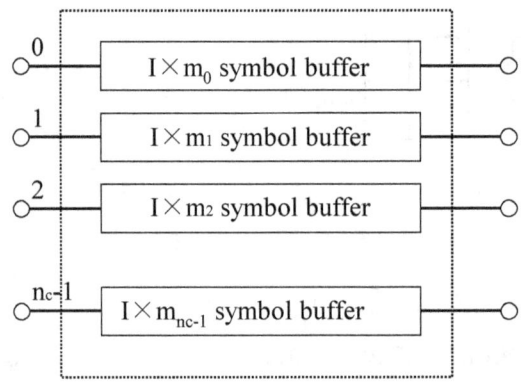

Provided that $m_i = (i \times 5) \bmod 96$
nc is 96, 192, and 384 in modes 1, 2, and 3, respectively.

Figure III-27   Configuration of the Intra-segment Time Interleaving Section[54]

Since "I" can be selected for each hierarchy as described above, a delay difference occurs between each hierarchy. This delay difference needs to be adjusted by adding delay that is equal to certain number of

---

[53] Quoted from Notification No. 303 of the Ministry of Internal Affairs and Communications (2011), Annexed Table 2, Annexed Statement 1
[54] Quoted from Notification No. 303 of the Ministry of Internal Affairs and Communications (2011), Annexed Table 2, Annexed Statement 2

symbols, as a result, a total delay both in transmission and reception is adjusted to be an integral multiple of the OFDM frame length. The relation between the Time Interleaving Length (I), the number of delay adjustment symbols and the number of delayed frames in total transmission and reception corresponding to each mode are shown in Table III-12.

Table III-12   Time interleaving lengths and delay adjustment values[55]

Mode 1			Mode 2			Mode 3		
Length (I)	Number of delay adjustment symbols	Number of delayed frames in transmission and reception	Length (I)	Number of delay adjustment symbols	Number of delayed frames in transmission and reception	Length (I)	Number of delay adjustment symbols	Number of delayed frames in transmission and reception
0	0	0	0	0	0	0	0	0
4	28	2	2	14	1	1	109	1
8	56	4	4	28	2	2	14	1
16	112	8	8	56	4	4	28	2

Note that delay adjustments on signals must be performed prior to time interleaving.

As described above, "Time interleaving" is intended to ensure further improvement on the robustness against channel fading interference by randomizing symbol data. The specifications of the interleaving length for each hierarchical layer allows for an optimal interleaving length for each reception modes.

III.3.8   Frequency interleaving

The frequency interleaving function was introduced in order to disperse burst errors present in the frequency domain due to multipath interference. Please refer to the section II.3.3 and III.4.2 for the principles of multipath interference.

A strong multipath interference generates serious signal level changes which results to burst errors in the frequency domain. To minimize this signal degradation effect, randomization, known as "Frequency interleaving" in the frequency domain is necessary.

The structure of "Frequency interleaving" is shown in Figure III-28. As described in the figure, the segment data 0-12 is divided into 3 parts: the partial-reception, differential modulation (segments for which DQPSK is specified for modulation), and synchronous modulation (segments for which QPSK, 16QAM, or 64QAM is specified for modulation). Please note that differential modulation is specified in ARIB STD-B31, but not adopted in ARIB TR-B14 (operational guidelines) so it is not used in the actual digital broadcast service.

---

[55] Quoted from Notification No. 303 of the Ministry of Internal Affairs and Communications (2011), Annexed Table 2, Annexed Statement 3

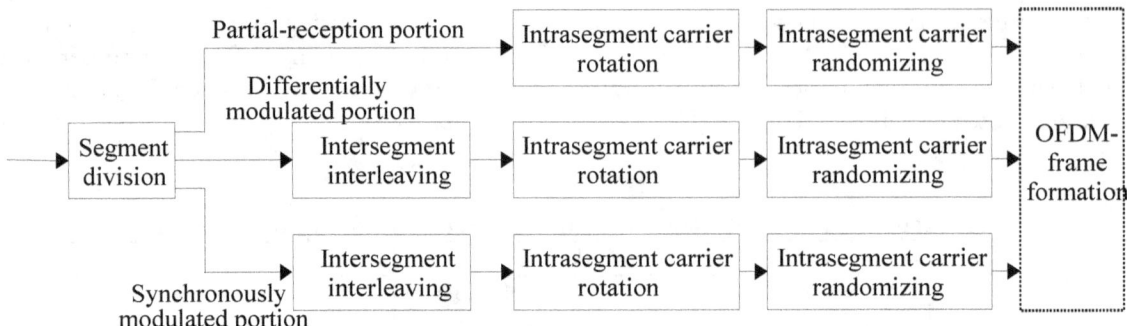

Figure III-28 Functional block of "Frequency interleaving"[56]

Why is it necessary to provide such configuration? For partial reception (One-Seg service), it is assumed that the partial reception receiver will be designed specifically to receive only the One-Segment service and not the other segments. Therefore, partial reception portion does not include other segment data and does not need Inter-segment interleaving process.

For differential modulation and synchronous modulation, different frame structures are assigned. See Figure III-30 for the segment structure of synchronous modulation portion, for differential modulation, see figure 3-29 in ARIB STD-B31. As shown in these figure, the segment structure is different, therefore, inter-segment interleaving is performed individually in each portion.

The intra-segment carrier rotation and intra-segment carrier randomizing are performed based on a table that defines how to mix the carriers in a pre-defined manner.. Please see ARIB STD-B31[57] for the details of intra-segment carrier rotation and randomizing.

III.4 OFDM Framing, Modulation

Figure III-29 illustrates the related functional block for OFDM framing & modulation[58].

---

[56] Quoted from Notification No. 303 of the Ministry of Internal Affairs and Communications (2011), Annexed Table 2, Annexed Statement 4
[57] ARIB STD-B31, Version 1.6 (Nov.30 2005), Transmission System for Digital Terrestrial Television Broadcasting, Section 3.11
[58] ARIB STD-B31, section 3-12, 3-13, 3-14, 3-15

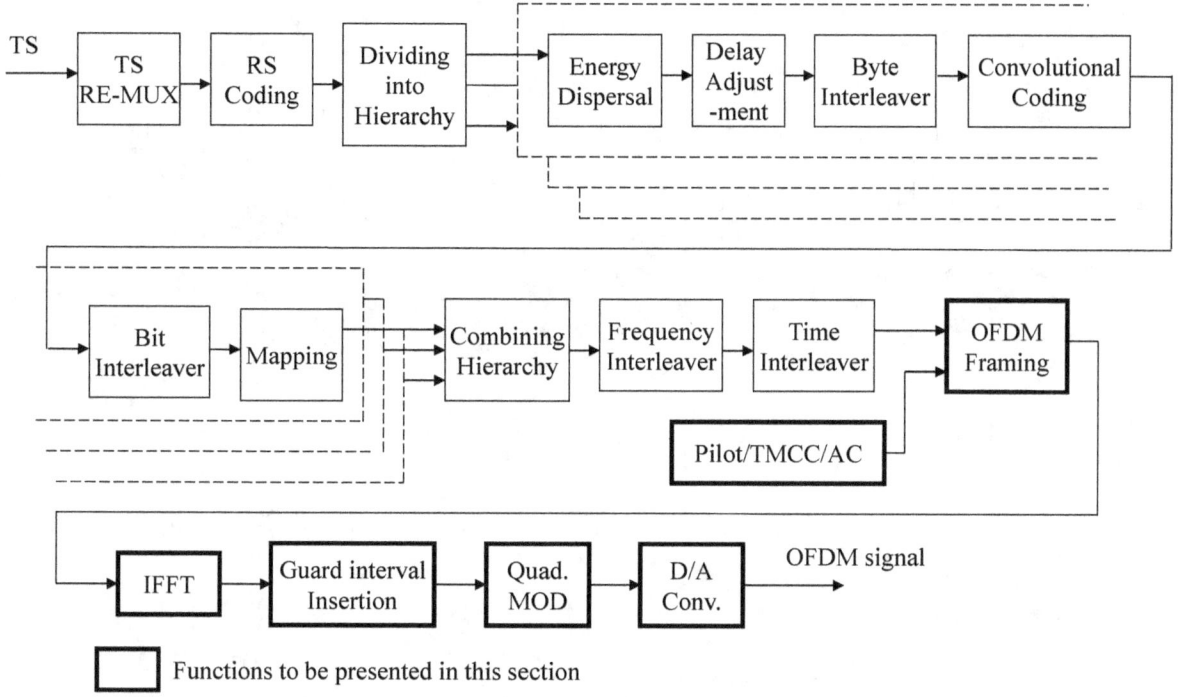

Figure III-29 Functional blocks of OFDM framing and modulation

III.4.1   OFDM framing

This section describes the OFDM framing structure which is composed of segment data, pilot signal and information for transmission.

As described in previous sections, two types of frame structure are specified, first is differential modulation and the other is synchronous modulation.

Since differential modulation is only defined in ARIB STD-B31, but not adopted in the ARIB TR-B14[59], that means this mode is not ready for the practical use.   This Chapter will skip the explanation of OFDM frame[60] for differential modulation and instead focus on the OFDM frame for synchronous modulation.

Figure III-30 shows an example of an OFDM segment frame structure (mode 1) for synchronous modulation.   As shown in the figure, the OFDM frame is composed of segment data (indicated as $S_{m,n}$), SP (Scattered Pilot signal), AC (Auxiliary Channel), CP (Continual Pilot) and TMCC (Transmission and Multiplexing Configuration Control signal).   The definition and details of SP, AC, CP and TMCC are described further in this Section.

CP is inserted into the left side frequency slot of each segment for differential modulation (DQPSK) segment.   While in case of synchronous modulation (QPSK, 16QAM and 64QAM), SP is inserted into the left side frequency slot instead of CP.   However, to make up for the entire transmission spectrum, a continuous carrier (CP) is provided at the right-hand end of the band.

SP locates in every 12 carriers, and 3 carriers are shifted for every symbols.   The positions of AC, TMCC in carriers are different depending on the mode[61].

---

[59] ARIB TR-B14, Version 2.8 (May 29,2006), Operational Guidelines for Digital Terrestrial Television Broadcasting
[60] ARIB STD-B31, Version 1.6 (Nov.30 2005), Transmission System for Digital Terrestrial Television Broadcasting, Section 3.12.1
[61] ARIB STD-B31, Version 1.6 (Nov.30 2005), Transmission System for Digital Terrestrial Television Broadcasting, table 3.15

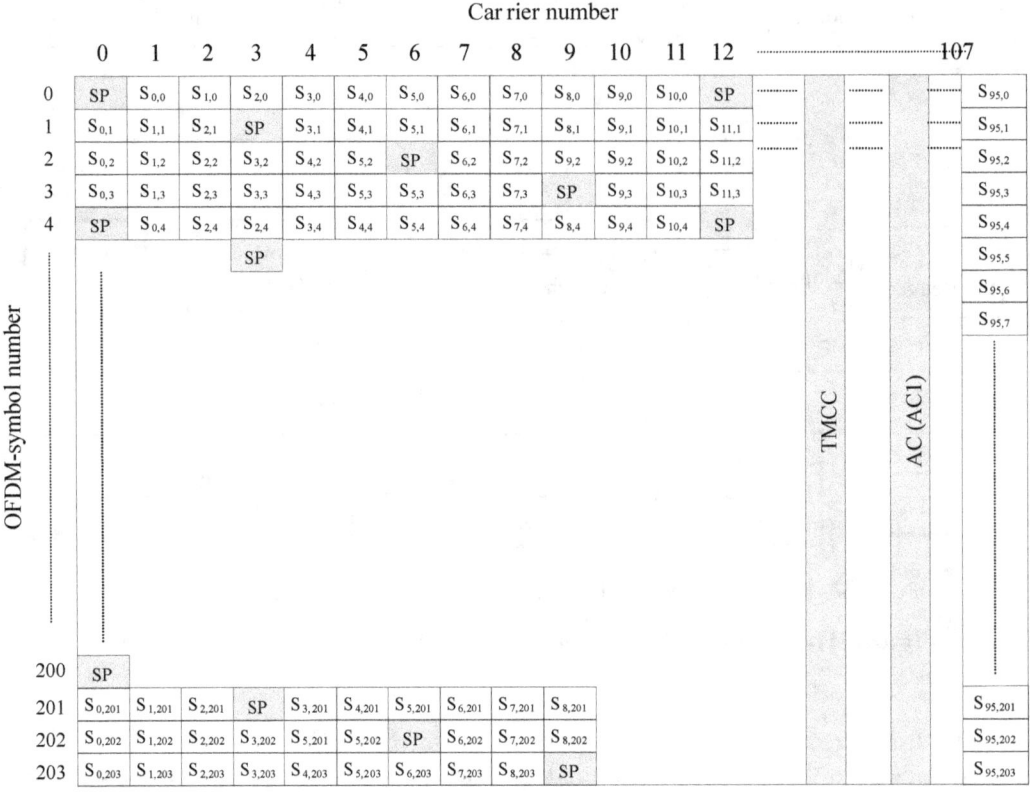

Figure III-30   OFDM segment structure of mode 1 for synchronous modulation[62]

This section discusses on each signals in OFDM framing.

III.4.2   Scattered Pilot (SP)
What is the purpose of SP?   This signal is used by the receiver for detecting multipath interference in the transmission path and for compensating the frequency distortion of the OFDM signal.

Figure III-31 shows an example of multi-path in a terrestrial transmission.   A direct path signal and a single multipath signal is shown in (b) of Figure III-31.

The phase difference of the direct signal and multipath signal is calculated as follows:

Phase difference(rad) = $(2\pi\lambda/\Delta L)$, $\lambda=1/f$

ΔL means the difference in physical distance between the multi-path (L_2) length and the direct path (L_1) length.   Accordingly, the phase difference between the direct path signal and the multi-path signal defined by above equation varies depends on the frequency.   Therefore, the received signal level which is a vector sum of direct path signal and multi-path signal varies according to its frequency (see Figure III-30 (b)).   Consequently, multi-path interference distorts the signal level received of each carriers of OFDM signal

---

[62] Quoted from Ordinance No.87 of the Ministry of Internal Affairs and Communications (2011), Annexed Table 7, Item 2.

because each OFDM frequencies are different. This is called the frequency distortion. Figure III-31 (c) shows the frequency distortion of an OFDM signal under multipath condition.

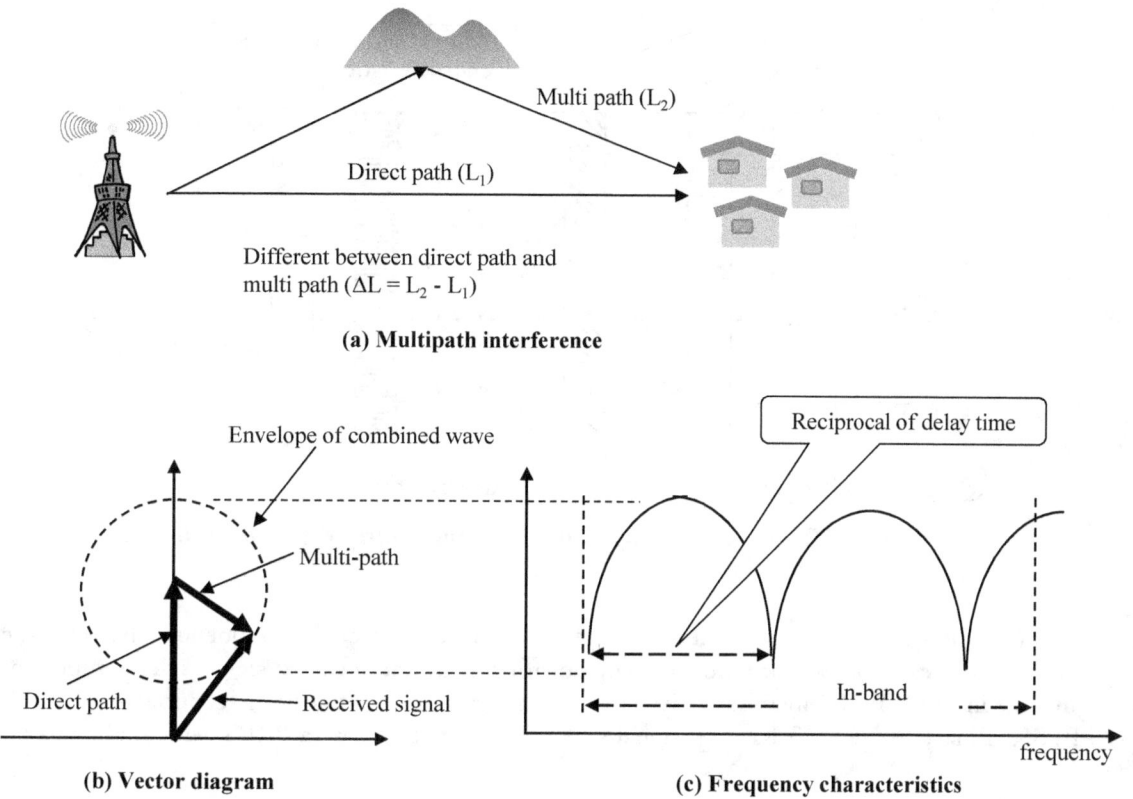

(a) Multipath interference

(b) Vector diagram

(c) Frequency characteristics

Figure III-31 Multipath and characteristics of the received signal

The scattered pilot (SP) signal is inserted as a reference signal whose signal level and phase are set to be known based on the standard so that the receiver is able to use it to compensate for the frequency distortion occurred by the multipath interference (see Figure III-32).

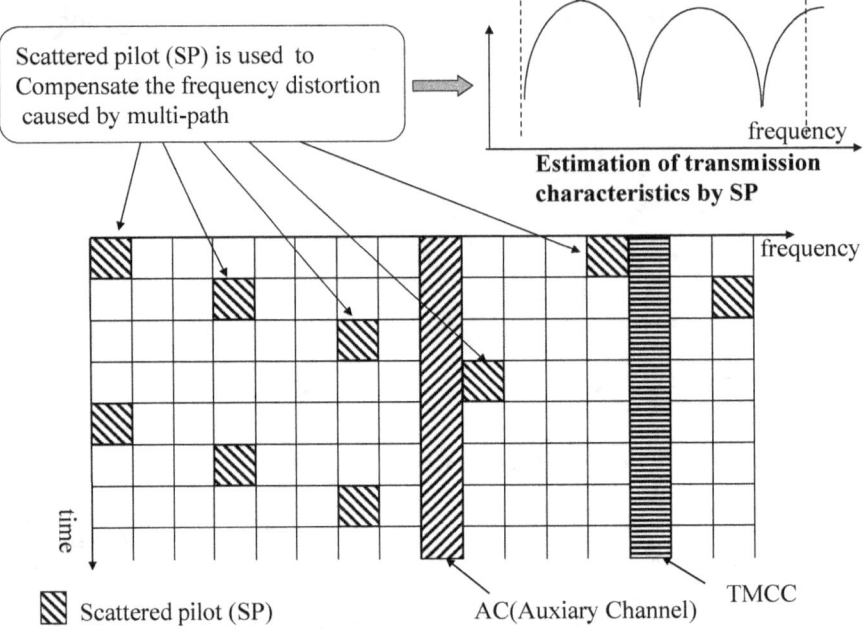

Figure III-32 Characteristics of SP under multipath condition

With this technology, by detecting the SP, receivers can calculate the frequency characteristics of transmission path and can compensate the frequency distortion of the received signal. This function plays an important role in an OFDM transmission system with the frequency interleaving technology.

SP is a BPSK (Binary Phase Shift Keying) signal which is modulated from a PRBS generated by its internal circuit[63].

III.4.3   Auxiliary Channel (AC)
AC is a channel designed to deliver additional information. The receiver does not use this information. Rather, it can be used as a container of transmission network information.

Two types of AC are specified in the ARIB STD-B31. One is named "AC1", which is common for differential modulation and synchronous modulation, and the other is named "AC2" which is exclusively used for differential modulation.

AC's additional information is transmitted by modulating the pilot carrier similar to CP through DBPSK. The reference for differential modulation is provided at the first frame symbol, and takes the signal point that corresponds to the Wi[64] value stipulated in Section 3.13.1 of ARIB STD-B31. Recently, usage of AC has been proposed such as transmitting network management information towards the relay station(s)[65].

III.4.4   Transmission and Multiplexing Configuration Control (TMCC)
TMCC is defined as the carrier frequency that delivers frame synchronization code and information for transmission/multiplexing parameters. These information are necessary to initialize a receiver's operation mode. Table III-13 shows the bit assignment of TMCC carrier.

---

[63] See ARIB-STD B31, Section 3.13 for details.
[64] Wi means the scattered pilot signal level and phase which are defined for each segment. For details, see ARIB STD-B31, setion 3.13.1
[65] ARIB STD-B31, Attachment Chapter 6.

Table III-13 Bit assignment of TMCC carrier[66]

B_0	Reference for differential demodulation
$B_1 - B_{16}$	Synchronizing signal (w0 = 0011010111101110, w1 = 1100101000010001)
$B_{17} - B_{19}$	Segment type identification (differential: 111; synchronous: 000)
$B_{20} - B_{121}$	TMCC information (102 bits)
$B_{122} - B_{203}$	Parity bit

(1) Synchronization signal
Synchronization signal is composed of 16 bits (B1 – B16) of TMCC carrier. Two types of synchronization word are specified, w0=0011010111101110 and w1=1100101000010001. They are transmitted alternatively for each frame.

(2) Segment type identification
Bits B17 – B19 are used to identify the segment type, "differential modulation" or "synchronization modulation".

(3) TMCC information
TMCC information[67] assists the receiver in demodulating and decoding various information including the system identification, the indicator of transmission-parameter switching, the start flag for emergency-alarm broadcast, the current and succeeding information.

It is here that the current information indicates the current hierarchical configuration and transmission parameters. The succeeding information contains the transmission parameters after a transmission parameter setting change.

The TMCC information are shown in Table III-14

Table III-14 TMCC Information[68]

Bit assignment	Description		Table No. in ARIB STD-B31
$B_{20} - B_{21}$	System identification		See Table 3-24.
$B_{22} - B_{25}$	Indicator of transmission-parameter switching		See Table 3-25.
B_{26}	Start flag for emergency-alarm broadcast		See Table 3-26.
B_{27}	Current information	Partial-reception flag	See Table 3-27.
$B_{28} - B_{40}$		Transmission-parameter information for hierarchical layer A	See Table 3-23.
$B_{41} - B_{53}$		Transmission-parameter information for hierarchical layer B	
$B_{54} - B_{66}$		Transmission-parameter information for hierarchical layer C	

---

[66] Quoted from Ordinance No.87 of the Ministry of Internal Affairs and Communications (2011), Annexed Table 11
[67] ARIB STD-B31, Section 3.15.6
[68] For details of each signals, see ARIB STD-B31 section 3.15.6

B_{67}		Partial-reception flag	See Table 3-27.
$B_{68} - B_{80}$	Next information	Transmission-parameter information for hierarchical layer A	See Table 3-23.
$B_{81} - B_{93}$		Transmission-parameter information for hierarchical layer B	
$B_{94} - B_{106}$		Transmission-parameter information for hierarchical layer C	
$B_{107} - B_{109}$	Phase-shift-correction value for connected segment transmission (Used for ISDB-T sound broadcast)		1 for all bits
$B_{110} - B_{121}$	Reserved		1 for all bits

III.4.5 OFDM modulation and spectrum

An example of the arrangement of OFDM segments is shown in Figure III-33. As shown in the figure, the segment number 0 is assigned for center segment, then segment numbers are assigned in the order of down, up, down ... on the frequency axis.

The figure is an example of the ISDB-T signal which is composed of three layers of partial reception, differential modulation, and synchronous modulation. In this case, the allocation of segment numbers are assigned to the partial reception layer (segment 0 in the figure), the differential modulation layer (segment 1 - 3 in the figure), and the synchronous modulation layer (segment 4 - 12 in the figure) in ascending order of segment numbers. Note that continuous carrier of which phase is defined in Table III-15 below is put at the right end on the frequency axis. This carrier is used as a pilot signal for carrier demodulation of the segment 12.

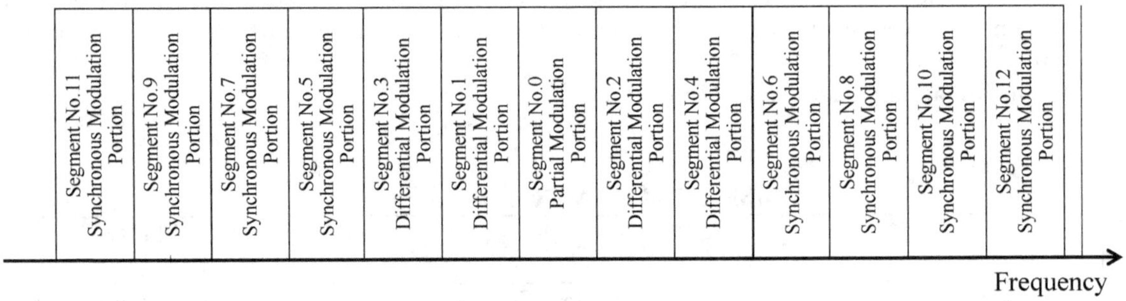

Figure III-33   OFDM-segment numbers on the transmission spectrum and an example of usage (same as Figure III-11)

Table III-15   Modulating signal for the rightmost continuous carrier

Mode	Modulating-signal amplitude (I, Q)
Mode 1	(-4/3, 0)
Mode 2	(+4/3, 0)
Mode 3	(+4/3, 0)

III.4.6   Guard interval

This section explains the signal processing and circuit for the guard interval.   Please refer to the purpose and effect of the guard interval explained in the section II.3.3.

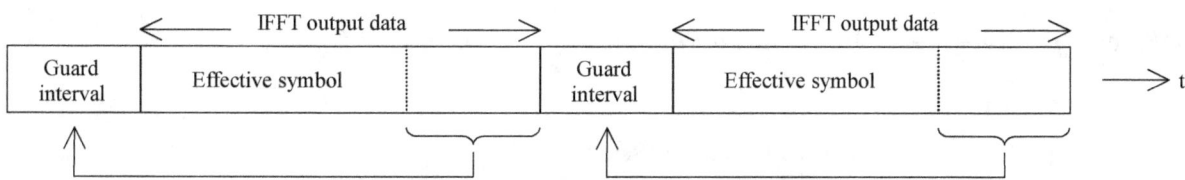

Figure III-34   Structure of an OFDM signal (time domain)[69]

As shown in above figure, the latter part symbol of the IFFT (Inverse Fast Fourier Transform) output is copied and inserted to the beginning of the effective symbol.   The guard interval length is specified as any of 1/4, 1/8, 1/16, 1/32 of the effective symbol length.

In Figure III-35, an example of "guard interval generation and insertion circuit" is shown.   A switch selects any of the non delayed signal and one effective symbol delayed signal during guard interval period shown in Figure III-34.   The switch selects the non delayed signal and during the effective symbol period, the switch selects the other delayed signal as an as the effective symbol.

As a result, an OFDM signal illustrated in Figure III-34 is generated.

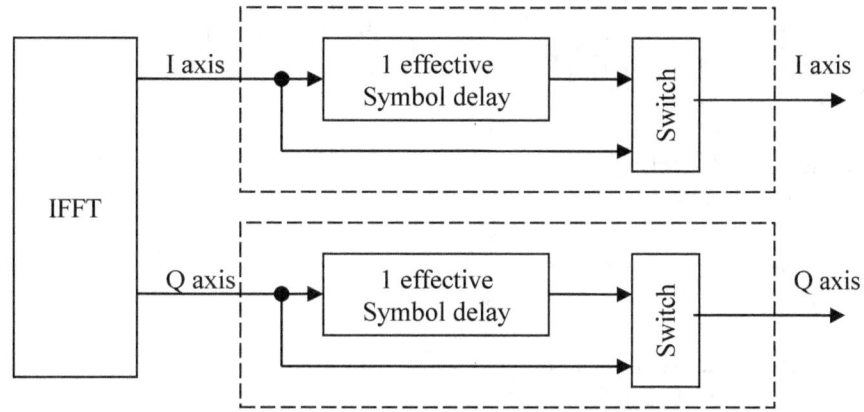

Figure III-35   An example of the Guard Interval generation and insertion circuit

III.5   Calculation of Data Rates

This Section will explain the calculation process of bit rates with various parameters[70] [71].

Step 1: Data rate calculation of a segment

---

[69] Quoted from Ordinance No.87 of the Ministry of Internal Affairs and Communications (2011), Annexed Table 5
[70] ARIB STD-B31
[71] ITU-R BT.1306-4 Error-correction, data framing, modulation and emission methods for digital terrestrial television broadcasting

An ISDB-T signal in one frequency band is composed of 13 segments, the first step to calculate total transmission data rate is to calculate the data rate of one (1) segment, followed by multiplying the number of segments of each layer to obtain total data rate of each layer.

The data rate of one (1) segment can be obtained with the following equation:

$$Data\ Rate = f_d \times N_f \times \frac{T_s}{T_s + T_G} \times M \times r \times RS\ coding\ Rate \times Effective\ Data\ Carrier\ Rate$$

Here, each parameter has following definition:

f_d: carrier spacing = effective symbol transmission speed

f_d (mode 1) = (6/14)/108*10³ kHz = 3.9682540 kHz   (for 6MHz bandwidth for 1 channel)

f_d (mode 1) = (7/14)/108*10³ kHz = 4.6296292 kHz   (for 7MHz bandwidth for 1 channel)

f_d (mode 1) = (8/14)/108*10³ kHz = 5.2910048 kHz   (for 8MHz bandwidth for 1 channel)

f_d (mode 2) = 1/2 fd (mode 1)

f_d (mode 3) = 1/4 fd (mode 1)

Here, (6/14)*10³ kHz, (7/14)*10³ kHz and (8/14)*10³ kHz is the bandwidth of one (1) segment.

N_f: Number of carriers in One-Segment
Nf$_{mode1}$ =108, Nf$_{mode2}$=216, Nf$_{mode3}$=432

$Ts/(Ts+Tg)$: ratio of total symbol length and effective symbol length.   This parameter is linked with the Guard Interval ratio.   If GI=1/16 then $Ts/(Ts+Tg) = 16 / (16+1) = 16/17$
any of 32/33, 16/17, 8/9, 4/5

M: modulation index (bit/ symbol)
QPSK=2,   16QAM=4,   64QAM=6

r: convolutional coding rate (depends on coding rate)
any of 1/2, 2/3, 3/4, 5/6, 7/8

$RS\ coding\ rate$: Code Rate for Reed Solomon Coding
188/204   (fixed value)

$Effective\ Data\ Carrier\ Rate$: (effective data carriers)/(total carriers)
= 96/108
(This is a fixed value common for mode 1, 2, 3)

The total carrier includes the effective data carriers, pilot carrier, TMCC, and scattered pilot symbols

Step 2: Total data rate calculation of each layer
The total data rate of each layer can be obtained from the data rate of a segment multiplied by number of segments (Nseg) that belongs to each layer.   The equation for this calculation is shown below :

$Total\ data\ rate\ of\ each\ hierarchy = (number\ of\ segment\ in\ same\ hierarchy)$
$\times (data\ rate\ of\ one\ segment)$

Some examples for this calculation are shown below.

III.5.1 Calculation examples of 6MHz system

[calculation example 1] of 6MHz system

The following case shows data rate calculation of one (1) segment of:
Mode 3, guard interval ratio=1/16, modulation =QPSK, coding rate(r) =2/3, 6MHz channel bandwidth

Data rate (1 segment)

$$= f_d \times N_f \times \frac{T_s}{T_s + T_G} \times M \times r \times RS\,coding\,Rate \times Effective\,Data\,Carrier\,Rate$$
$$= 0.9920635 \times 432 \times (16/(16+1)) \times 2 \times (2/3) \times (188/204) \times (96/108)$$
$$= 440.56\ kbps$$

[Calculation example 2] of 6MHz system

This example shows the data rate calculation of a channel (13 segments) with:
Single layer transmission, mode 3, single layer, G/I ratio =1/16, modulation index: M=64 QAM, coding rate: 3/4, 8MHz bandwidth

Data Rate (1 segment)

$$= f_d \times N_f \times \frac{T_s}{T_s + T_G} \times M \times r \times RS\,coding\,Rate \times Effective\,Data\,Carrier\,Rate$$
$$= 0.9920635 \times 432 \times (16/(16+1)) \times 6 \times (3/4) \times (188/204) \times (96/108)$$
$$= 1.4869\ Mbps$$

Data Rate (13 segment)
$$= Data\,Rate\,(1\,segment) \times 13$$
$$= 1.4869 \times 13$$
$$= 19.330\ Mbps$$

[Calculation example 3] of 6MHz system

This example shows a calculation of layered (Hierarchical) transmission data rate. The parameters shown are used for the actual DTTB service in Japan. In this case, the data rates for one (1) segment of each transmission mode (One-Seg or fixed reception) are calculated first, followed by summing up these data rates of segments.

This example is based on the following conditions:
(a) service configuration: One-Seg(portable reception) + HD/SD service(fixed reception)
(b) common parameters: Mode=3, G/I ratio=1/8, 6MHz bandwidth
(c) Layer A(portable reception) parameters: No of segments=1,   M=2(QPSK),   r=2/3
(d) Layer B(fixed reception) parameters   : No of segments=12, M=6(64QAM), r=3/4

Data Rate (1 segment of Layer A)

$$= f_d \times N_f \times \frac{T_s}{T_s + T_G} \times M \times r \times RS\,coding\,Rate \times Effective\,Data\,Carrier\,Rate$$
$$= 0.9920635 \times 432 \times (8/(8+1)) \times 2 \times (2/3) \times (188/204) \times (96/108)$$
$$= 416.09\ kbps$$

Data Rate (1 segment of Layer B)
 = *0.9920635×432×(8/(8+1)) ×6×(3/4) ×(188/204) ×(96/108)*
 = *1.40430 Mbps*

Data Rate (total)
 = *Data Rate (1 segment of Layer A)×1 + Data Rate (1 segment of Layer B)×12*
 = *17.2676 Mbps*

III.5.2 Calculation examples of 8MHz system

As described in III.5, this Section will explain the calculation process of bit rates with various parameters for 8MHz system.

As Step 1 and Step 2 are same as 6MHz system shown above, this section will skip these steps and show a couple of calculation examples of 8MHz system below.

[Calculation example 1] of 8MHz system
The following case shows data rate calculation of one (1) segment of:
Mode 3, guard interval ratio=1/16, modulation =QPSK, coding rate(r) =2/3, 8MHz channel bandwidth

Data rate (1 segment)

$$= f_d \times N_f \times \frac{T_s}{T_s + T_G} \times M \times r \times RS\,coding\,Rate \times Effective\,Data\,Carrier\,Rate$$

 = *1.322…(kHz) ×432 ×(16/(16+1)) ×2 ×(2/3) ×(188/204) ×(96/108)*
 =*587.4… kbps*

[Calculation example 2] of 8MHz system
This example shows the data rate calculation of a channel (13 segments) with:
Single layer transmission, mode 3, single layer, G/I ratio =1/16, modulation index: M=64 QAM, coding rate: 3/4, 8MHz bandwidth

Data Rate (1 segment)

$$= f_d \times N_f \times \frac{T_s}{T_s + T_G} \times M \times r \times RS\,coding\,Rate \times Effective\,Data\,Carrier\,Rate$$

 = *1.3227513 ×432 ×(16/(16+1)) ×6 ×(3/4) ×(188/204) ×(96/108)*
 = 1.982… Mbps
Data Rate (13 segment)
 = *Data Rate (1 segment) ×13*
 = *1.968… × 13*
 = *25.772… Mbps*

[Calculation example 3] of 8MHz system
This example shows a calculation of layered (Hierarchical) transmission data rate. The parameters shown are used for the actual DTTB service in Japan. In this case, the data rates for one (1) segment of each transmission mode (One-Seg or fixed reception) are calculated first, followed by summing up these data rates of segments.
This example is based on the following conditions:

(a) service configuration: One-Seg(portable reception) + HD/SD service(fixed reception)
(b) common parameters: Mode=3, G/I ratio=1/8, 8MHz bandwidth
(c) Layer A(portable reception) parameters: No of segments=1,   M=2(QPSK),   r=2/3
(d) Layer B(fixed reception) parameters   : No of segments=12,   M=6(64QAM),   r=3/4

Data Rate (1 segment of Layer A)

$$= f_d \times N_f \times \frac{T_s}{T_s + T_G} \times M \times r \times RS\,coding\,Rate \times Effective\,Data\,Carrier\,Rate$$

$= 1.322... \times 432 \times (8/(8+1))\ \times 2 \times (2/3)\ \times (188/204)\ \times (96/108)$
$= 554.78..\ kbps$

Data Rate (1 segment of Layer B)
$= 1.322... \times 432 \times (8/(8+1))\ \times 6 \times (3/4)\ \times (188/204)\ \times (96/108)$
$= 1.872...\ Mbps$

Data Rate (total)
$= Data\ Rate\ (1\ segment\ of\ Layer\ A) \times 1 + Data\ Rate\ (1\ segment\ of\ Layer\ B) \times 12$
$= 23.023 Mbps$

(This page is left intentionally blank.)

Chapter IV. Multiplexing and PSI/SI

In the analog terrestrial television systems, it can only accommodate a single video and audio in one channel assignment, Digital Television however, can transmit multiple video, audio and data casting in the same channel thanks to the digital transmission/compression technology. This chapter explains firstly a digital signal format, a multiplexing process and a synchronization process followed by the explanation on the control and service information for channel/program selection/indication etc.

IV.1 Structure of multiplexed signal

ISDB system including terrestrial, satellite and cable adopts the Transport Stream Structure in the MPEG-2 Transport Systems as defined in the ISO-13818-1 or ITU-T H.222.0. DVB-T and ATSC also use this transport stream structure. In addition to the definition of the international standard, each broadcasting standard adds its own details depending on their technical characteristics. ISDB-T defines its own detail based on the Directive No. 26, 2003, issued by the Ministry of Internal Affairs and Communications, Government of Japan and the corresponding MIC notifications as legally mandated specification designated. ARIB STD-B10 ARIB STD-B32 and ARIB TR-B14 serves as private industry standard in Japan to supplement legally mandated standard. Brazil adopts the same multiplex technology and defines its details in ABNT NBR 15602-3.

IV.1.1 Signal processing for TS generation

Figure IV-1 shows a general structure of multiplexing in digital broadcast site.

Video and audio signals from a camera and a microphone are encoded into digital video and audio, with the compression technology described in Chapter V and then fed into a multiplexer. Together with PSI/SI data, which defines what kind and how many programs are available, the video and audio streams are combined into one "Transport Stream (TS)." The Modulator converts the Transport Stream into Radio Frequency signal to be then fed into the Transmitter for transmission. Receivers follow the same path in the opposite order, where the received signal captured by an antenna goes to the tuner before going through the demodulation, de-multiplexing and decoding process to obtain video/audio data to be presented.

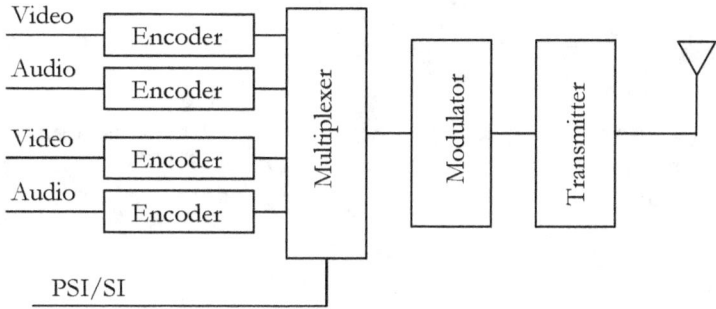

Figure IV-1 Multiplexing structure on a digital transmission site

Figure IV-2 illustrates how the audio and video signal is structured in the transport stream. Firstly, the video and audio signal are coded and compressed (to be discussed in the Chapter V), resulting in the output of "Elementary Stream (ES)" which is indicated as (a) in this figure. This ES are split by certain length (such as frame length) and headers are added for each split. This is called "Packetized Elementary Stream (PES)", which is indicated as (b) in the figure. This PES structure is common not only for digital broadcast but also for DVDs or Internet streaming that adopt MPEG-2 systems.

Then, each PES packet is split into 184 bytes and 4 bytes headers are added for each split. This packet is called Transport Stream Packet (TSP), which is indicated as (c) in the figure. This TSPs are combined with another TSPs for another broadcast service, this data stream is called Transport Stream (TS) which is indicated in the figure as (d).

Transport stream is unique for broadcast or communications media, very differently structured from Program Stream format that is for storage media such as DVDs. (Blue-ray disk uses MPEG2 Transport Stream) As TS is designed for transmission, its design bases on fixed and short length packet (188 bytes), while packet length of PS is variable. Also, TS is designed to carry multiple programs in one stream.

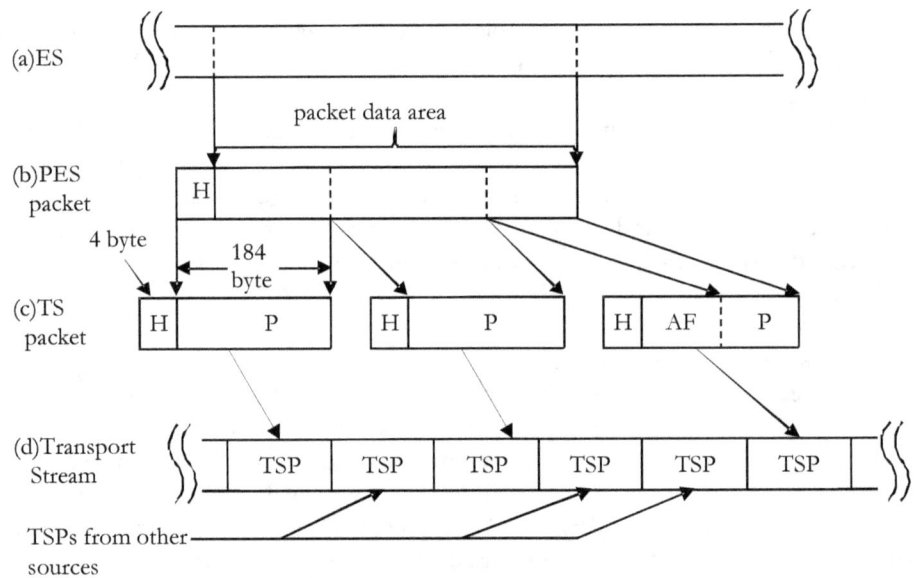

Figure IV-2   ES, PES and TS structure

IV.1.2   PES packet
(1) Structure of PES packet
Figure IV-3 shows the structure of PES packet. The PES packet format is common for both PS (Program Stream) which is mainly used for "Package Media" such as DVD, and TS (Transport Stream) which is used for broadcasting and communication media.

packet start code prefix	stream id	PES packet length	optional PES HEADER	stuffing bytes	PES packet data bytes
24 bytes	8 bytes	16 bytes			

Figure IV-3 Structure of PES packet[72]

As shown in the figure, PES packet is composed of the following data parts written in the box[73] below.

1. The packet start code prefix is a code representing the start of the PES packet and shall be set to 0x000001.
2. The stream id is used to identify elementary stream (coded signals; the same is true for the other signals) type and number. For the details for the assignment of elementary stream type and number, see Notification No. 88 of the Ministry of Internal Affairs and Communications (2009), Annexed Table 5, Annex note.
3. The PES packet length field indicates the number of bytes in the PES packet after this field. Value 0 indicates that the PES packet length is not specified and has no boundaries. The value 0 is permitted only for PES packets whose payloads are video elementary streams.
4. The optional PES header shall comply with the ITU-T Rec. H.222.0.
5. Stuffing bytes shall be set to 0xFF and shall not exceed 32 bytes in length.

In optional PES header, PTS (Presentation Time Stamp) and DTS (Decoding Time Stamp) field are included. These data are important for the presentation and display timing control in a receiver. As digital broadcasting system uses different encoders for video and audio, presentation timing is strictly controlled by PTS and DTS information that will be discussed on IV.1.3.

IV.1.3   TS packet

Figure IV-4 shows the structure of TS packet, which is composed of two parts, a header and a payload area. The length of TS Packet is fixed to 188 bytes which consists of 4 byte header and 184 byte data area.

(1) Header

The header is composed of a 8 bit sync_byte and information for TSP identification. The structure and each information are defined in ARIB STD-B32 part 3, Chapter 3.3, which is based on ISO-13818-1 or ITU-T H.222.0.

---

[72] Extracted from ISO/IEC 13818-1: 1996 (E), Annex F, Figure F.2 – PES packet syntax diagram
[73] Quoted from Notification No. 88 of the Ministry of Internal Affairs and Communications (2009), Annexed Table 5

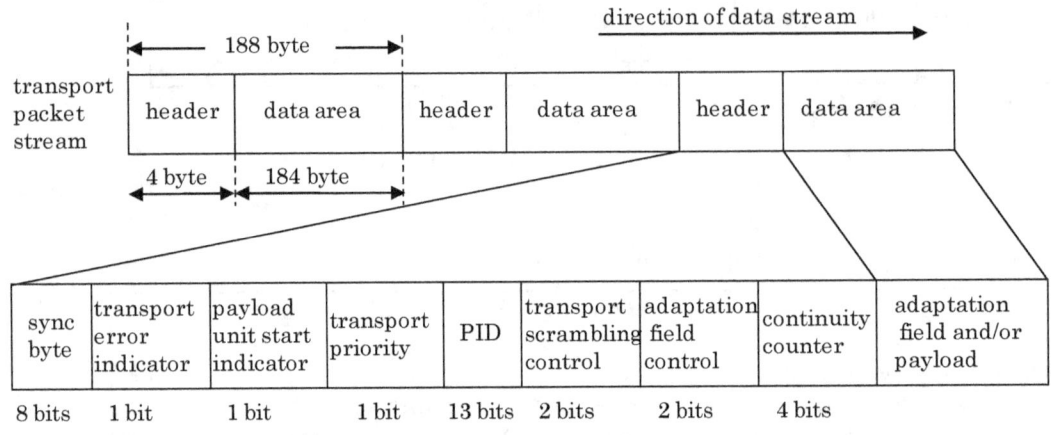

Figure IV-4　The structure of TS packet

For your reference, each control information are explained in the box[74] below.

1. The sync byte shall be 0x47.
2. The transport error indicator is a flag that indicates whether there is any bit error in the TS packet. If this flag contains '1,' it indicates that the TS packet has an uncorrectable error of at least one bit.
3. The payload unit start indicator indicates that the payload of this TS packet starts at the PES packet start or pointer when it contains '1.'
4. The transport priority is a flag that indicates priority among packets with the same PID. The packet is given priority if this flag contains '1.'
5. The PID is a field identifying the payload data type. Payload data type assignments shall be as shown in Notification No. 88 of the Ministry of Internal Affairs and Communications (2009), Annexed Table 7, Annex note 1.
6. The transport scrambling control is a field identifying the payload scrambling mode for TS packet. The value of this field shall be as shown in Notification No. 88 of the Ministry of Internal Affairs and Communications (2009), Annexed Table 7, Annex note 2.
7. The adaptation field control is a field indicating the configuration of the adaptation field/payload. Adaptation field/payload assignments shall be as shown in Notification No. 88 of the Ministry of Internal Affairs and Communications (2009), Annexed Table 7, Annex note 3.
8. The continuity index is a field specifying the sequence of TS packets with the same PID. The value of this field shall start with '0000' and be incremented by 1. The value shall change back to '0000' after '1111.'
   However, note that it shall be ensured that the same TS packet is transmitted only up to twice in a row and that in this case the value of this field shall not be incremented.
9. The adaptation field shall comply with ITU-T Rec. H.222.0.

Table IV-1 shows the assignment of PID. The details of PSI (Program Specific Information) will be

---

[74] Quoted from Notification No. 88 of the Ministry of Internal Affairs and Communications (2009), Annexed Table 7

explained in Chapter IV.2 later.

Table IV-1   Assignment of PID[75]

Value	Description
0x0000	PAT
0x0001	CAT
0x0002 – 0x000F	Reserved
0x0010	NIT
0x0011 – 0x1FFE	May be assigned to other than PAT, CAT, NIT, and Null packet
0x1FFF	Null packet

(2) Data area

The length of data area is 184 byte.   This area is used for carrying a payload and/ or adaptation field.   In order to carry variable length data with fixed length packets, the flag "adaptation field" is prepared as written in Table IV-2.

Table IV-2   Adaptation field control value[76]

Value	Description
'00'	Reserved
'01'	No adaptation field, payload only
'10'	Adaptation field only, no payload
'11'	Adaptation field followed by payload

(3) Relationship between PES packet and TS packet

PES packet are split into multiple chunks of fixed length and packed in the TS packets (see Figure IV-2).   If the PES packet size is larger than TS data area, the PES packet will be split into multiple TS packet data area. Figure IV-5 shows this mapping process from PES packet to TS packet.

As shown in the figure, three types of TS packets exist: (a)first TS packet, (b)intermediate packet, (c)last packet.   These types are identified by the combination of "payload_unit_start_indicator" and "adaptation_control_indicator" described in Table IV 3.

If the value is '01' or '11" and payload_unit_start_indicator=1 then 1byte pointer_field is placed immediately after the TSP header pointing where the payload starts.

---

[75] Quoted from ARIB STD-B32 Part 3, Chapter 3.3
[76] Quoted from ARIB STD-B32 Part 3, Chapter 3.3

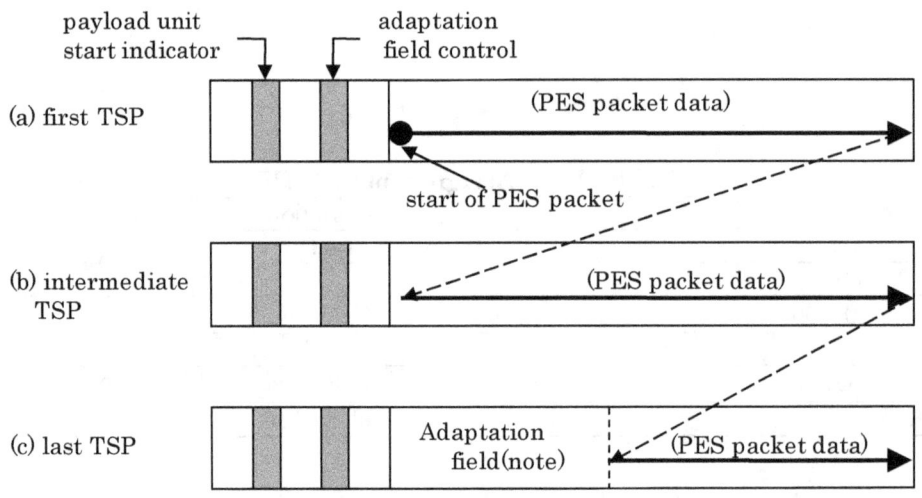

Figure IV-5 PES packet data distribution into TS packets

Table IV-3 Relationship between TSP type and the related header identifiers

TA packet type	payload_unit_start_indicator	adaptation_field_control
First TSP	1	01
Intermediate TSP	0	01
Last TSP	0	11

IV.1.4 Section data

(1) Structure of Section

Section is a syntactic structure that is used for mapping the PSI and SI tables specified in this standard, into the Transport Stream packets. The Section is also used for the DSM-CC carousel data transmission. These syntactic structures conform to ISO/IEC 13818-1. Sections may be vary in length. The sections within each table are limited to 1,024 bytes in length, except for sections within the EIT, which are limited to 4096 bytes.

Two types of data formats are defined: "general format" and "extended format." These formats are illustrated in Figure IV-6. Most of the information use "extended format" in ISDB-T.

What is written and in what format in the "Data" are defined as "tables" depends on information type called "PSI/SI" to be discussed in the Chapter IV.2 and IV.3.

Cyclic Redundancy Check (CRC) code, used in "extended format" is an error detection code. Please refer to (4) of this section for the details of CRC.

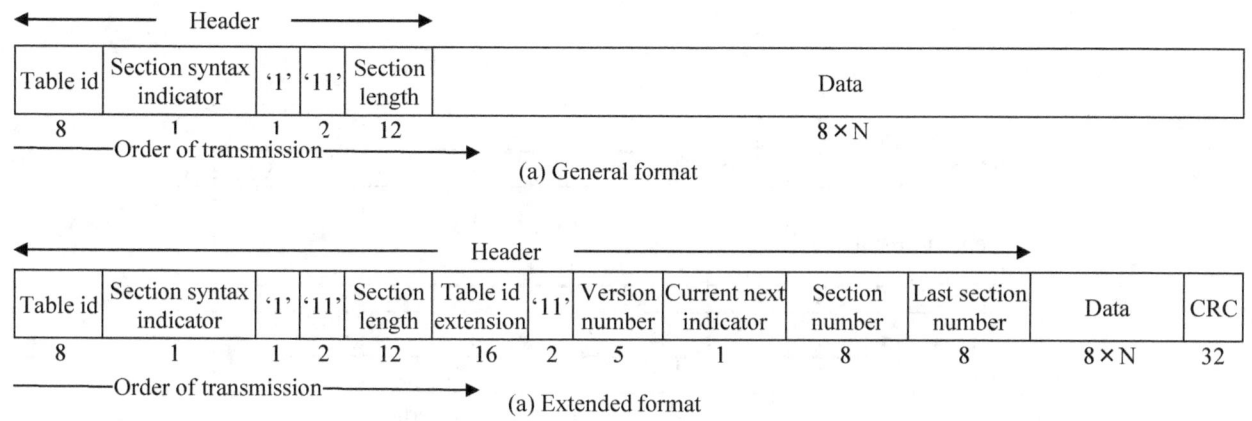

Figure IV-6 Formats of "Section"[77]

(2) Descriptors
As tables alone cannot completely describe the information to achieve actual broadcast service; rather, a mechanism called descriptor allows the delivery of additional information to tables. There are more than 50 descriptors defined in the standard. The use of descriptors will also be discussed in the Chapter IV.2 and IV.3.

(3) Mapping to TSP
Section data is placed in the TSP, as shown in Figure IV-7 (a). "1" is set to "payload_unit_start_indicator" of the first TSP and "0" (this number means the start byte number of Section data) is set at the first byte of payload which is assigned as a pointer. If the Section data size is smaller than TSP, Stuffing byte (0xFF) is inserted in the remainder. If Section data size is larger than TSP, Section data is placed in the following TSP as shown in Figure IV-7 (b). In this case, "0" should be set at payload_unit_start_indicator.

If the Section data continues further, it will be placed by the same method onto the following TSPs and Stuffing bytes are inserted into the remainder in TSP where Section data ends.

Figure IV-7(c) shows the case that Section data 1 ends in the middle and the next Section data (shown as section 2 in the figure) starts. In this case, payload_unit_start_indicator is set to "1" to indicate that new section data starts, and the starts byte point of Section 2 data is written at pointer (=N in the figure).

---

[77] Quoted from ARIB STD-B32 Part 3, Chapter 3.2

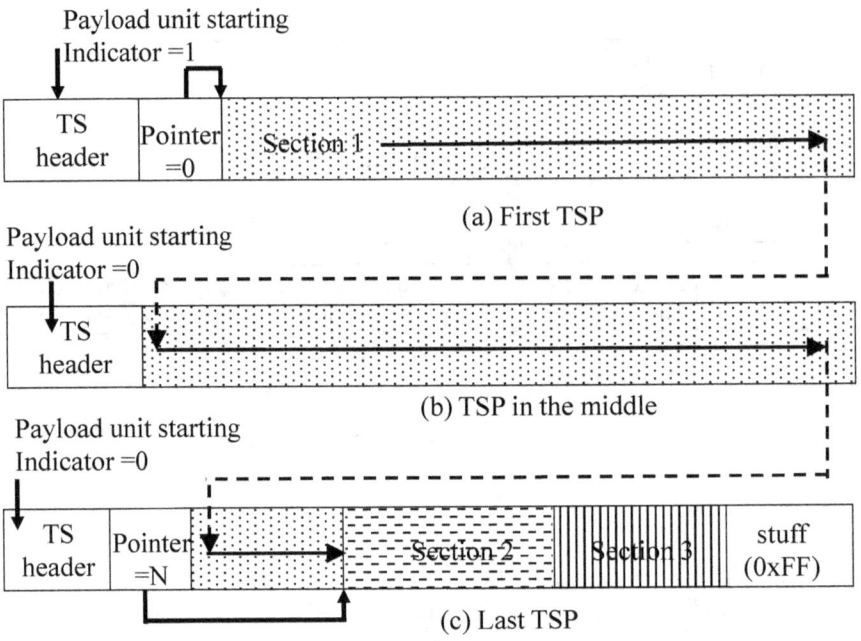

Figure IV-7 Mapping process of section data to TSP

(4) Cyclic Redundancy Check (CRC)
CRC is an error detecting code commonly used in digital communication medias. CRC attaches short check value based on the remainder of a polynomial division of the contents. CRC in ISDB-T is calculated by the following polynomial:

$$x^{32} + x^{26} + x^{23} + x^{22} + x^{16} + x^{12} + x^{11} + x^{10} + x^8 + x^7 + x^5 + x^4 + x^2 + x + 1$$

The CRC encoder applies XOR operation on the data field from the most significant bit to lower bits to obtain the remainder of the operation, which is the CRC check value. CRC field is encoded as such to guarantee the output integrity to its full capability.

IV.1.5   Synchronization control

In contrary to the analog system that have fixed amount of latency in both video and audio transmission, digital broadcasting system has variable output latency, as digital broadcasting system utilizes data compression technology. This means that synchronization mechanism of the timing between encoder input and decoder output should be necessary. Otherwise, if the encoder output is faster than that of decoder then the buffer that acts as the temporary storage of the TS in the receiver will overflow and loss of data necessary to the accurate reconstruction of the video or sound will likely occur, this is typified by blocking or frozen images. If the encoder output is slower than that of decoder, the buffer will ran out of the received and stored TS resulting to some glitches on the movements in the screen.

MPEG-2 system synchronizes the encoder and decoder clocks using a 27,000KHz Clock, so called System Time Clock (STC). To synchronize the decoder, the encoder sends the decoder a Program Clock Reference (PCR), which is multiplexed into an adaptation field of TS packet. The decoder compares the STC value contained in the received packet with its own STC counter value. If the values are not same and adjustment is necessary, decoders adjust its processing speed so that two values are always synchronized. To solve the fluctuations in the speed of the video and audio data rate, receivers usually incorporates an input buffer, and controls the overall latency of the controlling read-timing of the buffer.

Time-stamp is placed in each encoder by a 33-bit integer in 90 kHz clock period for every access unit. In this access unit, video and audio are encapsulated into PES and time stamp are placed in their headers. There are two kinds of time-stamps. One is Presentation Time Stamp (PTS), which indicates when the associated presentation unit should be displayed in regards to the System Time Clock. Other is Decoding Time Stamp (DTS), which indicates when the access unit should be extracted from the buffer and decoded in regards to the System Time Clock. Basically if PTS and DTS are the same, only the PTS is used.

This is because B-picture encoding requires the access units to be decoded in a different order of I-picture and P-picture as B-picture use past and future I or P-picture for decoding. Decoding timing can be swapped between B-picture and I or P-picture, therefore 2 time stamps are necessary.

IV.2   Program Specific Information (PSI)

Along with the actual video, audio and data casting information, the transport stream carries control information called Program Specific Information (PSI) and Service Information (SI).

PSI provides information necessary for receiver to traverse the structure of the transport stream, for example how many programs are in it or which video or audio belongs to what program, and for de-multiplexer to filter out necessary information out of the stream. SI provides information on the program, including Electronic Programming Guide.

The PSI data is structured as four types of table. The tables are transmitted in sections.

1) Program Association Table (PAT):
   For each service in the multiplex, the PAT indicates the location (the PID values of the Transport Stream packets) of the corresponding Program Map Table (PMT). It also gives the location of the Network Information Table (NIT).
2) Conditional Access Table (CAT):
   The CAT provides information on the Conditional Access (CA) systems used in the multiplex; the information is private (not defined within this standard) and depends on the CA system being utilized, this also includes the location of the EMM stream, whenever applicable.
3) Program Map Table (PMT):
   The PMT identifies and indicates the locations of the streams that make up each service and the location of the Program Clock Reference fields for a service.
4) Network Information Table (NIT):
   The location of the NIT is defined in this standard in compliance with ISO/IEC 13818-1 specification but the data formatting is outside the scope of ISO/IEC 13818-1. This is identified to provide information about the physical network, the syntax and semantics of the NIT are included in the standard.

IV.2.1   The PAT and PMT

The PAT and PMT are used to pick up desired video, audio and other information out of the transport stream.

Figure IV-8 illustrates how receivers sort out its input. Transport stream consists of fixed-length and various types of packets such as video, audio, data and PSI/SI etc. When a receiver receives a transport stream signal, the de-multiplexer of the receiver sorts out and picks up the necessary/desired information and then disregards unnecessary information inside the transport stream based on the PID value. PID is an id unique given to each type of stream.

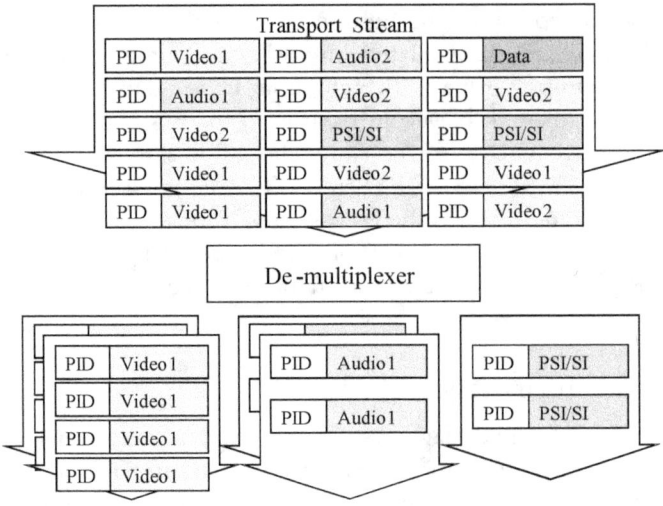

Figure IV-8 Concept of de-multiplexing

Figure IV-8 illustrates the basic structure and relations of PAT and PMT.

PAT lists all available programs in the transport stream by specifying the pair of information on the program number (or sometimes called "service ID") and PID of the programs (PMTs) in the stream. The first pair is on NIT with program_number: 0x0000 and PID of NIT. NIT is not a service, however, the MPEG standard defines to list NIT information in the PAT.

PMTs contain information on each program including program_number, list of elementary streams (audio, video or data) in the program and PCR (method used for synchronization). Optionally, PMT can carry information on the program as descriptors of the stream such as conditional access, copyright information, encoding format of audio/video and regional access definition.

Information contained in the table header (program_number, PCR_PID) and in the 1st loop is on the program as a whole. And information common to each of the elementary streams are listed in the 2nd loop. PCR_PID indicates the PID of the elementary stream that contains time reference of the elementary streams in the program. PCR can be common to other streams or can be independently set in each program. Also, if the program is encrypted, the decryption information such as Entitlement Control Message (ECM) will be referred based on the descriptors in the 1st loop.

Information in the 2nd loop describes on the each elementary streams in the program, such as stream_type and PID of the elementary stream. Values in the stream_type are predefined[78]. Major values used are as follows:

Table IV-4 Major values for stream_type

stream_type	Value
0x01	ISO/IEC11172 (MPEG-1 Video)
0x02	ISO/IEC13818-2 (MPEG-2 Video)
0x06	ISO/IEC13818-1 (MPEG-2 System) PES Packets containing private data
0x0D	ISO/IEC13838-6 (Data Carousel)
0x0F	ISO/IEC13818-7 (MPEG-2 AAC Audio)
0x1B	ITU-T H.264	ISO/IEC 14496-10 Video

---

[78] ARIB STB-B32, Part 3, 3.4, Transmission Control Signal (PSI)

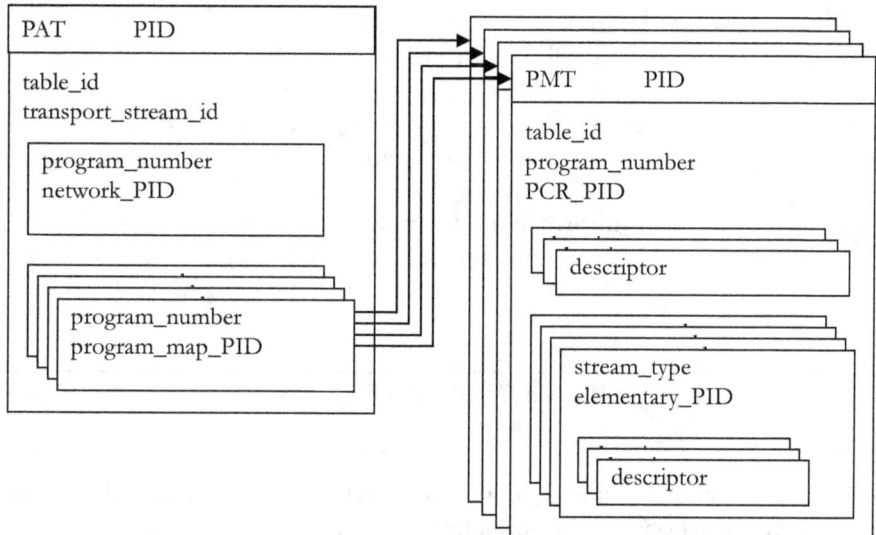

Figure IV-9   Basic structure of PAT and PMT

The PID of the PAT is assigned a fixed value 0x0000 so that the receivers can find the PAT and start analyzing the transport stream easily.   The following explains how the receivers analyze the structure of the transport stream.

1. Tune to a frequency with known NIT information to receive the transport stream (if there are no known NIT information, receivers have to perform initial scan to identify all available frequencies and find transport streams)
2. Receive PAT to obtain the PID's of the PMT
3. Receive PMT to obtain the PID's of a video/audio stream, PCR (and ECM if the stream is encrypted)
4. Obtain the necessary packets by filtering the transport stream based on the PID values
5. Descramble the video/audio stream from the decryption key obtained in the ECM packet)
6. Present the video/audio signal

IV.2.2   Network Information Table (NIT)
Network Information Table (NIT) describes the information pertaining to the physical structure of the multiplex/TS's of the network and its characteristics.   The syntax of the NIT is indicated in Figure IV-10. ARIB STD-B10 defines the outline, and ARIB TR-B14, Volume 4 defines specific details.

```
network_information_Section(){
        table_id
        network_id
        version_number
        network_descriptors
                (network_name_descriptor)
                (system_management_descriptor)
```

```
            for(i=0;i<N;i++){
                    transport_stream_id
                    original_network_id
                    transport_descriptors
                            (service_list_descriptor)
                                    - service_id, service_type x N
                            (terrestrial_delivery_system_descriptor)
                                    - area code, transmission mode, frequency
                            (partial_reception_descriptor)
                            (ts_information_descriptor)
                                    - remote key id, TS name, etc.
            }
            CRC_32
}
```

Figure IV-10   NIT syntax

Networks are assigned individual network_id values, which serve as unique identification codes for all networks. Networks here are synonymous to broadcast stations.

The 1st loop contains network_descriptors. Here, network_name_descriptor and system_management_descriptor will be inserted. network_name_descriptor shows human readable description on the information about this stream is broadcasted in what region and by which broadcasting station.

The 2nd loop contains transport_descriptors, which defines details (frequency, modulation, TS name, etc.) of the transport stream. While the MPEG standard allows to include information on multiple transport streams – not only the transport stream itself but also other transport stream – ISDB-T standard limits description on only one transport stream.

This transport_descriptors contain important information for services and reception, those are:
    service_list_descriptor,
    terrestrial_delivery_system_descriptor,
    partial_reception_descriptor and
    ts_information_descriptor.

service_list_descriptor lists pair of service ID and its service type. Service types are predefined value, and mostly used are 0x01: Digital TV service transmitted by type A (64QAM), 0xc0: Data broadcasting service transmitted by type C (QPSK). Please note that 0xc0 indicates this service is an One-Seg service as One-Seg is defined as data broadcasting with audio/video monomedia in the ARIB standard. Also service type 0xa4 is used for engineering service, which is mainly used for firmware updates of receivers.

Terrestrial_delivery_system_descriptor shows area code, guard interval, and mode (2k, 4k or 8k), as well as all the frequencies used to deliver this stream not only by the main station but also frequencies used by relay stations or repeaters.

If a transport stream is set for hierarchical transmission, especially for "One-Seg" transmission, the "partial_reception"_descriptor needs to be placed to indicate what programs are transmitted in what hierarchy.

Finally, ts_information_descriptor shows remote control key id, name of the transport stream (which is also implies name of the broadcast station), and which layers are in what modulation with service IDs in the layer.

Receivers have a function of "initial scan" and "re-scan." Receiver search through all channels in the

available frequency range at a location and gathers all NIT's received and store them in a Non-Volatile memory, which retains the information even if the receiver is turned off. The stored NIT information is used to tune into a program that basically identifies and assigns the Broadcaster station name. Information in NIT includes as follows:
- Assigned Number on remote control
- Frequency range, center frequency of the channel
- Network ID
- List of service ID

When a broadcaster multiplexes several programs into one transport stream, the first program in the TS table is identified as the primary stream, to which a channel button on a remote controller will be assigned (if there is one program in one TS, the program is identified as primary stream). The remote controller assignments of the programs may vary by region.

IV.2.3   Conditional Access Table (CAT)

Before explaining the Conditional Access Table (CAT), a brief explanation about the Conditional Access System (CAS) is needed. While there are several ways to encrypt contents for conditional access, ISDB-T adopts a block cipher called MULTI2. Details are explained in the ARIB STD-B25 and ARIB TR-B14, Volume 5.

Figure IV-11 indicates a model block diagram of a CAS system of the receiver unit. Please note that this Figure represents the model configuration and is provided only for the purpose of describing the specifications.

In this Figure, RF input from the antenna is fed into a tuner, which locks into the given frequency and extracts the transport stream through OFDM demodulation. The De-scrambler re-constructs the original contents of the packets based on the given descrambling method with the algorithm contained in the CAS or IC card. The succeeding stages will normally decode the video/audio signal from the transport stream.

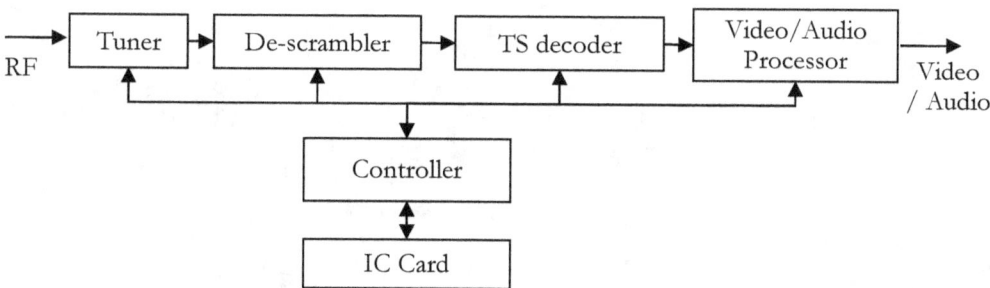

Figure IV-11   Block diagram on CAS system of example receiver unit

Figure IV-12 illustrates the basic concept of an implementation of a CAS system. The Left side is for the broadcast side. E1 encrypts the contents with an encryption key Ks, called "scrambling key" which is a common key among receivers. Then, E2 encrypts Ks with Kw, called "work key" which is a key to encrypt the scramble key, and is shared among the same conditional access category, such as per region, per subscription contract type, etc. Further, E3 Encrypts Kw with Km' called "master key," a key to encrypt the work key, and is a specific key for the security module. Packets that carry encrypted Ks are called

Entitlement Control Message (ECM); that carry encrypted Kw are called Entitlement Management Message (EMM).

In the receiver side, the right hand side of the illustration decrypts the keys and contents in the opposite process. Receiver stores Km, "master key" in order to identify the security module.

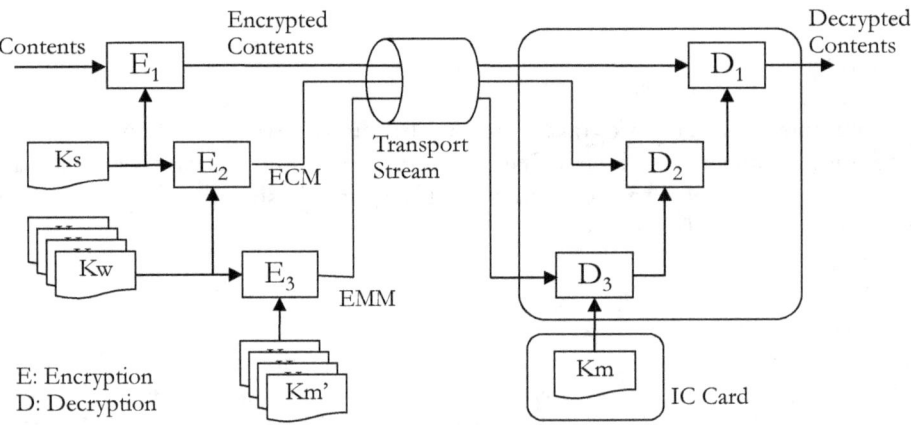

Figure IV-12   Conceptual drawing of a CAS system

Figure IV-13 illustrates a Flowchart example for the EMM/ECM Reception and Conditional Access. In this example, the receiver starts to receive the Transport Stream by acquiring and analyzing the PAT and NIT. Then, the receiver retrieves the CAT to extract the Entitlement Management Message (EMM). Here, EMM tend to be transmitted less often, therefore, receivers need to continuously acquire the transport information for certain period and to store the EMM information in a secure manner. Once the EMM is obtained, the receiver can start decoding the contents as illustrated in Figure IV-11.

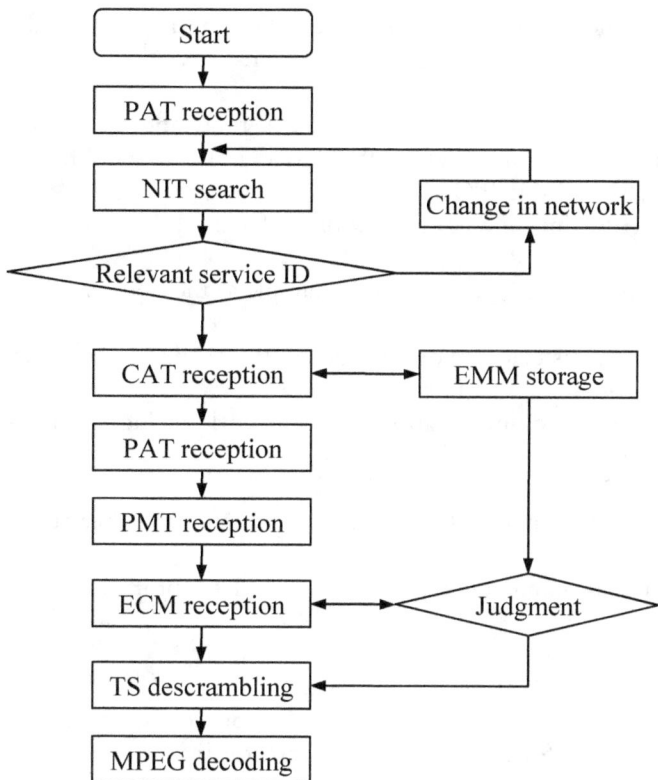

Figure IV-13　Example flowchart on EMM/ECM reception and conditional access[79]

IV.3　Service Information (SI)

In addition to the PSI, data are needed to provide identification of services and events for the user. The coding of this data is defined in this standard. In contrast to the PAT, CAT, and PMT of the PSI, which only gives information for the multiplex in which they are contained (the actual multiplex), the additional information defined within this standard can also provide information on the services and the events carried by the different multiplexes, and even the other networks. The data structure has eleven tables:

1) Bouquet Association Table (BAT):
    The BAT provides information regarding bouquets. As well as giving the name of the bouquet, it provides a list of services for each bouquet.
2) Service Description Table (SDT):
    The SDT contains data describing the services in the system e.g. names of services, the service provider, etc.
3) Event Information Table (EIT):
    The EIT contains data concerning events or programs such as event name, start time, duration, etc.
4) Running Status Table (RST):
    The RST gives the status of an event (running/not running). The RST updates this information and allows timely automatic switching to events.
5) Time and Date Table (TDT):

---

[79] Quoted from ARIB STD-B25, Part 1, 3.2.1 Types of Associated Information

The TDT gives information relating to the present time and date. This information is given in a separate table due to the frequent updating of this information.

6) Time Offset Table (TOT):
The TOT gives information relating to the present time and date and local time offset. This information is given in a separate table due to the frequent updating of the time information.

7) Partial Content Announcement Table (PCAT):
The PCAT includes starting time and continuing time of partial content in accumulated data broadcast.

8) Stuffing Table (ST):
The ST is used to invalidate existing sections, for example at the boundary of the delivery system.

9) Broadcaster Information Table (BIT):
The BIT includes the broadcaster's network or Transmitted SI parameter given to each broadcaster.

10) Network Board Information Table (NBIT):
The NBIT includes the board information in a network including the reference information for acquiring the board information.

11) Link Description Table (LDT):
the LDT includes various data collected to be used as reference to other tables.

Following sections will discuss on most commonly referred descriptors.

IV.3.1 Service Description Table (SDT)

Service Description Table (SDT) describes the information on service channels in the Transport Stream, including the service name, logo information, EIT transmission information. SDT's have multiple number of loops as many as the number of services in the transport stream. Each loop contain information on the service such as:

service_descriptor: service type, service provider name and service name
digital_copy_control_descriptor: generations of copy allowed, maximum transmission rate
logo_transmission_descriptor: type of the logo, pointer to CDT type logo or character instead of logos.

IV.3.2 Event Information Table (EIT)

Event Information Table (EIT) is a table that provides program data for the current and future shows that can be used to present on-screen program guides, such as title, time and description. The reason to choose EIT as an example of SI is to present how SI can carry additional information on the broadcast signal.

EPG is the user interface that helps the viewer to select the TV program currently being shown and the other programs that are lined-up for a certain period (i.e. day, week, month) of TV program schedule. Also, the EPG enables TV viewers to navigate through the TV program guide with the use of the remote control and access detailed information about the program such as the synopsis, schedule of airing, Channel/station, title, the cast or the abstract. If the receiver is equipped with a recording function coupled with the EPG, viewers can select and schedule the recorder to identify which program(s) is to be recorded by selecting it as listed in the EPG.

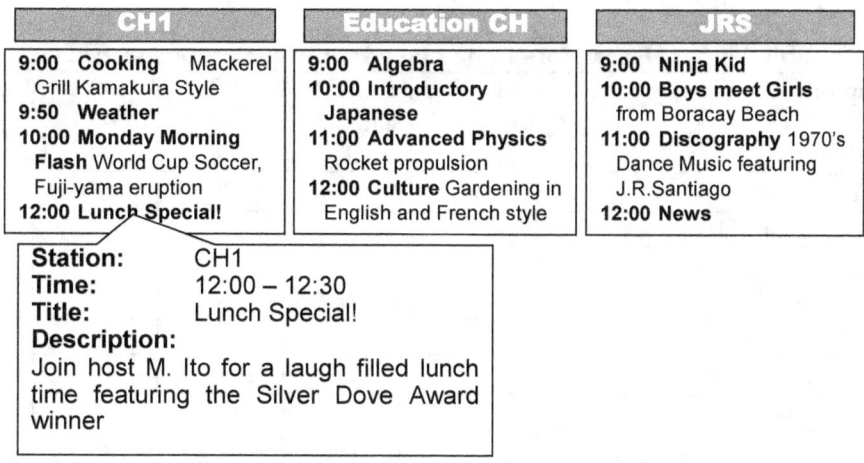

Figure IV-14   Example electronic program guide screen

EPG information is carried by the Event Information Table. The structure of the EIT is illustrated on Figure IV-15.

```
event_information_Section(){
        table_id
        service_id
        version_number
        for(i=0;i<N;i++){
                event_id
                start_time
                duration
                running_status
                free_CA_mode
                descriptors_loop_length
                (descriptors)
        }
        CRC_32
}
```

Figure IV-15   Structure of EIT

ARIB TR-B14 defines several types of EIT depending on the media type and timing information.
   H-EIT: for fixed receivers
   M-EIT: for mobile receivers (defined, but not used)
   L-EIT   for partial receivers (One-Seg receivers)

Also there are two types of EIT: EIT[schedule] (or EIT[p/f after]) and EIT[p/f]. EIT[schedule] contains the information of all programs within 8 days. (for One-Seg, EIT[p/f after] is used instead) EIT[p/f] contains the program information for the current and succeeding program from the schedule. When the current program ends and switches to the next, EIT[p/f] is refreshed, in which, the version number is incremented and the program information in the schedule are shifted. Through the detection of this change, the receivers will know that a transition has occurred and that the program has changed.

Descriptors contained in the EIT may vary depending on the type as described in the Table IV-5.

Table IV-5 Descriptors to be placed for event groups in EIT[80]

Tag	Descriptor	H-EIT [p/f]	H-EIT [schedule basic]	H-EIT [schedule extended]	L-EIT [p/f]	L-EIT [p/f after]
0x4D	Short Event Descriptor	✓	✓	×	✓	✓
0x4E	Extended Event Descriptor	○	×	○	×	×
0x50	Component Descriptor	✓	✓	×	×	×
0x54	Content Descriptor	○	○	×	○	○
0xC1	Digital Copy Control Descriptor	○	○	×	○	○
0xC4	Audio Component Descriptor	✓	✓	×	×	×
0xC7	Data Contents Descriptor	○	○	×	×	×
0xCB	CA Contract Info Descriptor	○	○	×	○	○
0xD6	Event Group Descriptor	○	○	×	×	×
0xD9	Component Group Descriptor	○	○	×	×	×
0xD5	Series Descriptor	○	○	×	×	×
0x42	Stuffing Descriptor	○	○	○	○	○

✓ : Should be inserted into the relative descriptor area in table
○ : Can be inserted into the relative descriptor area in table voluntarily
× : Cannot be inserted into the relative descriptor area in table

IV.3.3   Other tables (BIT, TOT)

Broadcaster Information Table (BIT) describes related information about the terrestrial TV operator such as station group information, which specifies stations belonging in the same broadcasting network group, and parameter of the SI being transmitted.

Time Offset Table conveys current date and time information in local time zone.

---

[80] ARIB TR-B14, Volume 4, 31.3.2, Table 31-47

Chapter V. Video and Audio Coding System

The purpose of applying coding on both audio and video is to allow better utilization of the available capacity. In analog systems, a lot of redundant information present in the audio and video, and are being transmitted which renders an inefficient utilization of the bandwidth. Digital Coding pertains to the application of compression and error correction which will provide a higher degree of data reduction while maintaining the perceptual information quality, this in effect will allow to use the freed up channel capacity for the delivery of additional information.

Compression techniques have been developed for the specific use of "humans", therefore, significant studies about the visual and aural perception has been made to ensure that the applied data reduction will not result in a perceived degradation of the processed image and sound.

The foundation of any data rate reduction technique utilizes either a lossy compression or a mathematically lossless compression system or a combination of both. The decision to use one or the other will entirely depend on the intended application and most significantly the cost. For practicality, the lossy compression technique prevails in most installations due to its practical benefit and higher compression ratio than the other.

Providing High Definition and/or several Standard Definition along with handheld video such as One-Seg in one channel (defined channel bandwidth) will not be possible in the analog system due to the limitations of the system to efficiently utilize the available channel capacity to support high data rate, multiple program service.

This chapter explains basic video and audio coding technology of MPEG-2 systems that structures the ARIB STB-B32[81].

V.1   Video Coding Technology

There are a series of processes involved in getting a video from light to its digital representation. First process is called digitalizing or sampling which converts analog visual information into a digital format. In this format, visual information is mapped into two-dimensional planes, generally a set of rectangles called pixels, and organized into discrete numerical values representing luminance or color value of the pixels. The digitized video information is then processed by the compression process.

V.1.1   Input video format

Video signals are composed of a combination of light called the luminance information and color that is called as the chrominance information. Luminance has been derived from the combination of the primary colors in the appropriate proportions and the Chrominance Cb and Cr are the In-Phase and Quadrature color difference signals.

In practice, video camera plays this role, capturing light and converting them into digital data representing the visual information. Figure V-1 illustrates an image of video camera mechanism and video signal processing. Following steps 1-4 corresponds to the (1)-(4) in the figure.
1. Visible Light passes through a beam splitter that filters the individual primary colors directing it to the desired optical pick-up device.

---

[81] ARIB STD-B32, Ver 2.1, Video Coding, Audio Coding, and Multiplexing Specifications for Digital Broadcasting

2. The primary colors (R,G and B) as a proportion derived from the visible light are then fed to the color matrix for the generation of the analog luminance and color difference signals.
3. The color matrix generates the luminance signal(Y), and the analog color difference signals(R-Y and B-Y), these are then fed to the Analog to Digital conversion.
4. After the A/D conversion, the Digital Color difference signals are now generated compliant with the Rec. ITU-R BT.601 standard that makes it suitable for the various MPEG-2 profiles.

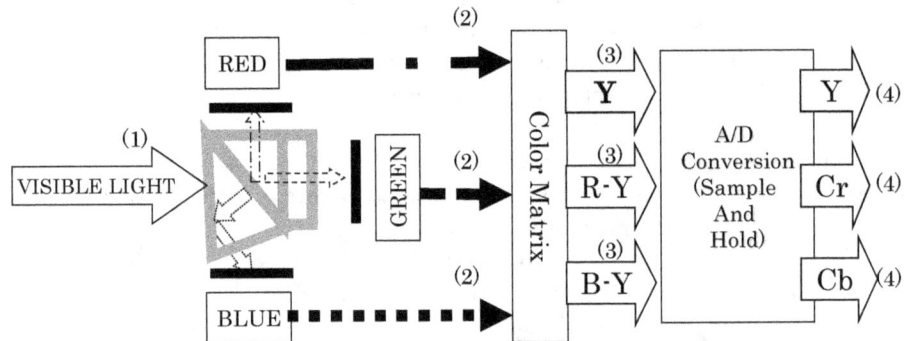

Figure V-1   An image of a video camera mechanism and video signal processing

For digitized video format, two types of video format are used. One is "Composite Video Baseband Signal" (CVBS) and the other is "component video signal". For composite type, the original analog TV signal with color subcarrier is directly sampled and digitized, while, for component type, the luminance component and the chrominance components of video signal are separately sampled and digitized. Component type has been specified to harmonize the different video signal formats such as, NTSC, PAL and SECAM. In addition, the component video signal is easy to handle in terms of digital signal processing. Therefore, component video signal became the mainstream digital video format.

Table V-1 shows the relationship between the component video signal parameters and the uncompressed video signal transmission rate.

Table V-1   Uncompressed component video signal transmission rate (in SD format)

Bit/sample	Chrominance format	Component	Sampling frequency (MHz)	Data rate of each component (Mbps)	Total data rate(Mbps)
8bit/sample	4:2:2	Luminance(Y)	13.5	108	216
		Chrominance(Cr)	6.75	54	
		Chrominance(Cb)	6.75	54	
	4:2:0	Two color components of 4:2:2 profile are transmitted alternately line by line.			162
	4:1:1	Luminance(Y)	13.5	108	162
		Chrominance(Cr)	3.375	27	
		Chrominance(Cb)	3.375	27	
10bit/sample	4:2:2	Luminance(Y)	13.5	135	270
		Chrominance(Cr)	6.75	67.5	

		Chrominance(Cb)	6.75	67.5	
	4:2:0	Two color components of 4:2:2 format are alternately transmitted by line.			202.5
	4:1:1	Luminance(Y)	13.5	135	202.5
		Chrominance(Cr)	3.375	33.75	
		Chrominance(Cb)	3.375	33.75	

For example in above table, let us use the 4:2:2 profile for a 10 bit system, the luminance, with a sampling frequency of 3.375MHz×4 will result into 135Mbps and each of the Chrominance Cr and Cb with 3.375Mhz x 2 will result to 67.5Mbps each. If we sum up the three then it will be 135Mbps + 67.5Mbps + 67.5Mbps = 270Mbps.

The resulting data rate should comply with the ITU.R BT601 standard for converting videos compliant with ISO/IEC 13818-1 known as MPEG Systems. Picture information in the real world contains a lot of redundant and irrelevant information that can be eliminated based on the model of the human eye in its ability to distinguish variations. This can be used as a baseline reference to attain data rate reduction allowing for a more efficient utilization of channel capacity. With this basic concept, more information can be delivered within the same channel allowing flexibility of choice to provide multiple SDTV or an HDTV program. Table V-2 shows the video parameters for the 525, 750 and 1125 line system which is used in Japan.

Table V-2   Parameters of video signal[82]

Title		525 Lines (i)	525 Lines (p)	750 Lines (p)	1125 Lines (i)
Number of Lines		525	525	750	1125
Number of active lines		483	483	720	1080
Scanning system		Interlaced	Progressive	Progressive	Interlaced
Frame Frequency		30/1.001Hz	60/1.001Hz	60/1.001Hz	30/1.001Hz
Field Frequency		60/1.001Hz	-	-	60/1.001Hz
Aspect Ratio		16:9 or 4:3	16:9	16:9	16:9
Line Frequency fH		15.750 /1.001kHz	31.500 /1.001kHz	45.000 /1.001kHz	33.750 /1.001kHz
Sampling Frequency	Luminance Signal	13.5MHz	27MHz	74.25 /1.001MHz	74.25 /1.001MHz
	Color difference signals	6.75MHz	13.5MHz	37.125 /1.001MHz	37.125 /1.001MHz
Number of samples per line	Luminance Signal	858	858	1650	2200
	Color difference signals	429	429	825	1100
Number of	Luminance	720	720	1280	1920

---

[82] Quoted from ARIB STD-B32, Section 2.4

samples per active line	Signal				
	Color difference signals	360	360	640	960

In the above given table, comparison among the 525, 750 and 1125 line system were shown. The major differences are on the selection of either interlaced or progressive scanning, the aspect ratio and sampling frequencies for the luminance and color difference signals.

V.1.2 Video resolution and aspect ratio

When analog television was designed and implemented in the early 1950's, the viewers have become accustomed with the 4:3 aspect ratio with either a 525 or a 625 line system. These system (PAL, NTSC and SECAM), aside from the line specifications actually give us a lesser number of visible lines that can be used, for example, in the 525 line system, there are only about 480 active lines.

In the late 1990's, several Digital Television standards envisioned the transition to a more spectrum efficient way of delivering audio/video/data to the receivers. Several offerings were made available such as Standard Definition Television which is comparable to the current analog video quality and High Definition Television which provides high resolution and realistic video quality comparable to current blue ray technology.

The introduction of higher resolution and "realistic" aspect ratio is one of the main considerations in the delivery of High Definition television. In the illustration provided (see Figure V-2), taking a 525 and a 1080 line system as an example, it provides a reference comparison to establish an idea of how picture resolution is increased in Digital Television.

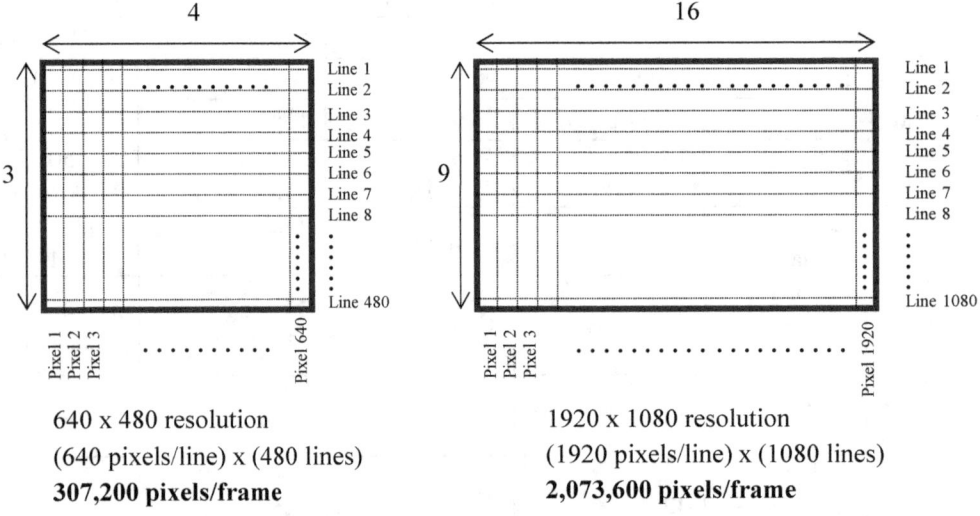

640 x 480 resolution
(640 pixels/line) x (480 lines)
307,200 pixels/frame

1920 x 1080 resolution
(1920 pixels/line) x (1080 lines)
2,073,600 pixels/frame

Figure V-2   Picture resolution of 525 and a 1080 line system

In Figure V-3, it shows the examples on how a 4:3 or a 16:9 image format will be made conformable to the desired aspect ratio. It is also a means to provide compatibility between the two methods of display formats.

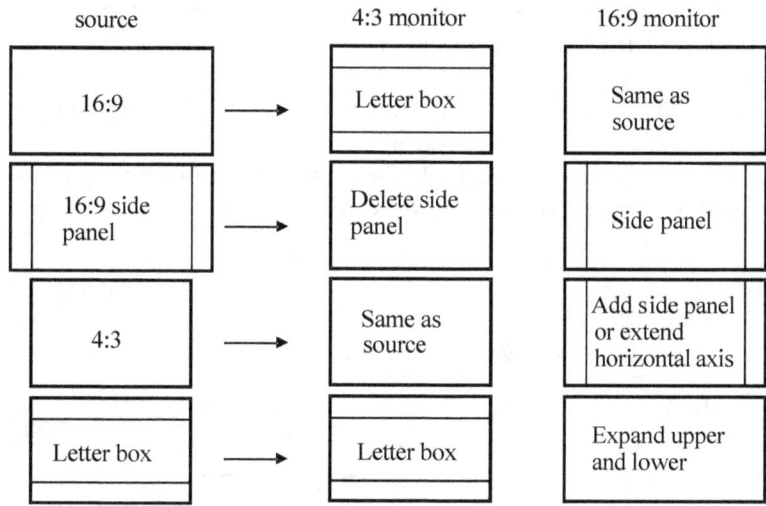

Figure V-3 Video format conversion

V.2 MPEG-2 Video Compression

Figure V-4 illustrates the general process of MPEG-2 compression.
There has been a significant effort made in the industry and academia to make video compression better. The goal of a video compression technology is to achieve efficient compression while minimizing the degradation of the video quality introduced by number of process.

Video information are handled as consecutive pictures that are referred as frames. There is substantial redundancy in these pictures, that is, most of the information contained in a given frame also exist in the previous frame. MPEG-2 Video compression technology uses this nature. By calculating where new information comes up and where old information remains, it is possible to reduce the data size of the frame.

Based on these fundamental concept, MPEG-2 video compression system are composed of four key technologies, "Motion compensation", "DCT", "Quantization" and "Variable length coding."

Among them, Motion compensation is a compression technique that utilizes the feature that the video signal consists of consecutive pictures.

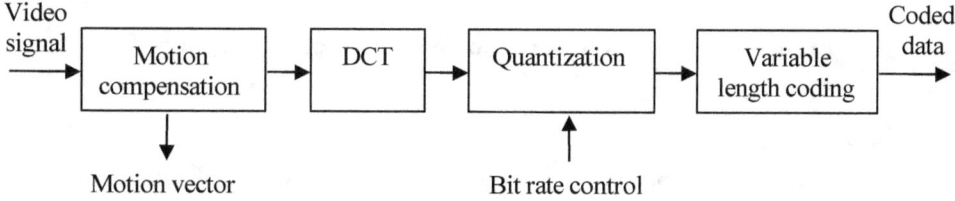

Figure V-4   General process of MPEG-2 compression

Before discussing each of technologies, an overall system block diagram is illustrated in Figure V-5. A motion vector is detected from the input Video signal to perform motion compensation. Motion prediction and motion compensation are described in V.2.2.

The difference components between the motion-compensated predicted picture and the input picture of video signal is converted into a frequency component by the two-dimensional DCT and quantized. The two-dimensional DCT is described in V 2.3.

The quantized signal is shortened based on the Huffman code, and output from the encoder. Quantization and variable length coding will be explained in V.2.4.

The quantized frequency component is fed to Inverse DCT to convert to a picture component. This picture component is added with the predicted picture component, and stored in the prediction memory.

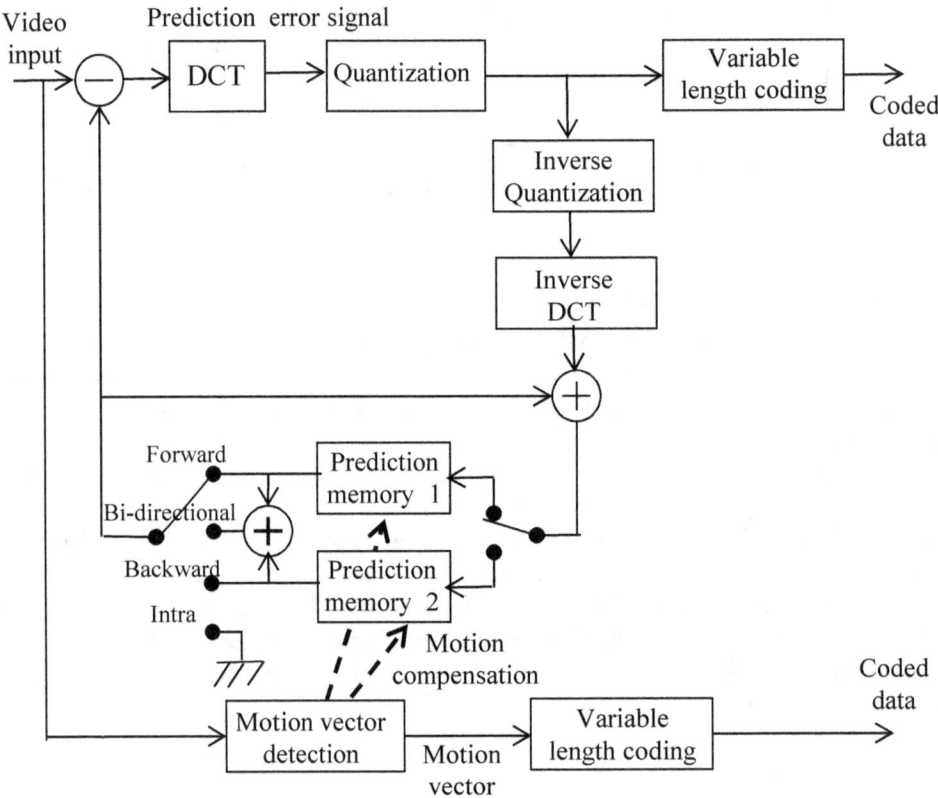

Figure V-5 Block diagram of MPEG-2 coding system[83]

V.2.1   Signal configuration

We need to understand the basic procedures in establishing a mathematical representation of the video information down to the pixel level in order to enable processing by computers. In Figure V-6 it illustrates how MPEG-2 analyzes the pixels and represents them in a mathematically efficient manner.

Figure V-6 below shows the structure of the signal processing in MPEG-2 video coding system.

---

[83] Quoted from ARIB STD-B32, Part 1, Chapter 4.1

The top of Figure V-6 shows the GOP (Group Of Picture, hereinafter call "GOP"), which consists of three types of picture, I-pictures (pictures encoded using only current picture information), B-pictures (pictures encoded using current, past and future picture information) and P-pictures (pictures encoded using current and past picture information) and contains at least one I-picture.

The I-Picture or I-frame represents the real part of the image, in itself, it is a whole image. The B-Picture or B-frame is the Bi-directional frame that exists in the previous and the current comparison of the I and B frame. The P-Picture of P-frame is the resulting predicted frame as a result of the I and B frame comparison. The basic difference between the B and P frame is that the B-frame exists before and after the I-frame, it contains the changes in the series of pictures with respect to the I-frame. The change (delta) then is used as a reference to construct the succeeding P-frame as a mathematical prediction of the likely image after gathering the information from the comparison.

In the GOP, starting from a reference I-frame, a slice from the picture will be taken with a 16 x 16 macroblock from which an 8 x 8 block will be taken and then subjected to the Discrete Cosine Transform.

A macroblock consists of a luminance signal of 16 x 16 pixels and two color difference signals spatially corresponding to an 8 x 8 or 16 x 8 pixels.

Knowing that the whole frame contains a lot of these blocks, isolating a single block will help us an elementary understanding of how it is dealt with by the MPEG systems in terms of the picture coding.

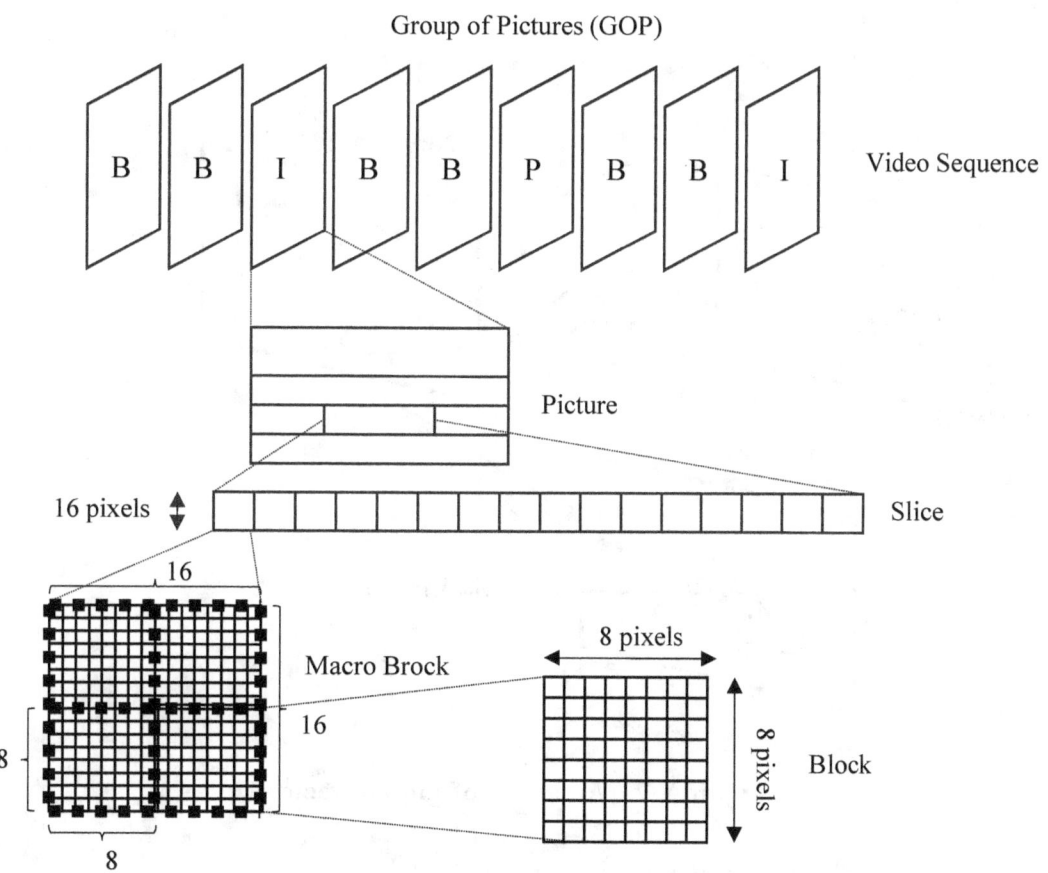

Figure V-6   Structure of signal processing in MPEG-2 video coding system

V.2.2 Motion estimation and motion compensation

"Motion Estimation and Motion Compensation" is a core technology for the bit reduction of moving picture to utilize a temporal correlation of a video signal. In some cases there are scenes in which only small part of the screen changes while the remaining part stays unchanged. This technique only codes the information on the moving part in the form of motion vector(s) so that coded data rates can be reduced significantly.

Figure V-7 shows an image of motion estimation and motion compensation. A moving object slightly changes its position in next video frame. For example of a scene of a soccer games, let's assume only a ball moves while picture of audience or ground stays the same. In case that the most part of the previous picture stays the same to the next picture, the difference between two parts is not significant.

The difference of the position (in this example, a soccer ball) in current and next frame is named "motion vector". In general, a quick scene change does not occur frequently, therefore, in most cases, it is possible to reproduce a motion picture by utilizing the information for the differentiation of the parts indicated by the soccer ball between current and next picture, the succeeding picture is called the P-frame (predicted) as a result of establishing the general direction of a uniformly moving "object" (in this case a block) with respect to its background. This general direction of motion is tagged with motion vector information to establish the likely position of the "object" in the succeeding frame.

This technique is called "inter frame prediction and bit reduction." Inter-frame refers to the correlation between succeeding frames.

In MPEG-2 video coding system, the motion estimation is done in macro block (16x16 pixel).

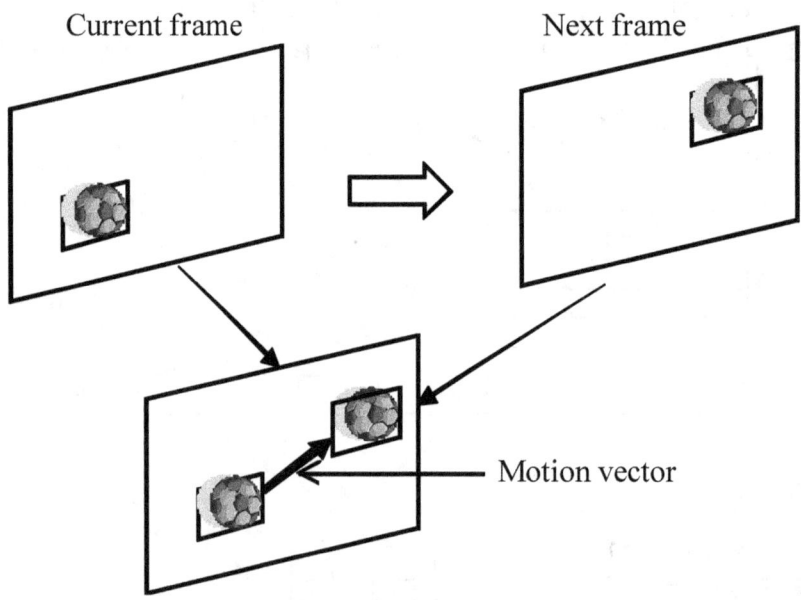

Figure V-7 An image of motion vector

V.2.3 Discrete Cosine Transform (DCT)

DCT is also one of the key technical elements for video compression. In general, neighboring pixels within an image tends to have high correlation. In other words, the energy of a picture tends to concentrate into the low frequency components rather than the high frequency components. Therefore, it is possible to reduce the bit rate by applying a function of transforming to the frequency component. The Discrete Cosine Transform (DCT) has been adopted in MPEG-2 video coding system as a key technology for compression.

Discrete Cosine Transform (DCT) represents a two-dimensional DCT coefficients F(u, v) for N × N pixels. The function f(x,y) represents the color or luminance value of the pixel, in this case, 8 pixels in the x-axis and 8 pixels in the y-axis.

$$F(u,v) = \frac{2C(u)C(v)}{N} \sum_{x=0}^{N-1} \sum_{y=0}^{N-1} f(x,y) \cos\left\{\frac{(2x+1)u\pi}{2N}\right\} \cos\left\{\frac{(2y+1)v\pi}{2N}\right\}$$

Provided that

$$C(u), C(v) = \begin{cases} \frac{1}{\sqrt{2}} & \text{for } u,v = 0 \\ 1 & \text{for } u,v \neq 0 \end{cases}$$

The two dimensional DCT is done in 8x8 pixels shown in Figure V-8 below. Pictures are divided into blocks of 8x8 pixels. Applying two-dimensional DCT on this block creates u, v matrix of coefficients.

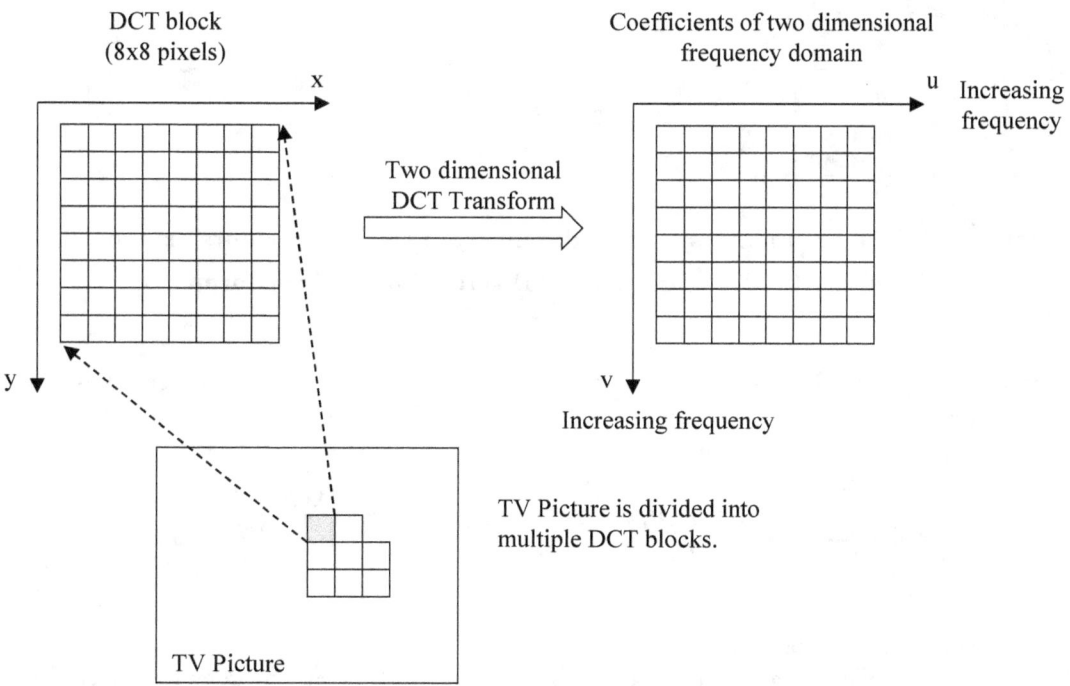

Figure V-8   An image of DCT processing of MPEG-2

Figure V-9 below shows sample coefficients of a DCT for an 8 x 8 ($N_1 = N_2 = 8$) two-dimensional DCT. As an example shown in the Figure V-9(a), from the reference pixel going to the right in the horizontal axis represents a vertical line pattern showing an increase in frequency (or variation of the pixel information) in the X-axis. In the vertical axis, the reference pixel going down in the Y-axis represents a horizontal pattern equivalent to the increase in pixel frequency. The top-left element will be the element with the least pattern variation representing the lowest frequency component. Each step from left to right and top to bottom is an increase in frequency by 1/2 cycle. The source data (8x8) is transformed to a linear combination of these 64 frequency squares.

By obtaining how much frequency component corresponding to each element of u, v matrix is in the block, we can apply quantization method to be explained later to reduce the data volume. Figure V-9 (b) shows an example of the energy of a picture from the low frequency to the high frequency. As illustrated in the Figure V-9 (b), the frequency component of the picture tends to be concentrated in a lower frequency region in general. With the knowledge of the energy concentration being present in the low frequency components, this has led to the significant reduction of the data volume.

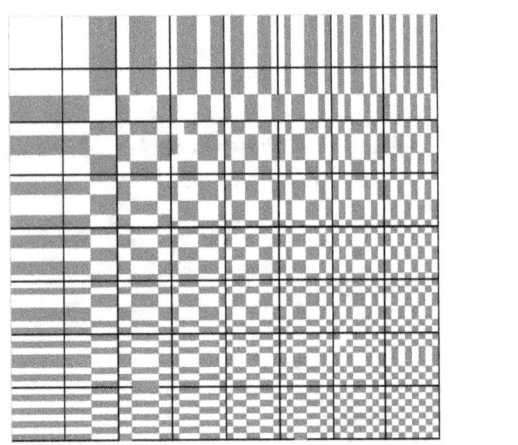 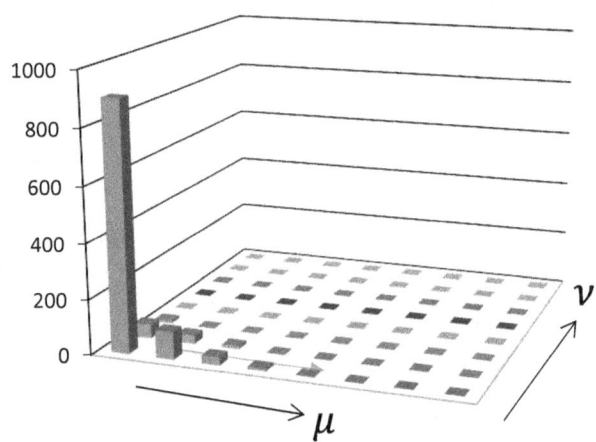

(a) Two dimensional DCT Frequencies   (b) 2-dimentional DCT coefficients example

Figure V-9 Two dimensional Discrete Cosine Transform

When decoding the picture Inverse DCT is used. Inverse DCT represents an Inverse Discrete Cosine Transform and is defined as follows:

$$F(x,y) = \frac{2}{N} \sum_{u=0}^{N-1} \sum_{v=0}^{N-1} C(u)C(v)F(u,v) \cos\left\{\frac{(2x+1)u\pi}{2N}\right\} \cos\left\{\frac{(2y+1)v\pi}{2N}\right\}$$

V.2.4   Quantization and variable length coding

The MPEG-2 encoder applies variable quantization to the DCT coefficients in order to reduce the number of bits required to represent it. In order to reduce the total bit rate of the stream, the high frequency coefficients will be quantized less than the lower frequency coefficients. See Figure V-10.

As described in section V.2.3, energy of high frequency components after DCT processing are very small with values, almost zero (0). In these scanning process, Figure V-10 (a) and (b) illustrates the scanning method taken for either a progressively scanned or an interlaced scanned system.

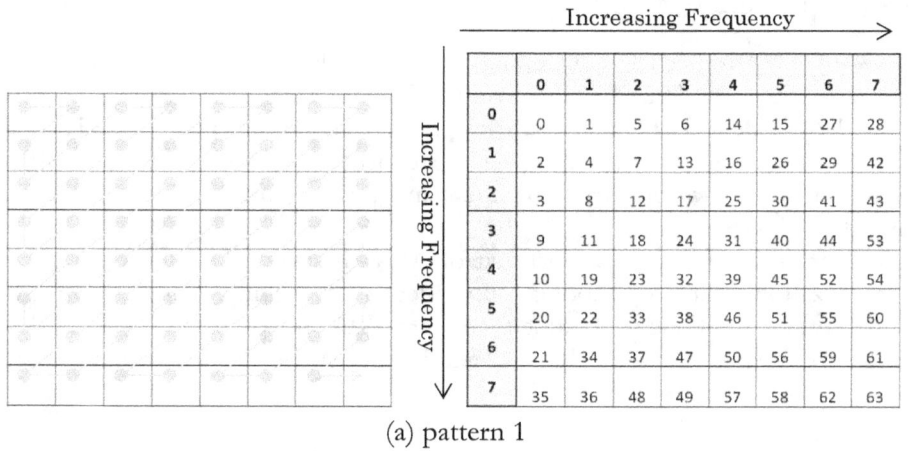

(a) pattern 1

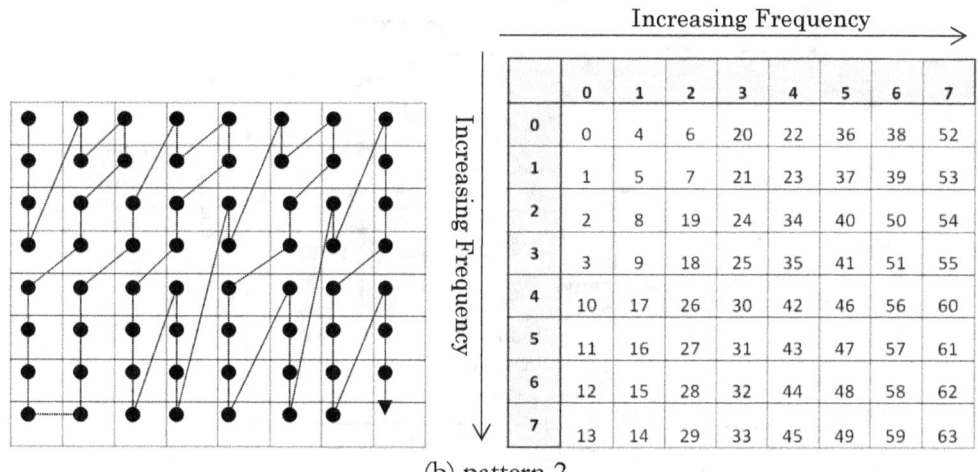

(b) pattern 2

Figure V-10   Order of quantizer output[84]

In order to apply further compression to the quantized bit-stream output, Huffman coding is used as a method of variable length coding. Huffman coding works by applying statistical values based on the probability of occurrence in the input bit-stream and applies a shorter bit representation of a more recurring character and a longer bit representation for a less recurring character.

In regular coding, code word and the value have fixed length, For example, character "A" is coded as "01100001" and "b" as "01100010". In contrast, Huffman coding changes the length of code words for "A" to be "0111" and "B" to be "10101100010" assuming that "A" occurs more frequently than "B". The code words are calculated based on a algorithm considering the probability of occurrence. As a result, Short code words are assigned to highly probable values and long code words to less probable values.

Huffman coding in MPEG-2 coding process is based on the same concept using binary values for both code word and values.

---

[84] Quoted from ARIB STD-B32, Part 1, Chapter 4.1

V.3 H.264 Coding System

H.264[85] or MPEG-4 AVC (Advanced Video Coding) is a standard for video compression, designed to enable high compression capability for a desired image quality. Standardization work was done by the Video Team formed by the Video Coding Experts Group of the Telecommunication Standardization Sector of International Telecommunications Union (ITU-T) and Moving Picture Experts Group (MPEG). Based on the naming convention of the two organizations, the standard was named H.264 in ITU-T convention and MPEG-4 AVC in MPEG convention.

The fundamental architecture is the same to that of MPEG-2 which consists of motion compensation, transformation, quantization and entropy coding. In order to achieve improvement for the coding efficiency and picture quality, various functions were added. Figure V-11 shows an outline of H.264 encoder block diagram. The major differences between MPEG-2 and H.264|MPEG-4 AVC are illustrated in Table V-3.

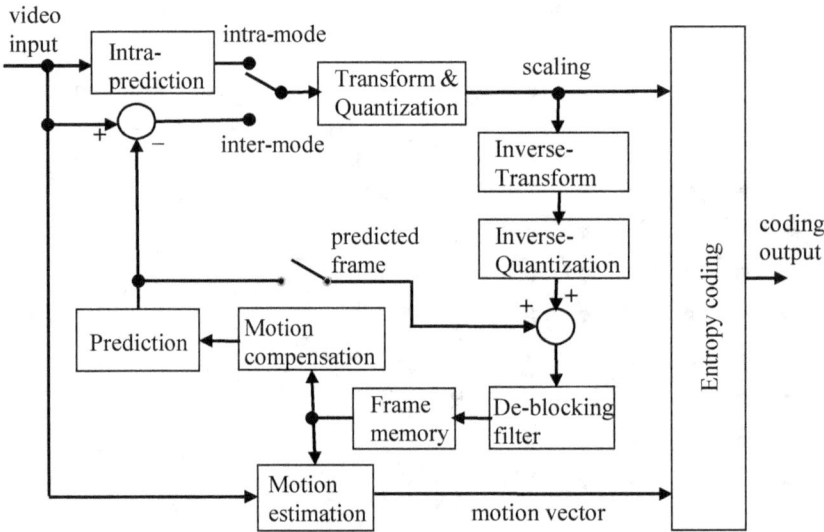

Figure V-11   Outline of the H.264/MPEG-4 Encoder

Table V-3   Comparison between MPEG-2 and H.264|MPEG-4 AVC

Technology	MPEG-2 (Main Profile)	H.264 / MPEG-4 AVC (High Profile)
Intra-picture prediction	None	Various modes of prediction in 16x16, 8x8, and 4x4 block size
Inter-picture prediction	Reference frame from immediately forward and after 16x16 or 16x8 block size 1/2 pixel precision	Allows up to 16 reference frames 16x16, 16x8, 8x16, 8x8, 8x4, 4x8, and 4x4 block size and their combination

---

[85] ISO/IEC 14496-10:2012, ITU-T H.264 (01/2012), Advanced Coding for Generic Audiovisual Services, http://www.itu.int/ITU-T/recommendations/rec.aspx?rec=11466

		1/4 pixel precision
Transform	8x8 Discrete Cosine Transform	Integer Transform in 8x8 or 4x4 block size
Entropy Coding	Huffman coding	Context-adaptive binary arithmetic coding (CABAC) or Context-adaptive variable-length coding (CAVLC)
Deblocking Filter	None	Yes

1) Intra-picture prediction

This is a new feature added in H.264|MPEG-4 AVC. Intra-picture prediction is a process in which, before the application of Integer Transform, subtraction is applied to the predicted value of the adjacent decompressed macroblock. In case of the luminance signal, 4x4, 8x8 and 16x16 block with several prediction modes can be utilized.

Compared to the MPEG-2 process in which pixel values of a block are directly fed into Discrete Cosine Transform, Intra-picture prediction can reduce the resulting volume of the coded data after quantization due to the subtraction of predicted values that reduce amount of information and high-frequency components with respect to the original picture.

2) Inter-picture prediction

The block sizes used for the prediction are not only 16x16 or 8x8 but also 16x16, 16x8, 8x16, 8x8, 8x4, 4x8, and 4x4 block size and their combination can be applied. Therefore, variable block sizing can be utilized depending on its applicability. If there is a complex and/or plain pixels in one 16x16, the encoder can flexibly choose the appropriate block size so that subtraction process will result to a reduced inter-picture variation.

In case of MPEG-2, the reference picture for P-frame can only be on a single direct adjacent I-frame or P-frame in the forward direction; and for B-frame, it is directly to the adjacent I-frame or P-frame in the forward or backward direction. It is always one frame for each direction. While H.264 can make references up to 16 frames depending on the level.

In case of level 4 for HDTV, 2 frames for each direction (up to 4 frames) can be utilized. Figure V-12 and Figure V-13 illustrates the difference.

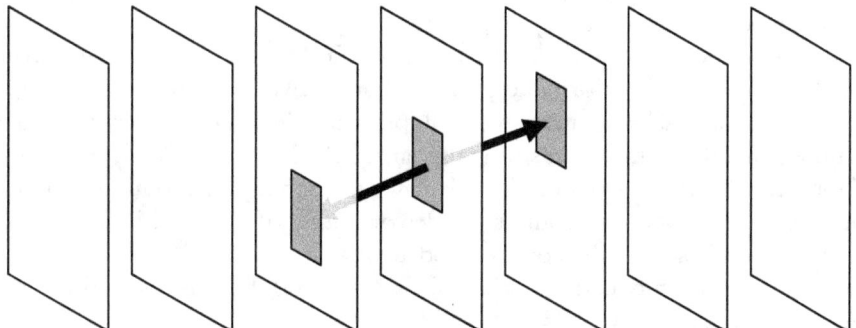

Figure V-12   Illustrates how MPEG-2 applies inter-frame prediction

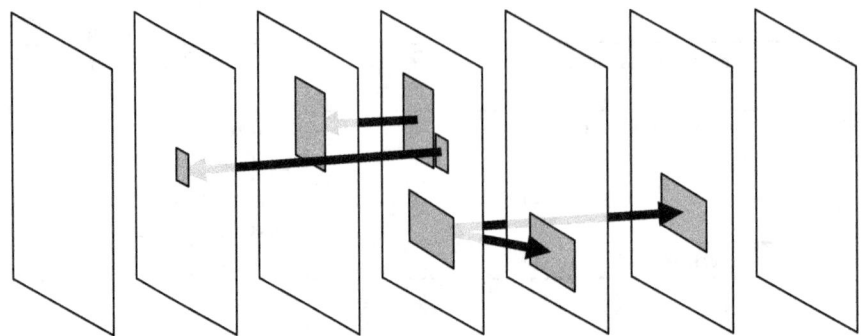

Figure V-13　Illustrates how MPEG-4 AVC/H.264 applies inter-frame prediction

In addition, precise motion compensation were improved to a Quarter-pixel dimension from what use to be a Half-pixel for MPEG-2, enabling accurate description of the displacements of moving areas. In motion compensation, prediction is "weighted," allowing an encoder to specify the use of scaling and offset which provide a significant improvement in the encoding quality for special cases such as brightness variations along with the time axis and fade-to-black or dissolve transitions.

3) Transform
H.264 adopts 4x4 or 8x8 block size Integer transform, whereas MPEG-2 adopts 8x8 Discrete Cosine Transform to convert video signal from the spatial domain to the frequency domain. H.264 transform algorithm uses four different matrices; 4x4 forward integer, 4x4 hadamard, 2x2 hadamard, and 4x4 inverse integer transform.

Since the calculations are done in integer precision, no round-off error will be introduced since the DCT has an in float precision calculation from which calculation can be faster than that of the DCT.

4) Entropy Coding
There are several types of entropy coding adopted in the H.264|MPEG-4 AVC standard which includes:
  a. Context-based Adaptive Variable Length Coding (CAVLC),
  b. Context-based Adaptive Binary Arithmetic Coding (CABAC)
  c. Variable Length Coding (Exponential-Golomb Coding).

Various coding are chosen to adaptively adjust depending on the syntax elements (header information, motion vector or transform coefficient values) of pictures.

In order to encode syntax elements, such as transform coefficients that shares a large volume in the coded data and require high compression performance, CAVLC and CABAC is used. CAVLC is used only for coding transform coefficients. CAVLC prepares multiple variable length codeword tables and chose the table depending on the contexts of the blocks near the target block. CABAC is more complex but it is an efficient coding algorithm. CABAC dynamically calculates probability distribution and generates codeword tables while performing the operation, in contrast, codeword table of CAVLC is fixed. Also, CABAC take correlations between symbols based on the context modeling.

In addition, Exponential-Golomb coding, which is simple and fast, is used for many of the syntax elements not coded by CAVLC and CABAC.

5) Deblocking Filter
With block coding technique, the sharp edges between blocks are sometimes visible, especially with motion compensated images in MPEG-2 technology which refers to the decoded picture with block noise, and the noise propagated between frames. In order to solve this, H.264|MPEG-4 AVC introduces deblocking filter.

The deblocking filter is incorporated in the in-loop filter in both the encoding and decoding path. It applies filtering not on the whole block, rather, it changes the filtering strength by adaptively considering the factors such as whether it is a macroblock boundary and the block coding (intra/inter) reference for motion prediction are different.

V.4 Audio Coding Technology

Audio Coding, much like that of video, takes into consideration the ability of the human ear to perceive sounds and therefore establishes a baseline reference that will determine what components of the sound should be given importance and components that should be discarded. This approach is sometimes called "Perceptual Coding" wherein irrelevant and redundant information can be reduced allowing for an efficient representation of the sound while preserving its perception as if data rate reduction has not been applied.

In the studies regarding audio, two coding techniques has been introduced:
1. Sub-Band Coding which deals with the defined band of frequencies for quantization and bit representation, the definition of the band of frequencies varies from the low end to the high end of the audible frequency which is referred to the Psycho-acoustical Model of the human ear.
2. Transform Coding which deals with the various types of sound that can immediately be discarded by qualifying its relevancy toward the existence of the property of its adjacent sounds, this lead to the significant reduction of the required data to represent the original information. (Please see auditory and temporal masking)

For both coding techniques, it has been used as the fundamental reference for the existence of the MPEG-1 and MPEG-2 audio system which can largely be attributed to the significant reduction of the amount of data required while maintaining a perceptually lossless quality.

V.4.1 Sampling and quantization of audio signals

The first step of the audio signal processing is digitization called sampling and quantization, in which audio signal is converted from analog to digital format. This process, as illustrated in the Figure V-14, measures the strength of the audio signal in every certain time period. The frequency, that is how many times does the encoder sample the signal, is called "sampling frequency." When the signal is sampled, the signal strength value is measured and rounded to the closest binary number. Here, the longer the binary number is, more accurately the number can be quantized. The length of the binary number is called sampling bit length.

The sampling frequency and the sampling bit length are the two most important Figures to measure how accurate the audio signal is sampled in the digital format.

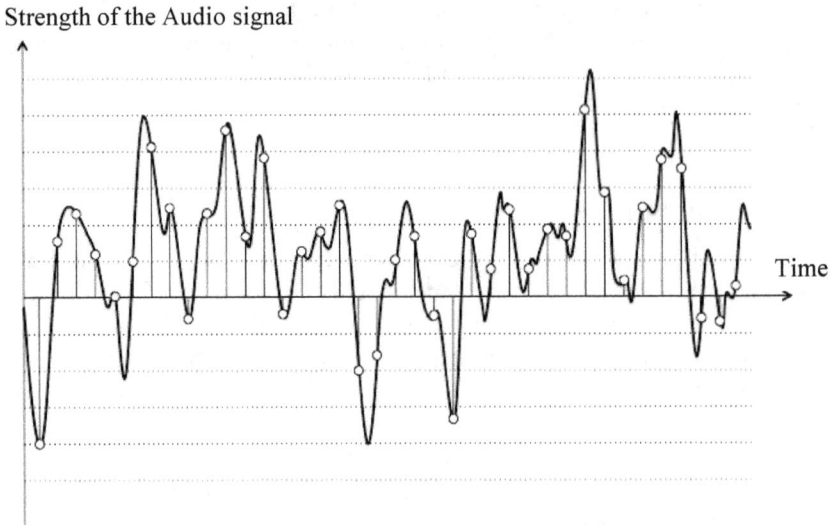

Figure V-14   Example of sampling and quantization of audio signal

The tendency can be observed that the more sampling frequency is and the more sampling bit length is the output bit rate glows up in the tradeoff of more accurate sampling.   Please note that the relevant data rates taken for example is nestled to the definition of the "CD quality audio" and therefore will be limited to approximately 1.5Mbps for both channels sampled at 44.1kHz without compression as against studio audio sampling frequency of 48kHz and 96kHz for audio mastering.

V.4.2   Audio coding algorithm

While sampling and quantization is a fundamental technique for digital audio processing and common for various coding standards, the audio coding algorithm to be described in this section differs among various standards.

This section focuses on Advanced Audio Coding (AAC) adopted in ISDB-T.   AAC is a standardized, lossy compression and encoding scheme for digital audio.   MPEG-2 AAC[86] has no backward compatibility with other MPEG audio standards (such as MPEG-1 audio layer III or MP3).

The fundamental concept of audio coding is quite similar to that of video coding in the sense that both system processes the signal in the flow of:
1. Sampling and Quantization
2. Conversion of the signal data into frequency domain (with MDCT etc.)
3. Quantization and Coding

The general outline of AAC compression process is illustrated in Figure V-15.

---

[86] MPEG-2 AAC (ISO/IEC 13818-7)

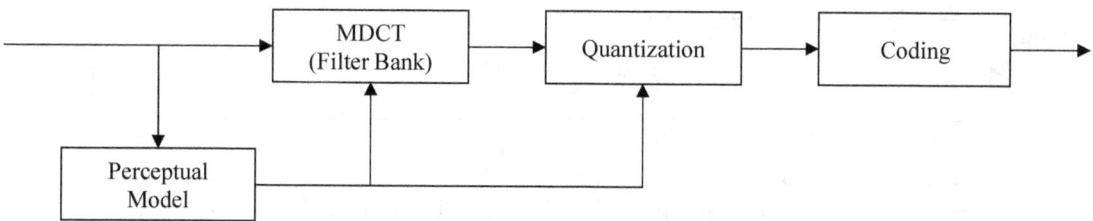

Figure V-15　General process of AAC compression

1) Perceptual Model
The audio signal input is firstly splitted in blocks in specified number of samples as a frame and fed to the Perceptual Model. The perceptual model calculates frequency spectrum of a frame by FFT in order to calculate masking effect of human perception, spectral shaping of quantization noise and parameter called "Perceptual Entropy."

Perceptial model also calculates the window size of MDCT. While long window is used for stable signal conditions in order to obtain high coding gain, short window is used for "attack" sound where sound level changes dramatically and longer window may result in mis-coding called "pre-echo."

2) Modified Discrete Cosine Transform
MDCT transforms input signal from time domain to frequeny domain based on a given window size. The output is passed to the various filtering processes.

While AAC use pure MDCT, other coding system like MP3 use combination of filters.

3) Quantization
The frequency domain signal in the form of MCDT coeffients is quantized here based on the parameters passed from the Perceptual model. Perceptual model is based on the phychoacoustic effect which provides in-audible limit of human ear so that the bit allocation of the inaudible informatin can reduced, to achieve efficient compression while maintaining degradation of preceptual quality minimum.

The phychoacoustic effect used in this coding system is as follows:
1. Auditory limit – the minimum sound level audible for human ear varies in frequency. Human ear can recognize sound around 1kHz clearly, whereas it is not sensitive for high and low frequency.
2. Temporal Masking - higher amplitude sounds tend to overwhelm lower amplitude sounds if their difference in frequency is not that significant, this exists in the pre and post existence of the higher amplitude sound.

4) Coding
The bit information generated in the quantization process is compressed by compression argorithm generally used for lossless binary data compression. In case of AAC, Huffman coding is used.

Figure V-16 illustrates an example of AAC encoder block diagram. The signal is inputted and processed from left to right to obtain the output.

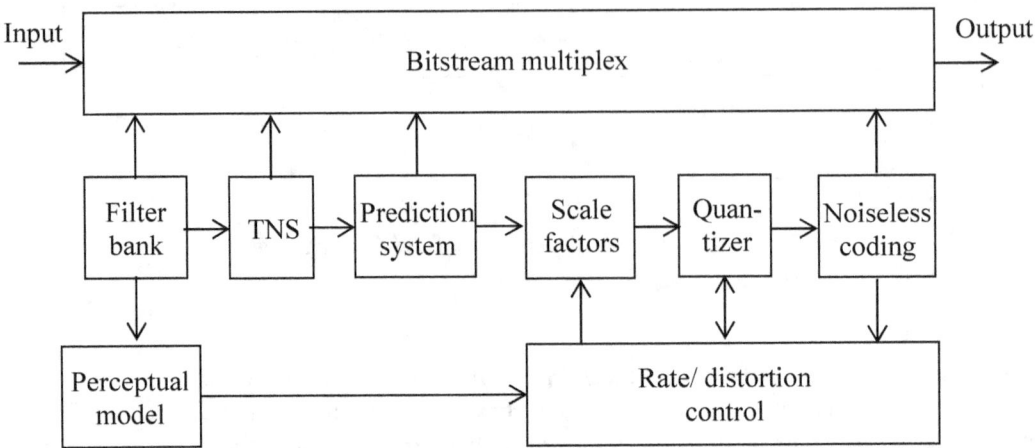

Figure V-16   Example of an AAC encoder block diagram

a) Filter bank
   A time domain input signal is converted to frequency domain by MDCT. AAC has adaptive window shaping function, which enables switching of frame size of 2048 samples or 8 frames of 256 samples. In addition, in order to reduce connection noise between frames, half size of block are overlapped next to each other and mixed like fade-in and fade-out style.

b) Temporal Noise Shaping (TNS)
   TNS flattens the frequency spectral by applying filtering to the MCDT coefficients.

c) Prediction system
   This function is optional for Main profile only. This system predicts MDCT coefficients of the current frame by the coefficients of the past 2 frames. Occurrence of statically present signals has longer transforms that enable the prediction in the backward structure for each frequency.

d) Scale factors
   The processed MDCT coefficients are grouped by certain spectral regions and amplified by scale factors, so that bit allocation is changed to increase signal to noise ratio.

e) Quantizer
   Processed MDCT coefficients are quantized to integer value here. AAC Quantizer assigns bits depending on the accuracy demands of the spectral bands with respect to the perceptual model of the human ear to effectively apply irrelevancy reduction.

f) Noizeless coding
   Huffman coding is applied, that assigns shorter code word for frequently occurring value and less occurring values for longer code word.

g) Rate/distortion control
   Quantization and coding are repeatedly applied so that generated output bit rate of a frame is shorter than the assigned bit length for the frame.

V.4.3   Specification of the audio coding in the standard

For audio coding system used in ISDB-T, MPEG-2 AAC (ISO/IEC 13818-7)[87] has been adopted in Japan and MPEG-4 AAC (ISO/IEC 14496-3, Subpart 4) has been adopted in Brazil[88]. As the concept of the two

---

[87] ISO/IEC 13818, "Generic coding of moving pictures and associated audio," http://mpeg.chiariglione.org/standards/mpeg-2
[88] ABNT NBR 15602, Audio and Video Compression

are almost the same, this Section will discuss on the MPEG-2 AAC, followed by the difference between the MPEG-2 AAC and MPEG-4 AAC.

The main parameters of audio coding system described in Table V-4 are defined in ARIB STD-B32 Part 2 Chapter 5.2.1

Table V-4 Main parameters of audio coding system[89]

Parameter	Restriction
Bitstream format	AAC Audio Data Transport Stream (ADTS)
Profile	Low Complexity (LC) profile
Max. number of coded channels	5.1 channels per ADTS (5.1: 5 channels+LFE)
Max. bitrate	Compliant to ISO/IEC 13818-7

MPEG-2AAC has three profiles described below:
(i) Main profile: high complexity
(ii) Low Complexity(LC) profile: reduces hardware complexity
(iii) Scalable Sampling Rates (SSR) profile: selectable coding 4 band

In ARIB STD-B32 Part 2 Operational Guidelines, it states the reason why these main parameters have been chosen for audio coding system. For reader's reference, the interpretation regarding the main parameters described in ARIB STD-B32 Part 2 Operational Guidelines are quoted in the following box.

> The ADTS format — a format with a header in each frame — has been adopted as the bitstream format, since it will be used for broadcasting purposes. Restrictions on ADTS header will be given later.
>
> The LC profile was initially adopted for use with BS/broadband CS digital broadcasting based on the following factors:
> (a)   As a result of the AAC audio quality assessment test conducted by ARIB in June 1998, we found that the LC and SSR profiles met the ITU-R broadcasting quality criteria or the criteria required by BS/broadband CS digital broadcasting at 144 kbps/2 channels or more.
> (b)   It was pointed out that SSR profile-specific features were not effective for BS/broadband CS digital broadcasting.
> (c)   It was pointed out that the LC profile could improve audio quality as a result of optimization and technical advance of encoders beyond year 2000 when BS digital broadcasting would begin.
> (d)   Based on the premise that BS digital broadcasting shall begin in 2000, it was pointed out that it would be possible to develop encoders and receivers for the LC profile, but would be difficult to do so for the MAIN profile.
> (e)   There is a significant difference in chip costs between MAIN and LC profiles.
> (f)   There are technical problems to be solved for MAIN profile.
>
> We have decided to adopt the LC profile for terrestrial digital television broadcasting and terrestrial digital audio broadcasting as well for the above reasons and in view of consistency with BS/broadband CS digital broadcasting.

---

[89] Quoted from ARIB STD-B32, Part2, Chapter 5.2.1

> No restrictions have been introduced in relation to the maximum bitrate. In terms of the standard, the maximum bitrate for AAC format is 288 kbps/channel when the sampling frequency is 48 kHz.

V.4.4 Comparison of the Japanese and Brazilian audio coding system

As described in the beginning of this chapter, both the Japanese and Brazilian ISDB-T system adopted AAC coding for audio coding technology. This is entirely different with the video coding. "Harmonization document"[90] published by ISDB-T International Forum" explains the differences between the Japanese and the Brazilian standard for whole technical detail.

In this text, only the outline of difference in the audio coding is given below, for further details, please refer to the Harmonization document.

(1) Coding system

ABNT NBR 15602-2 refers to MPEG-4 AAC[91], while ARIB STD-B32 part 2, refers to MPEG-2 AAC[92]. For further details, refer to Subsection 6.3 of "harmonization document Volume 3".

(2) Audio Coding Mode

ABNT NBR 15602-2 adopts "2/0 + LFE" as one of the audio modes, which ARIB STD-B32:part 2 does not. ABNT NBR 15602-2 recommends two modes, 2/0 and 3/2 + LFE, among several modes specified, while other modes are also recommended in ARIB STD-B32 part 2.

---

[90] DiBEG, ISDB-T Harmonization Documents (2009), Audio Coding, http://www.dibeg.org/techp/aribstd/harmonization.html
[91] MPEG-4 AAC (ISO/IEC 14496-3, Subpart 4)
[92] MPEG-2 AAC (ISO/IEC 13818-7)

Chapter VI. Datacasting System

One of the biggest features of digital broadcasting compared to analog broadcasting is Data Casting. With data casting system, broadcast signals can carry not only video and audio stream, but also high-bitrate data streams, which enable broadcasters to provide additional services.

VI.1 Data Broadcasting Service and Classification

There are two types of data broadcasting defined in the ISDB-T system; Multimedia Coding and Caption/Superimpose.

Applying multimedia coding technology, various new services have been introduced. Most common service is NEWS including society, market and sports news and weather information. Data broadcasting shows data broadcasting screen along with the main video content. By clicking the icons with push buttons on remote, viewers can follow the link just like internet surfing. The service is quite similar to that of the Internet.

In order to provide these multimedia services, Broadcasting Markup Language (BML) was defined for ISDB-T. Figure VI-1 shows screen image of BML and HTML. HTML assumes the use of computers. Generally, users watch screen in the distance of 30 to 50 cm; input devices are mouse and keyboard. On the other hand, BML assumes the use of fixed TV or mobile phone screen. Viewing distance is 1-3m for fix TV and around 30 cm for mobile phone. Both devices have cursor keys, whereas HTML can expect more sophisticated input device like keyboards.

a) BML screen   b) HTML screen

Figure VI-1 Example of BML and HTML screen

Incorporating these differences of the user interface, BML is specially designed for the TVs. While HTML screen generally have many hyper-links and text documents, BML is designed for presentation of a few movies/images and small number of characters with much fewer hyperlinks as viewers watch screen in farer distance and input device have limited capacity depending on the focus move by cursor key. Also, with limitation of input device, BML assumes no scrolling function.

In addition, it is desirable to present contents as contents creators intended regardless from BML browser manufactures or TV models. As a result, video/images of BML screen have absolute positioning and sizing towards screen.

In addition, Synchronization of BML contents with TV programs is important factor for data broadcasting. When TV program changes, BML needs to change also. For example, when switching from soccer game to music program, BML needs to change from tournament score to hit-charts triggered by the change of the program. Furthermore, multiple plane model was incorporated in order to display video/sound and BML screen simultaneously (see for section VI.2.6 for details).

Table VI-1   Comparison between BML and HTML

	BML	HTML
User environment	Viewing distance: 1-3 m Focus display: Focus of Hotspot Input device: Remote Controller with color key	Viewing distance: 30-50cm Focus display: free cursor Input device: mouse + keyboard, touch panel + keyboard
Feature of presentation	Fewer hyper-links in one screen Synchronization with TV programs No scrolling	Number of hyper-links in one screen Character-centric information presentation Assumes scrolling
Functionality	Synchronization with TV program Absolute positioning with CSS Multiple plane model including blending of multiple planes	No synchronization mechanism Relative positioning by browser Single plane model basis

Another type of data broadcasting service is Caption/Superimpose. Caption/Superimpose is a service to display characters or Figures embedded with the video contents.

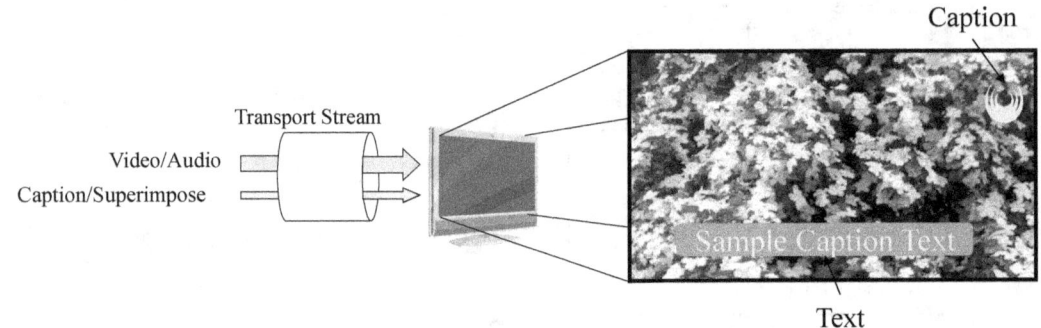

Figure VI-2   Image of caption/superimpose

VI.1.1   Profiles

Data Broadcasting are defined in ARIB standard (ARIB STD-B24 Chapter 1 Part 3, ARIB STD-B10), and ARIB operational guideline (ARIB TR-B14). While the standard defines the flexible functionality for future expansion, the operational guideline defines the minimum required functions for the moment.

As various types of receiver equipment are possible in ISDB-T services, ARIB operational guideline (ARIB TR-B14) categorizes receivers into three profiles based on the nature of mobility and screen size. The parameters for Data Broadcasting, such as screen or font size are defined for each profile. Accordingly, available functionality of the BML and scripts is also limited based on the profile. Table VI-2 shows the definition of profiles based on the receiver type in ISDB-T. Please note that it defines only on fixed and mobile receivers. At the beginning of the discussion for standardizing the ISDB-T in Japan, there were three profiles prepared for receiver types. However, as the discussion proceeded into details and considered the tradeoff between total bitrate and number of services (HD, SD or portable), most of the Japanese broadcast operator choose one HD and one One-Seg (portable). As a result, Profile B for car TVs were left undefined (To Be Determined) in the ARIB standard.

Table VI-2   Profiles for different types of receiver units

Profile	Contents
Profile A	Basic operation profile mainly targeting fixed receiver units (Stationary TVs, STBs, Portable TVs, etc.)
Profile B (T.B.D)	Basic operation profile of data broadcasting services mainly targeting transportable receiver units(car TVs, portable TVs, PDAs, etc.) (This profile is defined here but not actually used.)
C-profile	Basic operation profile of data broadcasting services mainly targeting portable receiver units (mobile devices, etc.)

VI.2   Broadcast Markup Language (BML)

Broadcast Markup Language (BML) is based on the tag sets defined in XHTML 1.0. In addition to the original tag sets, BML incorporates broadcasting unique functions such as video/audio play outs and synchronization with TV programs on air. Screen layout and display function adopts CSS, which is also common for website layout-design such as positioning and defining appearance of characters and images. Scripting language adopted is ECMAScript, which is based on JavaScript, also with some addition of broadcasting unique functionalities. In order for scripting language to dynamically access and update the content, the API called Document Object Model (DOM) – W3C recommendation - was adopted.

These language specification defined in ARIB B24 are very common also for the Internet use, and defined flexibly for the future use. However BML limits its functions, based on profiles of receivers, tailored for the receivers which generally have less computing power and resources than computers. These subsets for terrestrial broadcasting are defined in operational guideline TR-B14.

VI.2.1   Language structure of BML

BML is quite similar to HTML.[93] Internet engineer who is familiar with HTML or JavaScript will easily understand basic concept of the BML. A sample code is indicated in Figure VI-3.

```
<?xml version="1.0" encoding="EUC-JP" ?>          ←XML declaration

<!DOCTYPE bml PUBLIC "+//ARIB STD-B24:1999//DTD BML Document//JA"
    http://www.arib.or.jp/B24/DTD/bml_1_1.dtd">   ←Document Type Definition
```

---

[93] ARIB STD-B24 Vol.2, A2-4.8

```
<?bml bml-version="3.0" ?>                              ←BML version

<bml>
<head>
<style><![CDATA[
   #test { top: 260px; left: 330px; width:300px; height:60px;
           font-size:30px; color-index: 7; }
]]></style>

<script><![CDATA[
function catchEvent(){
   if(document.currentEvent.status ==0){
      browser.launchDocument("EventCatched.bml", "cut");
   }
}
]]></script>
<link href="basic.css" />
<bevent>
   <beitem type="EventMessageFired" es_ref="/89" message_id="1"
      Subscribe="subscribe" onoccur="catchEvent();" />
</bevent>
</head>
<body>
   <p id="test">Hello World!<br/>
</body>
</bml>
```

Figure VI-3　Sample BML code for A profile

(1) Declaration and Tag Sets
BML starts with declaration followed by BML body.　For A profile, Declaration consists as follows.

XML declaration:
　　XML version should be "1.0" and character encoding should be "EUC-JP."
Document Type Declaration:
　　The document type declaration names the document type definition (DTD) in use for the document.
　　IT should name DTD, "bml_1_0.dtd"
BML version
　　BML version should be "3.0" for the terrestrial broadcasting in Japan

　Declaration differs between profiles.　For C profile declaration indicated in Figure VI-4 is used, whereas A profile declaration is indicated in Figure VI-3.

```
<?xml version="1.0" encoding="Shift_JIS"?>
<!DOCTYPE html PUBLIC "-//ARIB//DTD XHTML BML 12.0//JA"
   "http://www.arib.or.jp/B24/DTD/bml_12_0.dtd">
<?bml bml-version="12.0" ?>
```

Figure VI-4　Document declaration for C profile

BML document follows the declaration. BML element must be consisted of head element and body element in this order. (In order to indicate start and end of the BML document, A profile use <bml></bml>; C profile use <html></html>).

Child elements of the "head" element are: "title", "meta", "style", "script" and "bevent" element. These elements appear only once in the "head" element, specifically in this order. Elements other than "title" are optional. "bevent" element has "beitem" element as child element.

Child elements of "body" element are: "div" and "p" elements. "Div" element has "div", "p" object and "input" elements as child elements. "P" element has "br", "span" and "a" elements as child elements. "A" element has "br" and "span" as child elements.

Table VI-3   BML tag sets

TAG			Description
Bml			(for C profile, html is used)
	head		
		title	defines the title of the document
		meta	supplies additional information about a document
		link	links the current document to related documents. Reference is limited only for a stylesheet.
		script	adds scripts into documents
		bevent	(for C profile, bml:bevent is used
		beitem	(for C profile, bml:beitem is used
	body		
		br	a forced line break
		div	logical divisions within the document
		p	paragraphs
		pre	preserves both spaces and line breaks of enclosed document (A profile only)
		span	allows localized formatting within documents for enclosed document
		a	designates the start or destination of a hypertext link
		form	defines a container for form controls and elements (C profile only)
		input	specify a field in a form for input
		textarea	specifies a multiline plain text edit control for the element's raw value (C profile only)
		img	specifies a image to display(C profiles only)
		object	specifies an object to be presented
		style	contains information on the stylesheet (A profile only)

(2) Event Synchronization

Tag sets of BML incorporate extension for event synchronization, the functionality of synchronous control between main video/audio stream and BML contents.[94] For example, BML can switch the contents from tournament score to hit charts, triggered by event message transmitted when program of soccer game ends and music show starts.

BML has "bevent" and "beitem" element in its tag sets to implement this functionality. "bevent" is a set of "beitem" elements. "beitem" defines nature of event to be watched and function to be fired when the events are caught..

Table VI-4   Representative attributes of the beitem element

Attribute	Value
Id	(ID) identifiyng this beitem element
type	EventMessageFired, EventFinished, EventEndNotice, Abort, ModuleUpdated, ModuleLocked, TimerFired, DataEventChanged, CCStatusChanged, MainAudioStreamChanged, NPTReferred, MediaStarted, MediaStopped, MediaRepeated, DataButtonPressed, IPConnectionTerminated, PeripheralEventOccured
onoccur	(Script) identifying a procedure that is executed when an event has occurred
es_ref	(URI) identifying an elementary stream (ES)
message_id	(Number) identifying each of the event messages distributed in the general event message descriptor
subscribe	"subscribe" (only "subscribe" can be taken)

VI.2.2   Monomedia

Monomedias, such as pictures and movies are presented in the object element. Object element cannot be placed directly below body element but below div element.

Table VI-5   Example of monomedias for BML

Monomedia Type	Encoding	Media Type
Video	MPEG-2 Video	video/X-arib-mpeg2
	MPEG-1 Video	video/X-arib-mpeg2
	H.264|MPEG-4 AVC Video (baseline)	video/X-arib-H264-baseline
Still Picture	JPEG	image/jpeg
Text/Graphic	PNG	image/png image/X-arib-png
	MNG	image/mng image/X-arib-mng
	8-bit character code (including EUC-JP)	text/arib-bml; charset="euc-jp" (text/X-arib-jis8text)
	S-JIS	text/arib-bml; charset="Shift_JIS"
Audio	MPEG-2-AAC	audio/X-arib-mpeg2-aac
	AIFF	audio/X-arib-aiff
	System Beep	audio/X-arib-romsound

---

[94] See ARIB STD-B24 Vol.2, 5.3.20.1

Video/Audio	TS with timestamp	application/X-arib-tts
Other	External Font	application/X-arib-drcs
	Binary Table	application/X-arib-btable

Figure VI-5 shows an example of an object element coded to process an image. Here, objects are to be identified based on the following naming convention.

```
<object data="image.jpg" type="image/jpeg" height="200" width="200" />
<object id="HD_video"
   style="left:0; top:0; width:960px; height:540px;"
   type="video/X-arib-mpeg2" data="/-1" remain="remain"/>
```

Figure VI-5 Sample code for object element

The namespace for a resource transmitted through a data carousel is defined as follows. The concept of the data carousel is explained in VI.5.2.

arib-dc://<original_network_id>.<transport_stream_id>.<service_id>
  [;<content_id>] [.<event_id>]/<component_tag>/<moduleName>

The AV streams and subtitle component is uniquely identified in the network with the following naming convention.

arib://<original_network_id>.<transport_stream_id>.<service_id>[;<content_id>]
  [.<event_id>]/<component_tag>[;<channel_id>]

In the example shown in the Figure VI-5, data="-1" is abbreviation of data="arib://-1,-1,-1/-1", meaning the video stream of a broadcasting service currently selected on a receiver is specified.

(1) Video Presentation
When a video or a still picture or graphic form is presented using an object element, a scaling operation shown below is performed. In the diagram, W indicates the number of horizontal pixels of an image, H indicates the number of vertical pixels of an image. Also, x and y indicate horizontal and vertical coordinates in the presentation screen. Sx and Sy indicate a horizontal and a vertical scaling ratio respectively. No value that places a video outside of a corresponding video plane must be specified. Also, no value that places a still picture outside of a corresponding still picture plane must be specified.

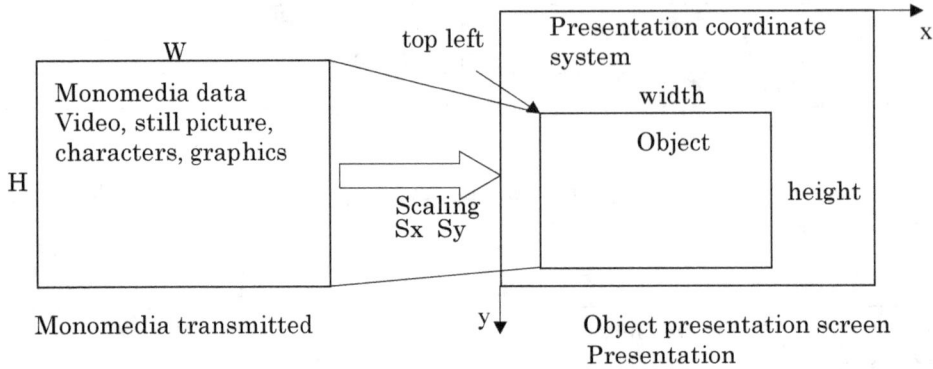

Figure VI-6 Video Presentation and coordinates

(2) Remain property
When this value is specified, the documents share the objects. During transition between documents, the presentation states of the shared objects are retained. The attribute values of the object element and the CSS properties in a document before transition are inherited to the object element after the transition. This feature enables continuous playback of video and audio between documents.

When pages switch, objects such as sounds or movies can be interrupted or redrawn from the start. BML retains the property for object tag to indicate the object stays and continues to play between the transitions of pages. This retention property can be used under rather strict conditions.

The same set of the id attributes must be applied to both a document before transition and a document after transition.

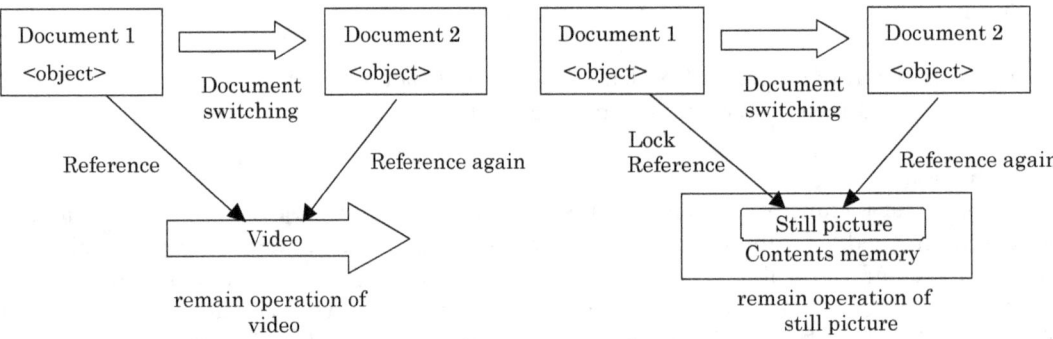

Figure VI-7 Remaining property and transitions

(3) Restriction on the use of Object
There are some restrictions on the use of object. It is desirable that the object element showing a video/still picture is the first element to be presented in the document. If it does not appear first, however, the element must be placed so that a character or graphic object is not on top of it.

The clipping operation is not performed by the parent div element when a video, still picture or MNG image is specified by the object element. In other words, monomedia image specified by object element must not be specified out of area with parent div element.

When a div element has as its child element, the object element associated with a GIF/MNG graphic item, the background color (background-color) of the div element must be colored other than the colors for rendering.

VI.2.3 Remote controller operation

Unlike a PC with a keyboard and mouse, TV is more of a low-interaction device, whose user interface is merely a simple remote control. At the same time, the users are already get used to the computers and smartphones where they can navigate and choose the desired service at the click of a button. BML are designed to enable interaction solely using buttons on simple remote controllers.

While BML browser presentation is very similar to that of Internet, BML browser cannot assume user interface like mouse which depicts arbitrary special coordinates of cursor movements and click. Rather BML is designed so that the user could use exclusively the arrow keys to move the focus within and between objects on the screen.

In order to describe focus movement operation with a remote control, BML equips various extensions.[95]

nav index:
> This property specifies an index for an element to which the focus is applied. Any element with this property set to none, is given no focus. The focus is initially applied to the element with this property set to 0. Any value of this property must be unique in a BML document.

nav-up, nav-down, nav-right, nav-left:
> These properties specify the value of the nav-index property describing an element to which the focus is applied when the arrow key (up, down, right, left) is pressed.

To indicate what object has focus, dynamic pseudo-classes is provided in the style sheet. Focus class specifies attribute of object presentation when a focus is coming over with an operation of a remote control or other triggers.

The nav-* property example is shown in the Figure VI-8. Initially focus is on the object with nav-index:0. If the down key is pressed when a focus is on object 1 (id=a), the focus is moved to object 3 (id=c) according to the specification of "nav-down:3."

```
<body>
 <div id="div1" style = "..." >
  <object id="a" style="... nav-index:0; nav-up:3; nav-down:1" />
  <object id="b" style="... nav-index:1; nav-up:0; nav-down:2" />
  <object id="c" style="... nav-index:2; nav-up:1; nav-down:3" />
  <object id="d" style="... nav-index:3; nav-up:2; nav-down:0" />
 </div>
</body>
```

Figure VI-8　Sample code on the "nav-index"

VI.2.4 Layout design with CSS

BML adopts CSS2 (and CSS1 in some part) style sheet that is a popular design representation for website coding, with certain restriction tailored for the capacity of receivers.

---

[95] Refer to ARIB STD-B24, Vol.2, 5.4.13.3

In order for BML to display precisely as the original design, all object needs the coordinates and size parameters including the foreground and background color parameters in typical cases.

In CSS, pattern matching rules, which is called "selector," determine which style rules apply to elements in the document tree. BML also adopts selector concept with certain limitations tailored for the capability of receivers.

The selectors are used only in the default style sheet and in the content of the style element. Only the following selectors defined in CSS 1 are operated:

For all elements	Ex.:	* { color-index: 1 }
Type selector	Ex.:	div { color-index: 1 }
Class selector	Ex.:	div.class1 { color-index: 1 }
	or	.class1 { color-index: 1 }
ID selector	Ex.:	#chap1 { color-index: 1 }
:active and :focus pseudo class	Ex.:	:focus { color-index: 1 }
	or	E:focus { color-index: 1 }

Please refer Figure VI-3 also for the example of use of style sheet.

Table VI-6 Elements to which CSS properties are applied

	div	p	br	span	a	input	object	body
Box model								
margin	△	△	-	-	-	△	△	-
padding-top	△	√	-	-	-	√	△	-
padding-right	△	√	-	-	-	√	△	-
padding-bottom	△	√	-	-	-	√	△	-
padding-left	△	√	-	-	-	√	△	-
border-width	√	√	-	√	√	√	△	-
border-style	√	√	-	√	√	√	△	-
Visual formatting model								
position	△	△	△	△	△	△	△	-
left	√	√	-	-	-	√	√	-
top	√	√	-	-	-	√	√	-
width	√	√	-	-	-	√	√	-
height	√	√	-	-	-	√	√	-
z-index	△	△	△	△	△	△	△	△
line-height	-	√	△	△	△	√	-	-
display	△	△	△	△	△	△	△	△
Other visual effects								
visibility	√	√	-	△	△	√	√	△
overflow	△	△	-	-	-	△	△	-
Background								

Property									
	background-image	-	-	-	-	-	-	-	√
	background-repeat	-	-	-	-	-	-	-	△
Font									
	font-family	-	√	-	√	√	√	-	-
	font-size	-	√	-	√	√	√	-	-
	font-weight	-	√	-	√	√	√	-	-
Text									
	text-align	-	√	-	-	-	√	-	-
	letter-spacing	-	√	-	△	△	√	-	-
	white-space	-	△	-	-	-	△	-	-
Extended property									
	clut	-	-	-	-	-	-	-	√
	color-index	-	√	-	√	√	√	-	-
	background-color-index	√	√	-	√	√	√	△	√
	border-top-color-index	√	√	-	√	√	√	-	-
	border-right-color-index	√	√	-	√	√	√	-	-
	border-left-color-index	√	√	-	√	√	√	-	-
	border-bottom-color-index	√	√	-	√	√	√	-	-
	resolution	-	-	-	-	-	-	-	√
	display-aspect-ratio	-	-	-	-	-	-	-	√
	grayscale-color-index	-	√	-	√	√	√	-	-
	nav-index	√	√	-	√	√	√	√	-
	nav-up	√	√	-	√	√	√	√	-
	nav-down	√	√	-	√	√	√	√	-
	nav-left	√	√	-	√	√	√	√	-
	nav-right	√	√	-	√	√	√	√	-
	used-key-list	-	-	-	-	-	-	-	√

√: implemented
△: partially implemented.   Receivers use default value regardless of this property being set
- : not used

(1) Color Look Up Table (CLUT)
In order to specify colors such as character, background and border colors, BML uses index color defined in the Color Look Up Table (CLUT) – a table used to convert an index color value to a physical value – with YcbCr color space with 8-bit depth (256 colors maximum).

If the receiver cannot find the CLUT property for body element, it uses the default CLUT. It should be noted that the default value for the background-color-index "0" is black and for body element; "8" is transparent for div and p element.

VI.2.5   Coordinate system and z-index

Screen resolution is 960×540 to be used in combination of HD screen (1080i or 720p) and 720×480 in combination to SD screen (480i/480p). In order to place objects in the screen, it is mandatory to indicate its vertical and horizontal coordinate and the size. Position of the elements – p, div, input, img, object, pre and form – is absolute. Default value of left, top, width, height and margin are 0. The positioning method uses relative coordinates from a parent element are as follows:

    left: Distance from the left of box of parent element (number of pixels);
    top: Distance from top of box of parent element (number of pixels);
    width: Width (number of pixels);
    height: Height (number of pixels);
    visibility: "visible", "hidden" or "inherit";

The objects are displayed in the order of appearance in the BML source. Therefore the objects appear first in the source is drawn first. This object is overdrawn when proceeding objects in the source with the same x-y position are drawn.

An example is illustrated on Figure VI-9. a) illustrates the structure of a BML document. An object is logical structure of a document – such as <tr>...</tr> or <div> ... </div>. (Details are explained in 0.) In this tree, object are written in the BML source cord in the order of up to down and left to right. b) illustrates resulting presentation screen. Objects are drawn in first-bottom and last-top basis.

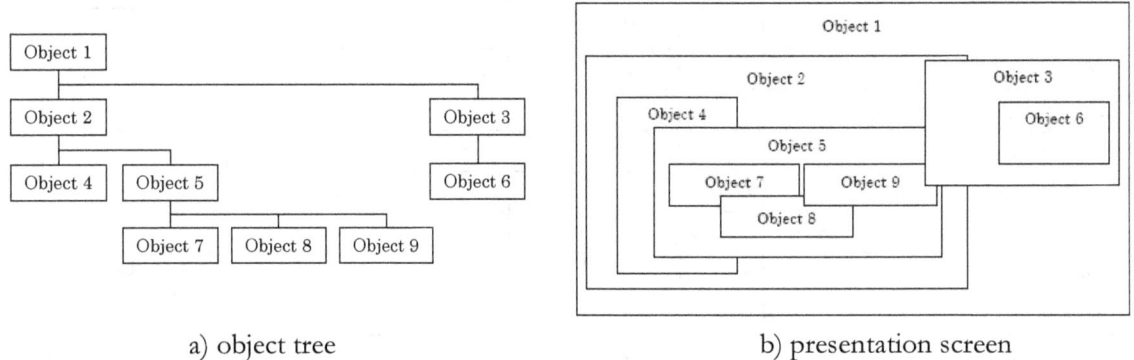

a) object tree   b) presentation screen

Figure VI-9   Object structure and presentation

VI.2.6   Plane model

Movies, pictures and text/shapes are mixed based on the drawing plane model, illustrated on Figure VI-10. Movies and pictures are switched by rectangular area, and composited into video/still picture plane. Further, video/still picture plane and text/graphic plane are mixed based on the α-value. When text graphic plane is mixed, the contents on the plane are colored based on the CLUT.

α-value represents the proportion of the color when mixed. α-value of 0 means transparent. α-value of 255 means non-transparent.

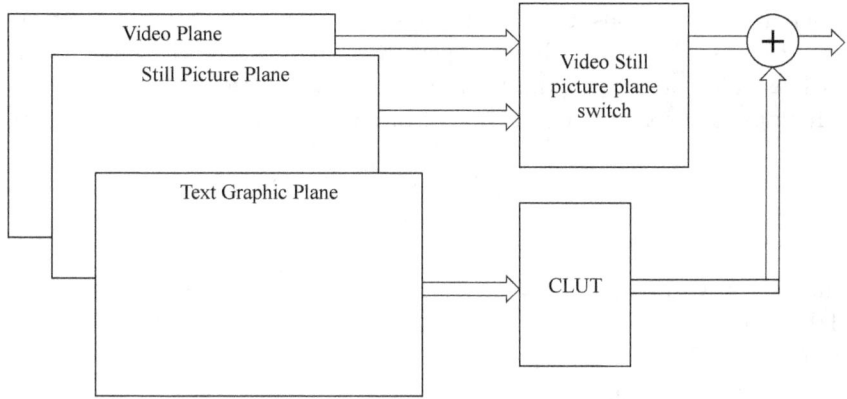

Figure VI-10   Model for drawing planes[96]

VI.2.7   Scripting

For BML browser to dynamically respond to the user operation, scripting language – ECMAScript - is defined, in which the same description in ECMA-262[97] is used with certain restrictions for certain receivers.

Further, DOM (Document Object Model) - An API also known as DOM-API is equipped that defines the logical structure of an XML or HTML document and the way to access or manipulate an XML or HTML document.

The Syntax, semantics, and built-in objects of the scripting language conform to ECMA-262.   However, the following exceptions are allowed in real operation.

Using EUC-JP for character encoding.
Setting the Number size smaller than 64 bits.
Not implementing Float.

Broadcasting Extended APIs including date/time, table operations, external characters, EPG, operation control, event acquisition, and timer are added.   Conventions for Broadcasting Extended Object Group and Browser Pseudo Objects are also added.

- Broadcasting Extended Object Group
    This object group handles table data.   It includes CSVTable object and BinaryTable object.   Its operation is the same as the native objects of ECMAScript.
- Browser Pseudo Objects
    These objects have been added to implement functions specific to broadcasting.   Unlike the native objects, they do not inherit functions.   They have global scopes.
- Navigator Pseudo Objects
    A string identifying a BML browser.   The available values are defined in an operational standard regulation.

(1) Extended Functions for Broadcasting

While BML incorporates ECMAScript basically as it is, to handle the broadcast of unique features such as timer recording of TV program, or memory management, "Extended Functions for Broadcasting" are

---

[96] Excerpt from ARIB STD-B24, Volume 2, Appendix 2, Chapter 5
[97] Standard ECMA-262, ECMA International, https://www.ecma-international.org/publications/standards/Ecma-262.htm

implemented.[98] When these functions used to bind ECMAScript, they are treated as Browser Pseudo Objects. A Browser Pseudo Object is a global object that provides broadcast of extended functions as a method. It is accessible with a browser.method name or a browser.Ureg.

The Extended Functions for Broadcasting defined in the ARIB STD-B24[99] are listed as follows:
- EPG functions
- Event group index functions
- Series reservation functions
- Subtitle presentation control functions
- Non-volatile memory functions
- Extended APIs for Storing
- Interaction Channel functions
- Operational control functions
- Receiver sound control
- Timer functions
- External character functions
- Functions for controlling external devices
- Functions for controlling bookmark areas
- Other functions (random number generation, data and time)
- Ureg pseudo object properties/ Greg pseudo object properties
- Functions for Printing
- Server-based broadcasting functions

(2) Document Object Model

DOM (Document Object Model): An API, also known as DOM-API, that defines the logical structure of an XML or HTML document and the way to access or manipulate an XML or HTML document. This API is an independent interface of platforms and languages. The APIs conform to W3C Recommendation DOM Level 1.

The definition of the DOM standard consists of the next two parts: the Core DOM and HTML DOM. Further, the Core DOM is divided into the following two parts: fundamental interfaces and extended interfaces. The fundamental interfaces of the Core DOM represent basic functions for an XML document, and work as the base of HTML DOM. It is preferred that the DOM must be applied by implementing all of the fundamental Core interfaces before incorporating the HTML Section and the extended Core interfaces.

This interface is an extended DOM interface for operating elements and attributes defined in BML. It is HTML DOM Interface defined in ARIB STD-B24[100] with an extension for BML.

---

[98] ARIB STD-B24, Vol.2, 7.6
[99] ARIB STD-B24, Volume 3, 7.6 Extended Functions for Broadcasting (Browser Pseudo Object)
[100] ARIB STD-B24, Section 7.1.3 HTML DOM Interfaces

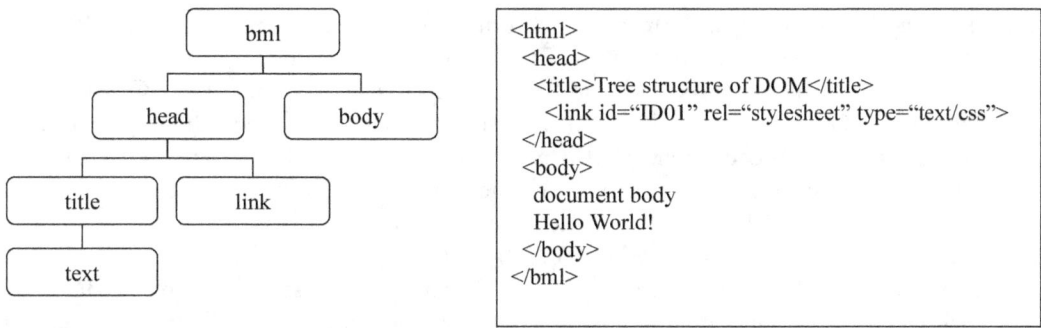

Figure VI-11 DOM document tree and sample BML code

VI.3 Ginga

Ginga[101] is the middleware adopted by the ISDB-T standard in Brazil and Latin American countries. Ginga architecture defines two types of applications: declarative and procedural applications, whose protocol stack is shown in the Figure VI-12.

Declarative application, Ginga-NCL[102][103][104], is similar to the HTML documents. Ginga-NCL is a derivation of XML language, incorporating ECMAScript, CSS layout design, DSM-CC stream events and "Lua" a scripting language. NCL formatter functions as a language interpreter and presentation engine.

Procedural application, Ginga-J is based on Java technologies with Java Virtual Machine and special APIs implementing functions unique to broadcasting receivers. Java software can be broadcasted from the station via the broadcasting signal or they can be placed inside the receiver on factory out basis (resident application) or transferred via a external input (USB portable, network port, memory card, etc.). These software are interpreted by the Ginga execution engine.

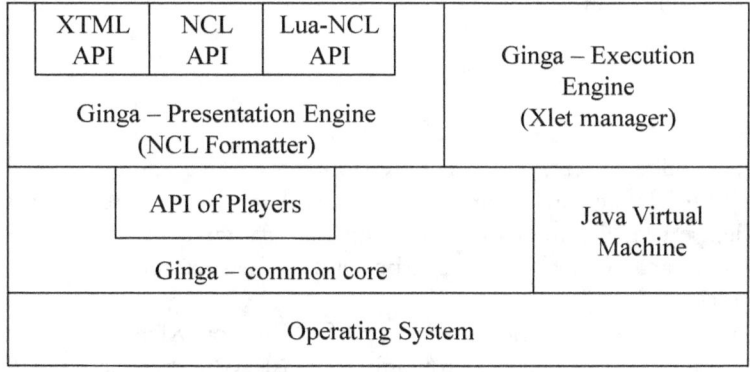

Figure VI-12 Ginga architecture

Both declarative and procedural application engine are placed on the placed on the Ginga-common core

---

[101] L. F. G. Soares, G. L. S. Filho, "Interactive Television in Brazil: System Software and the Digital Divide"
[102] ABNT NBR 15606, "Data Coding"
[103] ITU H.761, "Nested context language (NCL) and Ginga-NCL"
[104] L. F. G. Soares et.al., "Ginga-NCL: the Declarative Environment of the Brazilian Digital TV System"

that provides functions for decoding and presenting common contents types such as PNG, JPEG, MPEG etc. and functions for obtaining these contents from MPEG-2 Transport Streams.

In general, implementation of GINGA middle ware consists of Ginga-CC (Common Core), the presentation environment Ginga-NCL (declarative) and the runtime environment Ginga-J (procedural).

Contents developers can choose their platform of interactive (and non-interactive) contents from Ginga-NCL declarative application environment or Ginga-J procedural application environment, depending on their requirement. Ginga-NCL offers high degree of abstraction, where programmers simply need to provide the set of tasks to perform without addressing the details of the hardware behavior.

The Brazilian Association of Technical Standards (ABNT) defines the specification. ABNT NBR 15606 "Digital terrestrial television — Data coding and transmission specification for digital broadcasting" defines the specification of the GINGA system. ABNT NBR 15606 is structured as follows:

- Part 1: Data coding specification
  Part 1 specifies the reference model enabling data broadcasting, the monomedia supported by the data broadcasting system and code of caption and superimpose characters.
- Part 2: Ginga-NCL for fixed and mobile receivers – XML application language for application coding
  Part 2 specifies an XML application language, named NCL (Nested Context Language), the declarative language of the middleware Ginga, and the data coding and transmission for digital broadcasting
- Part 3: Data transmission specification
  Part 3 specifies data transmission for the data broadcasting scheme, part of the digital broadcasting scheme specified as the standard in Brazil.
- Part 4: Ginga-J – The environment for the execution of procedural applications
  Part 4 specifies Ginga-J, a Java-based script language designed for Television use. Details will be discussed in section VII.3.1.
- Part 5: Ginga-NCL for portable receivers – XML application language for application coding
  Part 5 also specifies Ginga-NCL. While Part 2 specifies Ginga-NCL for fixed and mobile receivers that receives full segment of one frequency channel, Part 5 specifies for portable receivers that receives only One-Segment of the channel. "One-Seg" is categorized in portable receivers.

VI.3.1 Ginga-J

Application of Ginga-J is called "Xlet", which is very similar to a Java applets embedded on HTML pages. The Xlets are generally broadcasted in a digital TV stream and fed to a receiver, where the middleware identify and run the Xlets automatically or manually with the Java virtual machine. Alternatively, Xlets can be put in the receivers on a factory out basis or can be send to the receivers via external inputs such as USB or communication channel.

The Ginga-J platform provides a standardized set of APIs for Xlets (transmitted applications). In addition to that, resident applications can use not only Ginga-J APIs, but also access to functions such as closed captions, conditional access (CA) messages, resident receiver menus and resident electronic programming guides (EPG).

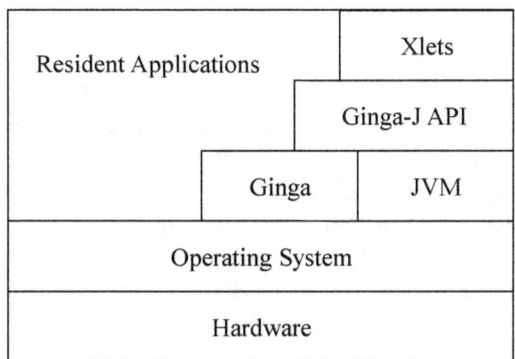

Figure VI-13 Ginga-J architecture and the execution environment

Life cycles of the Ginga-J applications are controlled by the Javax.microedition.xlet.Xlet interface. This interface allows Application Manager to change the running state (loaded/paused/active/destroyed) of the Ginga-J applications when the Application Manager receives:
- Flags originating from the broadcasters
- Selection based on a property menu with a list of applications
- Order originated in another Ginga-J application by means of the "application lifecycle management and control api" (JavaDTV)
- Order originated in NCL documents deploying one or more Xlets

Ginga-J API consists of Java DTV APIs and Java TV APIs on top of the common Java Runtime components including the Connected Device Configuration with Foundation Profile and Personal Basis Profile.

(1) Foundation Profile (FP) and Personal Basis Profile (PBP) on Connected Device Configuration
Connected Device Configuration (CDC) was developed as part of Java Platform, Mobile Edition (Java ME) family. The targets for CDC-based technology are mainly for consumer and embedded devices with resource-constrains but with network connection, including Set Top Boxes, high-end personal digital assistants (PDAs) or smart communicators. CDC provides the most basic set of libraries and virtual machine capabilities for these devices. Further, the Foundation Profile (FP) and Personal Basis Profile (PBP) are included. A profile here is defined as "a set of standard APIs that support a narrower category of devices within the framework of a chosen configuration." Foundation Profile provides very basic functions like I/O or networking without Graphical User Interface. Personal Basis Profile further supports lightweight components and Xlets.

(2) Java TV
Java TV provides APIs for digital TV-related capabilities for set-top boxes, Blu-ray Disc players, and other digital media devices. Java TV is designed to sit on atop the CDC, FP and PBP in order to provide control over functionality unique to broadcast- and media-related equipments, especially television receivers, in the form of API for access to the Service Information database, content selection, media control and access to the broadcasted data over the television signal. Java TV intends to be run on different platforms for digital TV reception independent of the nature of the transmission network.

Please note that the Xlet API was originally introduced in Java TV specification and now re-defined as part of PBP in javax.microedition.xlet package. While the javax.tv.xlet package is considered obsolete for Ginga-J, in order to accommodate legacy applications, both interfaces needs to be included for a platform. Newer interface is preferable for software development.

(3) Java DTV

Further, Java DTV environment is build atop Java TV to provide APIs for additional User Interface APIs, Access to broadcast data, or Access to receiver hardware and peripherals (smart cards), etc.

Java DTV adopts the Lightweight User Interface Toolkit (LWUIT) for its user interface APIs similar to Swing for java application running on PCs. LWUIT is commonly used in the Java ME platform especially for mobile phone application. LWUIT was developed to provide UI functionality unique to the hardware architecture so that UI will look and behave the same on all devices with LWUIT. LWUIT offers a set of pre-made graphical functions and features, without handling low-level programming, so that developers can quickly build a user interface.

(4) Protocol dependent service information API in Ginga-J

Ginga-J APIs absorbs protocol-dependent parameters especially Service Information and bridging function between Ginga-J and Ginga-NCL by providing br.org.sbtvd package. This package bases on the Java DTV package and provides modifications and additional functions for parsing and handling protocol-dependent parameters.

The br.org.sbtvd.bridge and br.org.sbtvd.bridge.ncl package bridge between Ginga-J and Ginga-NCL. This API allows Ginga-J application to present Ginga-NCL document through NCLPlayer class. Ginga-NCL document is also capable of including Ginga-J Xlets in the code as <media> element. The transition in the Ginga-NCL document can invoke javax.microedition.xlet.Xlet interface or NCLAnchorEvent class events in order to provide means for Ginga-J and Ginga-NCL to be synchronized.

(5) Sample codes of Ginga-J

The following sample code demonstrates the use of Ginga-J by simply displaying a text field on the screen.

```java
import java.awt.*;
import java.awt.event.*;
import javax.tv.xlet.*;
import org.havi.ui.*;
import org.havi.ui.event.*;

public class XletExample implements Xlet, KeyListener {
   private XletContext context;
   private HScene scene;
   private HStaticText label1;
   public XletExample() { /* empty function */}
   public void initXlet(XletContext xletContext) throws XletStateChangeException {
      this.context = xletContext;
   }
   public void startXlet() throws XletStateChangeException {
      HSceneFactory hsceneFactory = HSceneFactory.getInstance();
      scene=hsceneFactory.getFullScreenScene(HScreen.getDefaultHScreen().getDefaultHGraphicsDevice()); scene.setSize(640, 480);
      scene.setLayout(null); scene.addKeyListener(this); //el propio Xlet es el oyente
      label1 = new HStaticText("Hola Mundo Java", 35, 45, 660, 50, new Font("Tiresias", 1, 36), Color.red, Color.white, new HDefaultTextLayoutManager());
      scene.add(label1);
      scene.setVisible(true);
```

```
      scene.requestFocus();
   }
   public void pauseXlet() {
   }
   public void destroyXlet(boolean unconditional) throws XletStateChangeException {
      if (scene!=null) {
         scene.setVisible(false);
         scene.removeAll();
         scene = null;
      }
      context.notifyDestroyed();
   }
   public void keyTyped(KeyEvent keyevent) {
      /* empty function */
   }
   public void keyReleased(KeyEvent keyevent) {
      /* empty function */
   }
   public void keyPressed(KeyEvent e) {
      String mensagem = "Hello World!";
      }
      scene.removeAll(); scene.add(label1); scene.repaint();
   }
}
```

VI.3.2 Ginga-NCL

Ginga-NCL was developed by the Pontifícia Universidade Católica do Rio de Janeiro (PUC-Rio) and the Universidade Federal da Paraíba (UFPB), designed to provide a platform for the applications written in the declarative Nested Context Language (NCL). Ginga-NCL is a XML (extensible Markup Language) based language focuses on the aspect of interactivity, time and special

Ginga-NCL is also a declarative application environment of the Ginga middleware focusing on structure and relations of objects in time and space axis. Ginga-NCL was designed strictly separating between context and structure. Ginga-NCL does not define the media itself, rather Ginga-NCL document only defines how objects of the media are structured in relation to time and space. The medias used often used in Ginga-NCLs include: The video (MPEG, etc.), Audio (AAC, etc.), Image (JPEG, GIF, etc.) And text (TXT, HTML, etc.). Also, Ginga-NCL incorporates concept of synchronization in which timing information of medias (ex. 10 seconds elapsed in video play out) or signals from broadcasting stations can trigger new actions.

In addition to the declarative environment, Ginga-NCL enables interactivity of its contents by offering supports of two procedural languages: LUA and Java. Lua is the NCL script language and Java follows the Ginga-J specification.

Lua is an powerful but light programming language designed for general procedural programming and object-oriented programming. Lua works as a set of functions to be invoked by the program container (or a host client) to execute Lua codes, to read/write Lua variables and to execute C functions called from Lua codes.

It will be easier for readers to take a glimpse of the Ginga-NCL code rather than descriptive explanation. The Figure VI-14 shows a sample Ginga-NCL code, which will display a movie and two buttons. When you push the left button, the program will end.

```xml
<?xml version="1.0" encoding="ISO-8859-1"?>
<ncl xsi:schemaLocation="http://www.ncl.org.br/NCL3.0/EDTVProfile
http://www.ncl.org.br/NCL3.0/profiles/NCL30EDTV.xsd"
xmlns:xsi="http://www.w3.org/2001/XMLSchema-instance"
xmlns="http://www.ncl.org.br/NCL3.0/EDTVProfile" id="newDocument1">
<head>
  <connectorBase>
    <importBase documentURI="myConnectorBase.conn" alias="connBase"/>
    <causalConnector id="composerOnSelectionStop">
      <connectorParam value="xs:string" name="keyCode"/>
      <simpleCondition key="$keyCode" role="onSelection"/>
      <simpleAction role="stop"/>
    </causalConnector>
  </connectorBase>
  <regionBase>
    <region width="320" height="180" id="region0"/>
    <region left="61" top="197" width="40" height="40" id="rgButton1"/>
    <region left="183" top="198" width="40" height="40" id="rgButton2"/>
  </regionBase>
  <descriptorBase>
    <descriptor region="rgButton1" id="dButton" focusIndex="1" moveRight="2" focusBorderWidth="-2" focusBorderColor="black"/>
    <descriptor region="rgButton2" id="dButton2" focusIndex="2" moveLeft="1" focusBorderWidth="-2" focusBorderColor="black"/>
    <descriptor region="region0" id="node0"/>
  </descriptorBase>
</head>
<body>
  <port component="video" id="port_newDocument1_node0"/>

  <media descriptor="node0" src="../exemplo01/media/video1.mpg" type="video/mpeg" id="video"/>
  <media descriptor="dButton" src="../btn.gif" type="image/gif" id="btn1"/>
  <media descriptor="dButton2" src="../btn2.jpg" type="image/jpeg" id="btn2"/>

  <media id="luaSample" type="application/x-ginga-NCLua" src="main.lua" descriptor="dImages"/>

  <link xconnector="connBase#onBeginStartN" id="lnStart">
    <bind role="onBegin" component="video"/>
    <bind role="start" component="btn1"/>
    <bind role="start" component="btn2"/>
  </link>

  <link xconnector="connBase#onKeySelectionStopN" id="lnStopButton">
    <bind role="onSelection" component="btn1"/>
```

```
        <bind role="stop" component="video"/>
    </link>

    <link xconnector="connBase#onEndStopN" id="lnEnd">
        <bind role="onEnd" component="video"/>
        <bind role="stop" component="btn1"/>
        <bind role="stop" component="btn2"/>
    </link>
</body>
</ncl>
```

Figure VI-14　Sample Ginga-NCL Code

(1) Language Structure of Ginga-NCL

The Ginga-NCL code is structured by xml declaration and root element: <ncl> which has <head> and <body> element as child elements, that also has various elements as their sub elements (Table VI-7).

Table VI-7　Tag sets of Ginga-NCL

TAG			Description
ncl			
	head		
		importedDocumentBase	specifies a set of documents to import
		ruleBase	defines rules based on property, operator and values
		transitionBase	defines transition effects
		regionBase	defines set of regions which is rectangular presentation areas of objects
		descriptorBase	defines optional attributes, timing and special presentations of objects
		connectorBase	defines relations which is cause and resulting action of nodes
		meta / metadata	contains information which is not used by the NCL player for the application presentation, but can be used for other NCL user agents
	body		
		port	defines a interface point of a context
		property	defines object properties (local variables) or a group of object properties as one of the object interfaces
		media	defines a media object containing: video/audio/text/etc. content; code content; UTC content; or global variables for the application.
		context	defines set of objects (media, context or switch) and links
		switch	allows for defining alternative objects to be chosen in presentation time
		link	defines a relationship among media and other objects.
		meta / metadata	same as above (meta/metadata)

(2) Layouts and presentation

The Ginga-NCL adopts concept of region to define where presentation of objects take place. Region is a square which have height, width, top, left and z-index attributes where the square will be drawn. These attributes names are self explanatory: height, width, top, left shows the relative position to its parent region. Z-Index means smaller z-Index will be drawn lower with objects with greater z-Index value to be stacked on top and over painted.

Within these regions, media objects are binded by the <descriptor> elements in the <descriptorBase> element. Descriptor element indicates what media object (id) is drawn in what (region), also it specifies other temporal and special information needed to present each document component including timing and key-navigation attributes.

Media objects are placed in the body element, whose attributes include id, src (URI of the object) and type. URI (Uniform Resource Identifier) is commonly used on the Internet to identify where is the resource like "http://server_identifier/file_path/fragment_identifier". URI can take both absolute URI and relative URI. Type attribute follows MIME Media Types format and defines the classes of media and a media encoding type. For example, type can be "text/html", "image/png" or "video/mpeg4". Refer ABNT NBR 15606-2, 7.2.4 for complete set of definition.

The NCL Formatter and the Private Base Manager consists the Ginga NCL presentation engine. The NCL formatter receives an NCL document from the broadcast stream and controls its presentation, whereas the Private Base Manager also receives NCL document editing commands and maintains the active NCL documents.

The NCL Document files are transmitted through an object carousel or through an interactive channel protocol, and the document editing commands are transmitted via DSM-CC protocol.

(3) Event and transition handling

Behavior of objects in Ginga-NCL is best understood as state-machines. The <port> element gives an initial state for the objects in the code. Transitions of state of objects will be defined by <link> element. For example, when a video ends, it will fire "onEnd" message, that will trigger "start" or "stop" role of other objects. This binding is represented by <bind> element that has role (such as "start" or "onBegin") and component. The type of state transition (for example, when a object begins other objects starts, or when an object ends other objects starts) is defined in the xconnector attribute of the <link> element, which refers to the URL of a connector. <causalConnector> element in the <conectorBase> in the <head> element or another document refered by <importBase> element defines the set of link behaviors.

(4) Scripting

Ginga-NCL includes LUA Scripts as an imperative scripting language run by LUA engine. LUA Script in a NCL code is reffered from "<media>" elements with the "application/x-ncl-NCLua" (or "application/x-ginga-NCLua") type. Sample codes of LUA script is shown in Figure VI-15.

```
local text = "Hello World"
canvas:attrColor("silver")
canvas:attrFont("vera", 24)
canvas:drawText(20, 30, text)
canvas:flush()
```

Figure VI-15 Sample code of LUA Script

The lifecycle of an NCLua object is controlled by the NCL player, which is responsible for triggering the execution of an NCLua object and handling communication between an NCLua object and other nodes in an

NCL application. The LUA player is initialized by the code specified by the NCLua object code. At the same time, modules shown in Table VI-8 are connected. After the initialization, the execution of the NCLua object becomes event oriented: any event created by the NCL player reaches the registered event handler, or any NCL event state change notification is sent as an event to the NCL player.

Table VI-8 Default modules of NCLua object

Module	Functionality
canvas	offers an API to draw graphical primitives and manipulate images
event	allows NCLua applications to communicate with the middleware through events (NCL, pointer and key events)
settings	exports a table with variables defined by the NCL document author and reserved environment variables contained in an "application/x-ncl-settings" node
persistent	exports a table with persistent variables, which may be manipulated only by imperative objects
security	provides basic functions such as the generation and verification of digital signatures, message digest generation, and data encryption

VI.4 Caption and Superimpose

ARIB standard defines separately for Caption and Superimpose. Among the services that display characters on video screen, Caption refers to a service whose context relates to the main movie, sound and data, for example: translated transcript of dramas or news. In contrast, Superimpose refers to a text service whose context does not relate to the main movie, sound and data, for example: weather alert or news ticker.

Captions and Superimpose can be differentiated also where these texts are inserted, from the broadcaster's side or receiver's side. Open caption is inserted on the broadcasting side, where text data is combined with video signal and the resulting video signal is then transmitted. For closed caption, text data are not combined with video signal in the broadcaster's side, rather, text data and video signal are transmitted on the same broadcasted transport stream in the separate packet stream.

Viewers can operate with the remote to switch on/off the closed caption or change the display language of the closed caption. On the other hand, for the open caption, the viewers have no such choice to activate /deactivate.

Caption and Superimpose are drawn in accord to the Logical Structure of Screen Display indicated in the Figure VI-16. First, Video and Still picture plane are blended whose output is drawn on video and still picture switching plane. Then, Text and Graphic plane and Subtitle plane are drawn over top of the video and still picture switching plane.

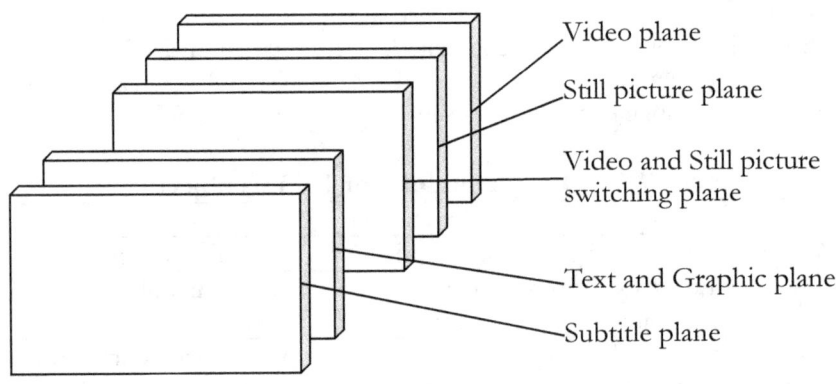

Figure VI-16 Logical structure of screen display

The usable display format is 1920x1080, 960x540, 1280x720 and 720x480 in the standard, however the operational guidelines limit the format to only 960x540 and 720x480 for practicability. It is applicable to both in the horizontal and vertical scanning. The starting point (0,0) of the coordinates of display area is the upper left of the "caption plane" regardless of the direction (vertical or horizontal) of scanning. Caption and Superimpose have only one display area for each. Characters are drawn in the same direction of the program content: for right hand side presentation for the horizontal writing; and downwards for vertical writing.

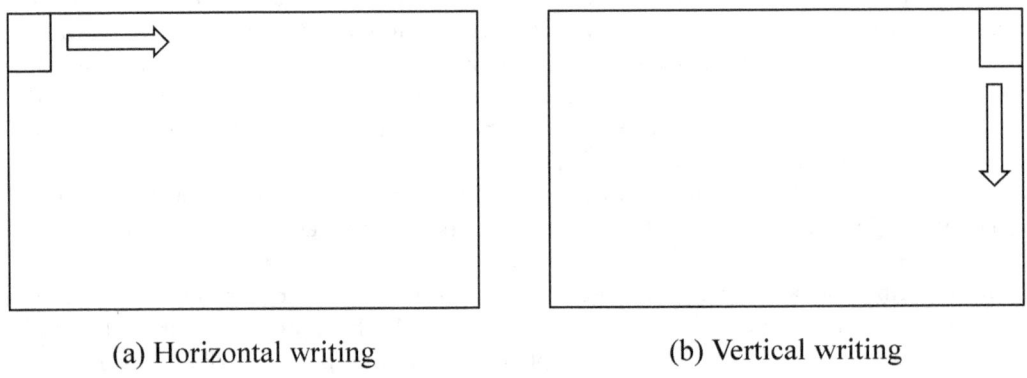

(a) Horizontal writing   (b) Vertical writing

Figure VI-17 Performance direction for horizontal and vertical writing

The character encoding method used for caption/superimpose is 8-bit character codes or utf-8. 8-bit character code is a coding system to display not only symbol, alphanumeric characters and control codes but also Japanese characters such as Kanji, Hiragana, Katakana and other characters for other languages. In this coding system, typical symbols such as alphanumeric characters and control codes are represented in 8 bits, whereas Japanese characters are represented in two bytes or 16 bits. Also 8 bit character codes can incorporate external bitmap font in the 16 bit coding system, whose bitmap data is transferred in the broadcasting stream, as Dynamically Redefinable Character Sets (DRCS).

VI.5 Transmission of Data Broadcasting

Various Transmission protocols are used depending on the nature and usage of data types. Table VI-9 shows the relation between transmission protocols of Data Broadcasting and their major functions and usage.

Table VI-9 Types of transmission protocol[105]

Transmission Protocol	Major Functions and Usage
Independent PES transmission protocol	Used for streaming synchronous and asynchronous data for broadcasting services. Applied to subtitles and superimposed characters.
Data carousel transmission Protocol	Used to transfer general synchronous and asynchronous data for broadcasting services. Applied to data transmission for download services and multimedia services.
Event message transmission protocol	Used for synchronous and asynchronous message notification to an application on the receiver unit from the broadcasting station. Used in multimedia services.
Interaction Channel Protocols	Transmission protocols used in a fixed network such as a PSTN/ISDN network and a mobile network including a mobile phone/PHS network when bidirectional communication is also used in a broadcasting service.

VI.5.1 Independent PES transmission protocol

Independent PES transmission is transmission of streams of data like video and audio elementary streams. Here, data are transmitted in a block format.

The major difference of independent PES transmission from data carousel is that PES transmission can add Presentation Time Stamp (PTS), which indicates when the contents should be presented on the screen so that the presentation of the contents synchronize with those of video and audio. Also PES transmission is one transmission of data, where as carousel transmission transmits data repeatedly.

Sentences or Figures are packaged into data unit along with header indicating data type and size. These data unit are packaged into data group. This data unit is PES data in the PES packets.

VI.5.2 Data carousel transmission protocol

Resources such as BML documents, JPEG/PNG, binary tables are transmitted by the data-carousel, which bases on the Digital Storage Media Command and Control (DSM-CC) data carousel specification defined in ISO/IEC 13818-6. Data carrousel is one form of transmission in which data are repeatedly delivered.

In broadcasting, the transmission is uni-directional from broadcasting stations to receivers, timing for the receivers on start and stop reception are unknown, repeated retransmission of data allows receivers to retrieve necessary information in an unpredicted timing manner. In the event when necessary information changes, i.e. viewers change a channel or programs on-air change, receives redefines the necessary information set and waits until the required information can be transmitted via the data carousel.

In addition, event messages are defined to enable services in synchronized TV programs transmitted on

---

[105] ARIB STD-B24, Volume 3, Chapter 4

the air. Conditions are as follows :
  1) to switch all display receivers at certain time simultaneously
  2) to switch display of data broadcasting in synchronized videos

In order to satisfy these conditions and for broadcasting stations to control receivers, event messages are defined to enable broadcasting stations to send messages to the receivers instantaneously or at a predetermined time.

According to the data carousel transmission protocol, data is transmitted using the DownloadInfoIndication (DII) message and the DownloadDataBlock (DDB) message. DII message and DDB message are transmitted in Section format.

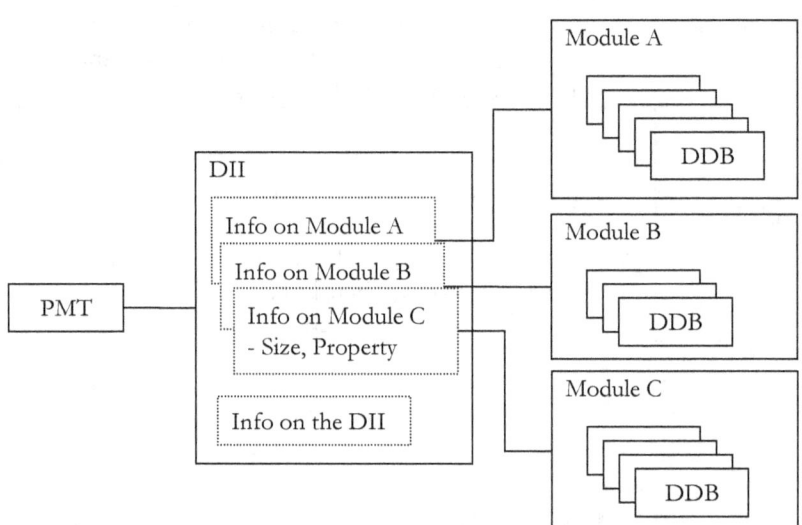

Figure VI-18 Organization of data carousel

DII message contains directory information on uniquely identification of the data carousel or the number of the modules. Also it contains information on property of the directory or modules in the format of descriptors.

Modules are the set of DDB messages. DDBs are sequentially numbered and sent in a fixed size except for the final numbered DDB. Receivers concatenate DDBs to form a original module.

As one screen of the contents are usually consisted with multiple resources (BML documents, images), multiple resources can be aggregated into one module in multi-part format to shorten the time needed to retrieve data and to save receiver resources.

VI.5.3   Event message transmission protocol

Event message is transmitted in DSM-CC Section containing NPT reference descriptor or General event descriptor. PID value of DSM-CC is stated in PMT.

The NPT reference descriptor is a descriptor to indicate the relation between NPT (Normal Play Time) and STC (System Time Clock). The general event descriptor is a descriptor to indicate information applied generally to event messages.

General_event_descriptor has mainly following fields.
  descriptor_tag
  event_msg_group_id

      reserved_future_use
      time_mode
      event_msg_MJD_JST_time or event_msg_NPT or event_msg_relativeTime
      event_msg_type
      event_msg_id

Among the fields of the General_event_descriptor, id and the time of the event is important. Id needs to be a value that a beitem element of BML states in order to be identified and caught. Time indicates when this message is going to be effective.

(This page is left intentionally blank.)

Chapter VII. ISDB-T receiver

ISDB-T service started in 2003 in Japan, 2007 in Brazil and thereafter in many South American countries, Philippines, Botswana and Maldives. As ISDB-T service become very popular, variety of receivers have been sold.

There are many types of receiver developed and sold based on user request. The variety include high quality HDTV receiver with large Flat Panel Display (FPD), medium quality reasonable price receiver with display, STB (Set Top Box) and One-Seg receiver.

As it is difficult to explain details on all receivers, this text limits its focus to common technical elements based on ARIB STD-B21 and other technical documents.

VII.1 ISDB-T receiver overview

Figure VII-1 shows an image of digital terrestrial broadcasting service and receiver type.

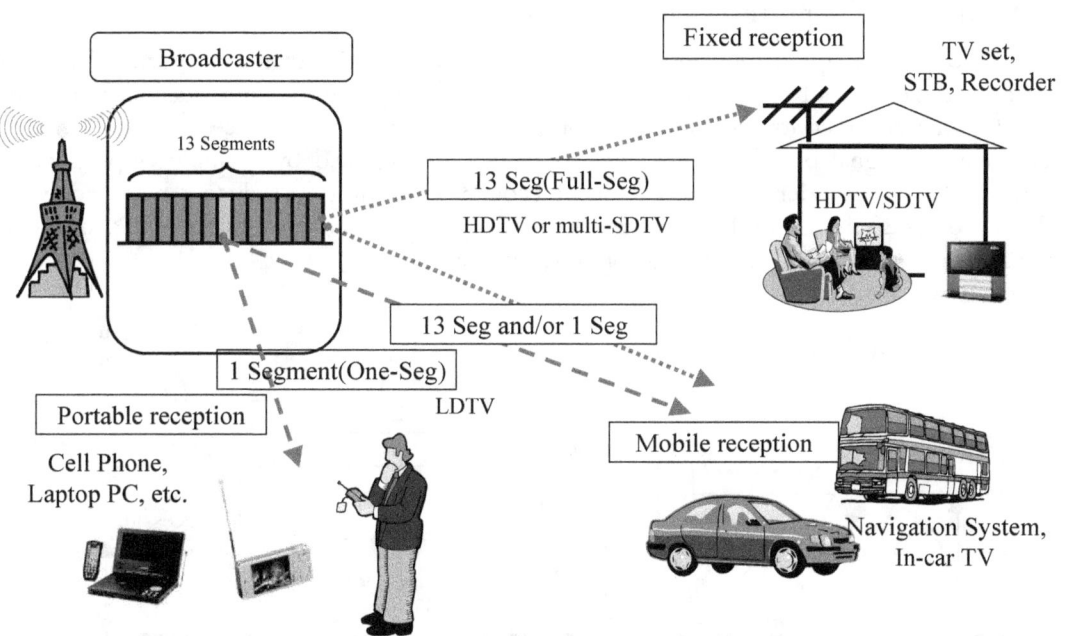

Figure VII-1   An image of ISDB-T service and receiver type

ISDB-T receivers can be classified by reception style as shown in Table VII-1.

Table VII-1   Classification of ISDB-T receivers

Classification (note 1)		Purpose	Remarks
In house receiver	With large size display	mainly for HDTV service	Antenna type:

	With medium size display	mainly for SDTV service	-fixed outdoor type
	STB only(note 1)	depend on display size	-indoor type
Mobile(mounted in car)	With car-navigation system	in car use	in general, diversity antenna system used
	TV receiver only		
Handheld (portable)	Mounted in mobile phone	outdoor use	Mainly One-Seg Receiver
	TV receiver only		
Other case	Mounted in PC	Use like as handheld service	
	USB type receiver	used with PC	

(note1) take note that 2 types of STB is commercialized, one is SDTV receiving only, the other is HD/SD. In general, HD reception needs large memory, so HD-STB is rather expensive.

As well known, ISDB-T has a unique service named "One-Seg" service, categorized in mobile or handheld type. In addition, as described in Chapter II.4.4, mobile reception by car is also unique reception type. Both "time interleaving" and "diversity reception" techniques are used for car-mounted receiver of ISDB-T system.

VII.2 Structure of digital receivers

Figure VII-2 shows an example of overall functional block diagram of digital receiver. As illustrated in figure, a digital receiver is mainly composed of tuner/demodulator portion, demux/decoder portion, display processing, CPU for receiver control and data-decoding, memory and I/O interface.

In earlier stage, the signal processing function is generally composed by multiple chips, such as tuner, demodulator, TS demux/decoder or Audio/Video Decoder. Recently, to reduce a manufacturing cost, 1 chip type LSIs (for example, OFDM demodulator, TS demux/decoder and Audio/Video decoder in one chip) are commercialized and used for more affordable receivers.

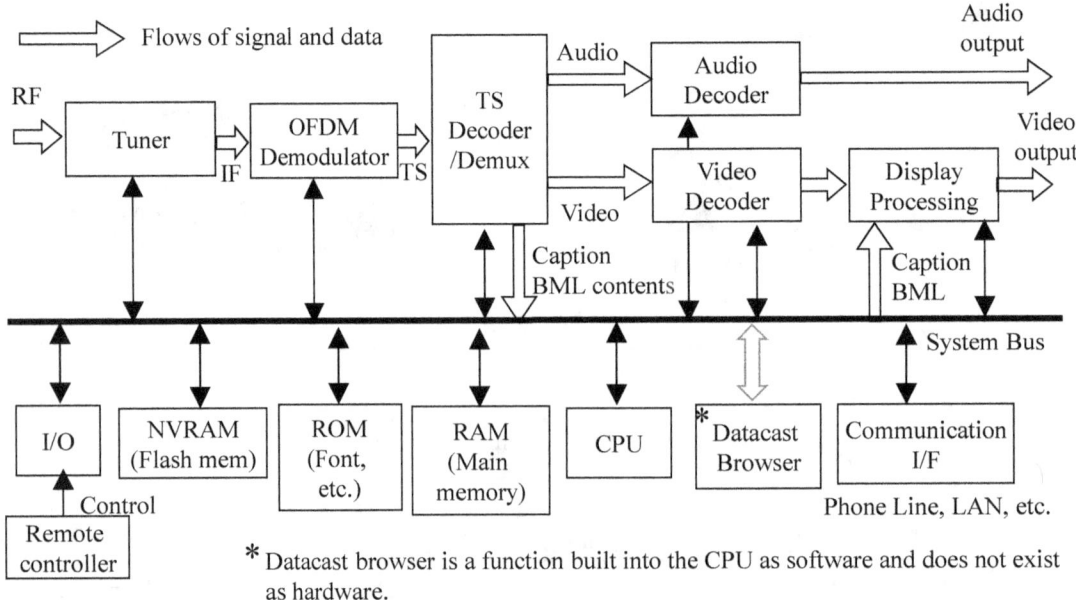

Figure VII-2 ISDB-T receiver structure

VII.3 Synchronization Technology in Digital Receiver

After the antenna receives the transmitted broadcast signal, the receiver synchronizes with the signal and extracts information carried on the signal. This section explains the synchronization method equipped in the receivers. In this section, the structure of tuner/demodulator portion and synchronization procedure are explained.

VII.3.1 Block diagram of tuner/demodulator

In Figure VII-3, an example of tuner/demodulator block diagram of full segment receiver is illustrated. For One-Segment receivers, see Chapter VIII.

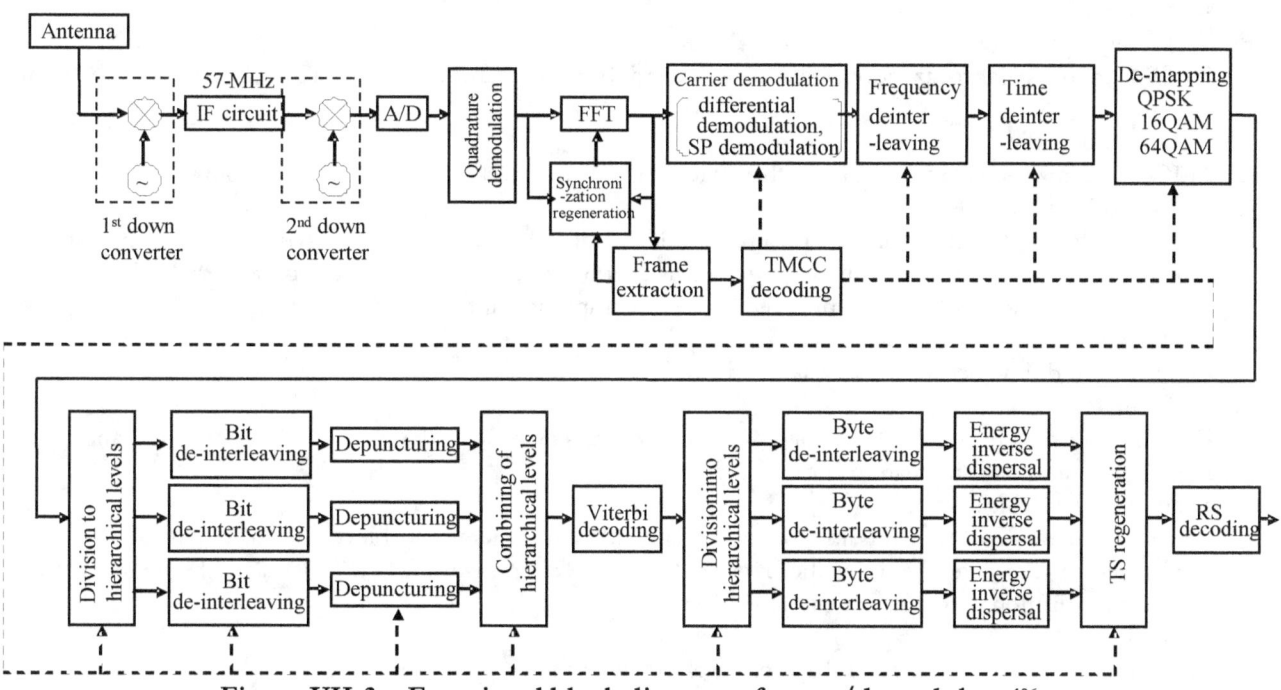

Figure VII-3 Functional block diagram of tuner/demodulator[106]

Figure VII-3 shows an example of Japanese case. Intermediate Frequency (IF) of 57 MHz is used in Japan. In other countries, different IF frequency may be used. Recently, a silicon tuner is also used for digital receivers. In this case, IF frequency can be selectable.

Demodulator function block diagram shown in the figure follows the process in the opposite order compared to ISDB-T transmission system block diagram illustrated in Figure III-2. The key elements of signal processing is described below. For details of each functions, see Chapter III.

Channel selection:
This function is composed of 1st down converter, IF circuit and 2nd down converter. Received RF

---

[106] Quoted from ARIB STD-B21, Section 5.2.6

signal is down converted to 1st Intermediate Frequency (IF, in case of Japanese receiver, 57 MHz IF is used), then amplified and fed to 2nd frequency converter.

Synchronization:

This function is composed of quadrature demodulation and synchronization regeneration. In this Section, a recovered carrier, a recovered clock and a recovered FFT window timing are obtained. As digital reception uses this synchronization signals in the following demodulation and decoding process, these synchronization processes are important. See Chapter VII.3.2 for details.

FFT:

A FFT function is to demodulate an OFDM signal, which is composed of multi-carriers, from time-domain waveform to each frequency component so that following process can extract information from them. FFT operation is executed during a period corresponding to an effective OFDM symbol duration. This symbol duration is named "FFT window". Under multipath condition, the position of FFT window should be adaptively set. For the best position of FFT window under multipath condition, see Chapter II.3.3.

Frame timing regeneration:

OFDM frame synchronization signal is extracted from the front end of TMCC signal (see Chapter III.4.4).

TMCC signal decoding:

TMCC information is extracted from the TMCC signal and used to conduct various controls.

Carrier demodulation:

Based on the TMCC information, differential demodulation for DQPSK or synchronous demodulation through the use of I-Q coordinates[107] which is made by referring the received scattered pilot (SP) for QPSK, 16QAM, or 64QAM is conducted, to detect amplitude and phase information.

De-interleaving:

Frequency and time de-interleaving is conducted.

De-mapping:

De-mapping of QPSK, 16QAM, or 64QAM is executed in accordance with the amplitude and phase information and bit information is extracted.

Division into hierarchical levels:

In case of hierarchical transmission, the signal is divided into hierarchical levels. Note that the division is performed of 204 bytes between the byte next to the synchronization byte (47 H) of the TS packet and the synchronization byte of the next TS packet.

Bit de-interleaving:

Bit de-interleaving is executed in the each level of hierarchy.

De-puncturing:

Bit-interpolation is executed for the each level of hierarchy, in accordance with the convolution coding rate indicated in the TMCC information.

Viterbi decoding:

Viterbi decoding with a coding rate of 1/2 is executed. In Viterbi decoding, a soft-decision algorithm is employed to improve performance. Further, to avoid error propagation due to the convolutional code, termination processing is conducted based on the fact that the synchronization byte (47 H) of the TS packet is already known. For this process, see section III.2.3.

Byte de-interleaving:

De-interleaving is executed on a byte-by-byte basis.

Energy inverse-dispersal:

Inverse dispersal is conducted by applying Exclusive-OR with the 15th M-sequence PN signal on a

---

[107] See Section II.2.1 of this book for details

bit-by-bit basis, except for the synchronization byte of the TS packet. Note that during the period of the synchronization byte, a shift register is in operation, and initialized at every OFDM frame.

TS regeneration:

Processing for regeneration of a transport stream is conducted. On this occasion, the order of the TS packets and the temporal location of the PCR should be the same as they are on the transmitting side.

RS decoding:

Shortened Reed-Solomon code RS(204,188) is decoded. During RS decoding, if an error is detected, transport_error_indicator, which is positioned at the 9th bit of the transport stream packet (specifically, MSB in the second byte), is set to "1."

VII.3.2 Synchronization techniques in ISDB-T receiver

As described in the previous section, the demodulator function blocks shown in the Figure VII-3 follows opposite process compared to the modulator function blocks illustrated in Figure III-2 except "synchronization" function. The Synchronization process is necessary to demodulate a received signal. In OFDM transmission system, several kinds of synchronization process illustrated in Figure VII-4 below are requested before demodulation.

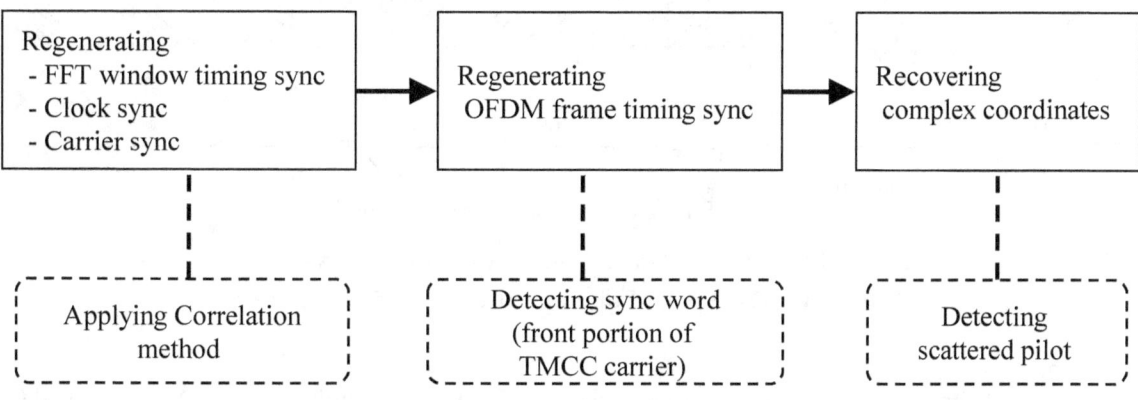

Figure VII-4 Synchronization for OFDM demodulation[108]

(1) FFT window timing detection and clock synchronization

To recover FFT window timing and to recover clock synchronization, a unique technique, correlation method, is used. Correlation method is one of the effective techniques for synchronization. The feature of this method uses the correlation between "guard interval" and "end of effective symbol."

As explained in Chapter II.3.3 before, "guard interval" is generated from a copy of the end portion of "effective symbol" and inserted in front of the effective symbol (see following Figure VII-5 (a). This means that the waveform of "guard interval" and "end portion of effective symbol" is identical, that is, both portion has strong correlation.

---

[108] See Section III.4.2 for the role of the scattered pilot signal

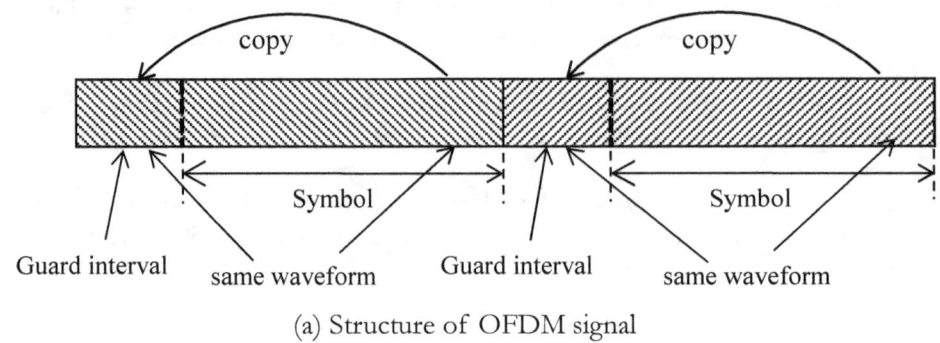

(a) Structure of OFDM signal

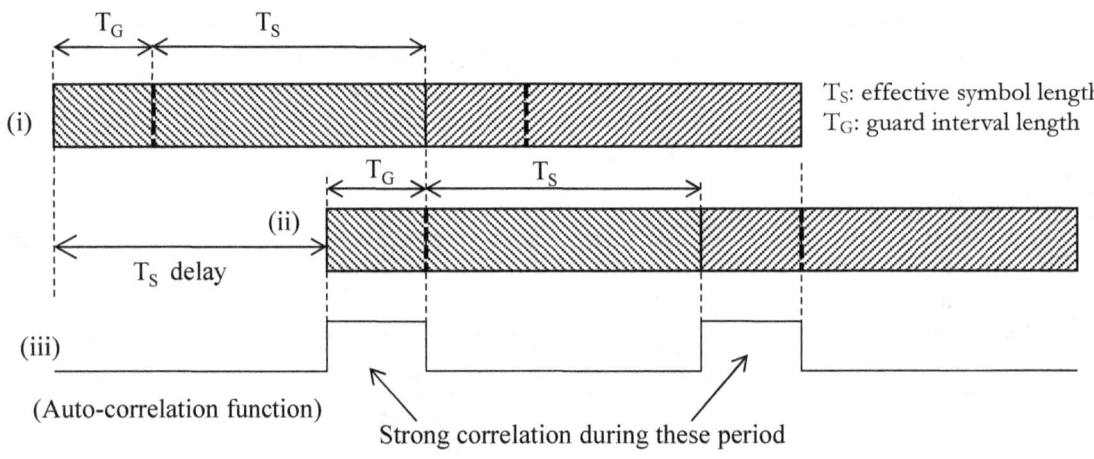

(b) Auto-correlation function of OFDM signal

Figure VII-5   OFDM signal structure and correlation

Figure VII-5(b) shows the relationship with an original OFDM signal (indicated (i)) and Ts delayed OFDM signal. As illustrated in Figure, the strong correlation portion appears in every (Ts+T_G) period. This period is just same as a symbol timing period of OFDM signal.

Therefore, the symbol timing (known as also "FFT window timing") can be easily detected by auto-correlation circuit shown in Figure VII-6 below.

The output of auto-correlation circuit is used as a FFT window timing for FFT signal processing.

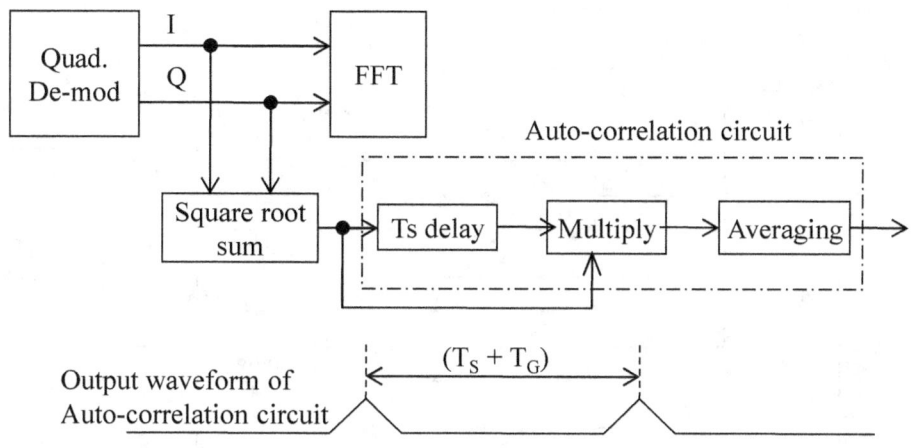

Figure VII-6　Symbol timing detection

The number of sampling clocks in FFT window (whose length is Ts) is identical to the number of FFT points.　From this relationship, the number of sampling clocks in the period of $(T_S + T_G)$ illustrated above is easily calculated by following equation:

Number of sampling clock in $(T_S + T_G)$ = (number of FFT points in a FFT window) × {$(T_S + T_G)/ T_S$}

For example, in case mode 1 and Guard interval ratio=1/8, the number of sampling clock in $(T_S + T_G)$ is calculated as follows:

Number of sampling clock in $(T_S + T_G)$ = (number of FFT points in a FFT window) × {$(T_S + T_G)/ T_S$}

= 2048 × 9/8 = 2304 clocks

For other cases of modes and Guard Interval ratios, please see Table III-5.

By keeping above relationship between the regenerated clock and the recovered "symbol timing," the sampling clock is exactly regenerated.

(2) Carrier synchronization
One of carrier synchronization measures, a cross-correlation method is used.

As explained in Chapter II.3, the effective symbol length, Ts, is reciprocal of carrier spacing (fd) of OFDM signal, that is, Ts=(1/fd).　Further, effective symbol length is integer multiple of carrier waves for all carriers. Because of above relation, the phase of carrier wave at the beginning of and at the end of effective symbol is coincident for all carriers of OFDM signal.　If recovered local signal at receiver side have offset in frequency at Δf (Hz), the received all OFDM carries have offset Δf (Hz).　Therefore, the phase difference, $\Delta\varphi = 2\pi(T_S*\Delta f)$ (rad), occurs between the beginning and the end of Ts period (= effective symbol length) for all OFDM carrier.

The following circuit as illustrated in Figure VII-7 can detect the frequency offset Δf by calculating phase difference of $\Delta\varphi$ from I, Q output of the Quad Demodulator.

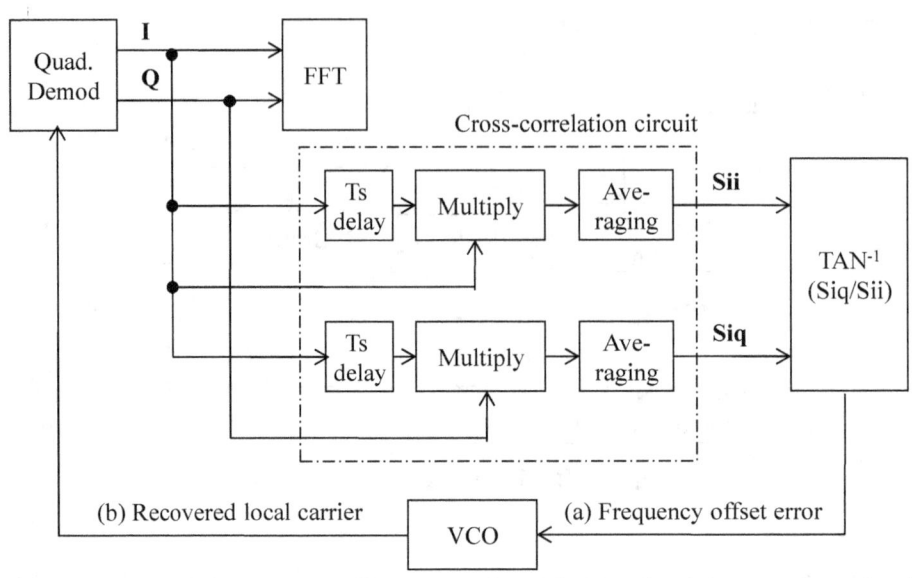

Figure VII-7　An example of frequency offset error detection and carrier synchronization circuit

The circuit shown above calculates a cross correlation of Ts period and an end portion of effective symbol. The symbol Sii means the cross correlation of I axis of Ts period and I axis of an end of effective symbol. The symbol Siq means the cross correlation of I axis of Ts period and Q axis of an end of effective symbol. The ratio (Siq/Sii) is equal to Tan($\Delta\varphi$).

Therefore, $TAN^{-1}(Sii/Siq)$ is proportional to frequency offset error, Δf, which is defined as the difference between the frequency of received signal and the frequency of recovered local carrier shown in Figure VII-7.

The Voltage Controlled Oscillator (VCO), illustrated in Figure VII-7, generates a local carrier signal for Quadrature demodulation of received signal. The frequency of local carrier is corrected by an error signal provided from $TAN^{-1}(Sii/Siq)$ circuit. That is, all functions as a whole illustrated in Figure VII-7 work as a local carrier signal generation circuit of which frequency is controlled by negative feedback.

Figure VII-8 shows the relationship between frequency offset error and $TAN^{-1}(Sii/Siq)$. The frequency of local signal converges to stable points of $TAN^{-1}(Sii/Siq)=0$, which is shown in Figure VII-8. The allows(→) illustrated in the Figure shows the direction of convergence of the frequency of regenerated carrier signal. As shown in Figure VII-8, many stable points exist in every frequency spacing of f_d (frequency spacing of each OFDM carrier). By monitoring VCO output, we can know if the center frequencies of local carriers and actual OFDM carriers are synchronized.

However, as described above, there are many stable points, VCO alone cannot identify that a VCO output frequency is converged to the center frequency of the received OFDM signal. Next step is to identify the location of AC and SP in order to know which stable point correspond to the center frequency of the received OFDM signal. By doing so, local carrier can synchronize with the OFDM carrier. AC and SP is pilot carriers placed in predefined locations for carrier synchronization. They are modulated in DBPSK and can be easily found out among other carriers. Details of AC and SP was discussed in Section III.4.

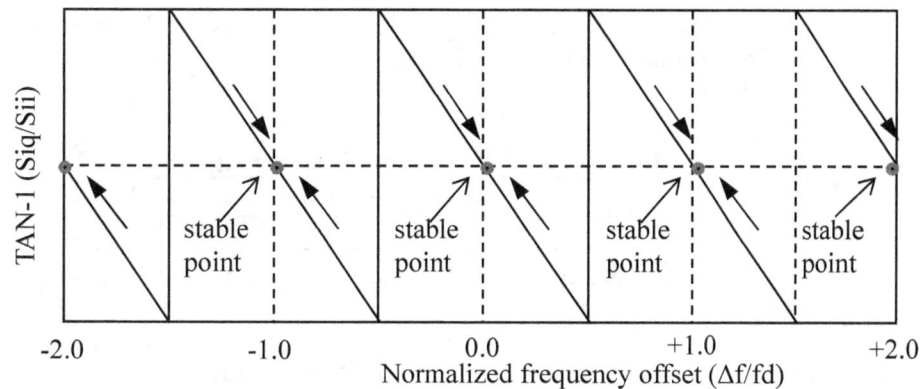

Figure VII-8 Relationship between cross correlation and frequency offset

Quadrature demodulator output of I-Q format are fed to FFT to be converted from time-domain waveform to frequency components which carries information. After FFT, AC carriers and CP carriers will be detected. These carriers are used as frequency reference. And finally, the frequency synchronization is completed.

As a summary, frequency synchronization has two process described below:
Process 1: the frequency of local carrier is moved into any stable point at which $TAN^{-1}(Siq/Sii) = 0$.
Process 2: AC & CP carrier are detected at output of FFT to decide the recovered frequency position.

(3) Frame timing synchronization
After going through FFT process, each carrier are demodulated as a stream of digital carrier symbols. A frame synchronization word which is a unique code to be used to detect the start of the OFDM frame is inserted into a top portion of TMCC signal. See Chapter III.4.4 for details.

By the circuit illustrated in Figure VII-9, OFDM frame timing synchronization can be obtained. After carrier synchronization, the position of TMCC carrier can be easily known. Next, the selected TMCC carrier is differentially demodulated (illustrated as "DBPSK demodulation", here, DBPSK means Differential Binary Phase Shift Keying). The demodulated data is fed to "Frame sync. word detection" circuit, which finds synchronization word pattern in the data stream.

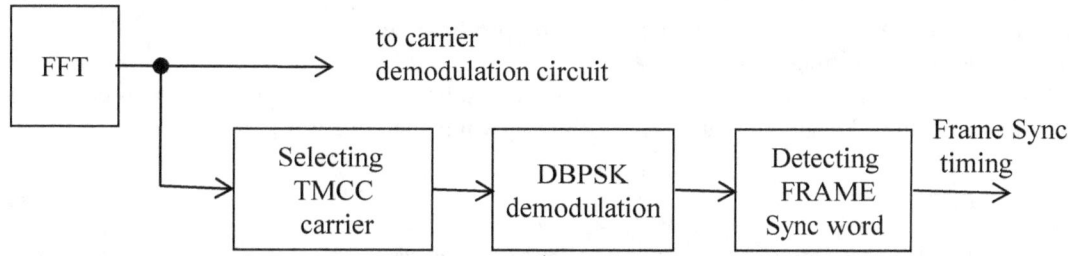

Figure VII-9 An example of frame synchronization timing detection

(4) De-mapping of each carrier
As described in Chapter III, ISDB-T system has 4 kinds of mapping, DQPSK, QPSK, 16QAM and 64QAM. Except for DQPSK, reference complex coordinate should be necessary for de-mapping.

Scattered pilot signal, described in previous Chapter III.4.2 is used as for the reference of I-Q coordinates. The value of Scattered Pilot signal is defined in following table.

These values are used for a reference of mapping.

Table VII-2 Wi and modulating signal[109]

W_i value	Modulating-signal amplitude (I, Q)
1	(-4/3, 0)
0	(+4/3, 0)

VII.4 Composition and Specification of ISDB-T receiver

In this Section, it mainly explains some key points for composition and specifications of digital receiver.

The composition and specifications of digital receiver explained in this Chapter are receivers sold in Japan, based on ARIB STD-B21. Countries adopting ISDB-T also defines their standard with minor modifications on ARIB standards, because of the difference of regulation, culture and another reason. Therefore, it should be necessary to check the regulation/guideline of each country.

VII.4.1 Specifications of tuner/demodulator

In ARIB STD-B21 Section 5.2.1- 5.2.5, the specifications of Digital Integrated Receiver and Decoder (DIRD) are defined as below:

For the receiver specifications of each country which adopts ISDB-T system, please read through the regulations and/or guidelines of each country.

(1) Input
    Impedance: 75 Ω
    Received frequency: UHF ch 13-62[110] [111]
    Center frequency[112]: 473 + 1/7[113] MHz (ch 13), 479 + 1/7 MHz (ch 14), . . ., and 767 + 1/7 MHz (ch 62)

(2) First intermediate frequency (1st IF)
    Center frequency: 57 MHz (frequency reversed)
    Local oscillator frequency: At the upper side of the received frequency

(3) Synchronization range of the received frequency
    Synchronization range of the received frequency: ±30 KHz or wider

(4) Synchronization range of the received clock
    Synchronization range of the received clock: ±20 ppm or wider

(5) Characteristics of the tuning unit
    A tuning unit for receiving 13 segments and a tuning unit for receiving 1 segment located in the central part of the 13 segments must satisfy the following specifications:

---

[109] Quoted from ARIB STD-B31 main body table 3-17
[110] Channel number of UHF band is specified for Japan (not US channel number)
[111] ITU resolved over 700 MHz band will be assigned to mobile communication. Please check the assigned bandwidth for digital terrestrial broadcasting in each countries
[112] If considering to use for cable network, it is desirable that the reception channel range thereof include the SHB band (ch C23– C63) in addition to the UHF band. It is also desirable that the reception channel range of the receiver includes the VHF band (ch 1 to ch 12) and the MID band (ch C13 to ch C22). These frequency assignment may vary in regions.
[113] 1/7 MHz offset of center frequency is adopted for 6MHz countries, but, for 8MHz countries ,this offset is not applied

Minimum input level: -75 dBm or lower (targeted value)
Maximum input level: -20 dBm or higher.

However, when the input level in a One-Segment receiver is measured in terms of electric power per segment, the level must be reduced by a factor equivalent to the bandwidth (i.e., one-thirteenth, or -11 dB).

Table VII-3　Protection ratios of the 13-segment receiver[114]

Undesired wave	Item	Protection Ratio
Analog television	From the co-channel	18 dB
	From the lower adjacent channel (undesired wave on the lower side)	-33 dB
	From the upper adjacent channel (undesired wave on the upper side)	-35 dB
Digital television	From the co-channel	24 dB
	From the lower adjacent channel (undesired wave on the lower side)	-26 dB
	From the upper adjacent channel (undesired wave on the upper side)	-29 dB

(Note)　The transmission parameters used for the measurement must be as follows: Mode 3, guard interval ratio of 1/8, no time interleaving, modulation of 64 QAM, and an inner-code of 7/8

VII.4.2　Functions and specifications of back end

As for an example, a Japanese ISDB-T receiver's backend portion is illustrated in Figure VII-10, from decoded TS input to video and audio output, of digital receiver.

---

[114] Quoted from ARIB STD-B21, table 5-2

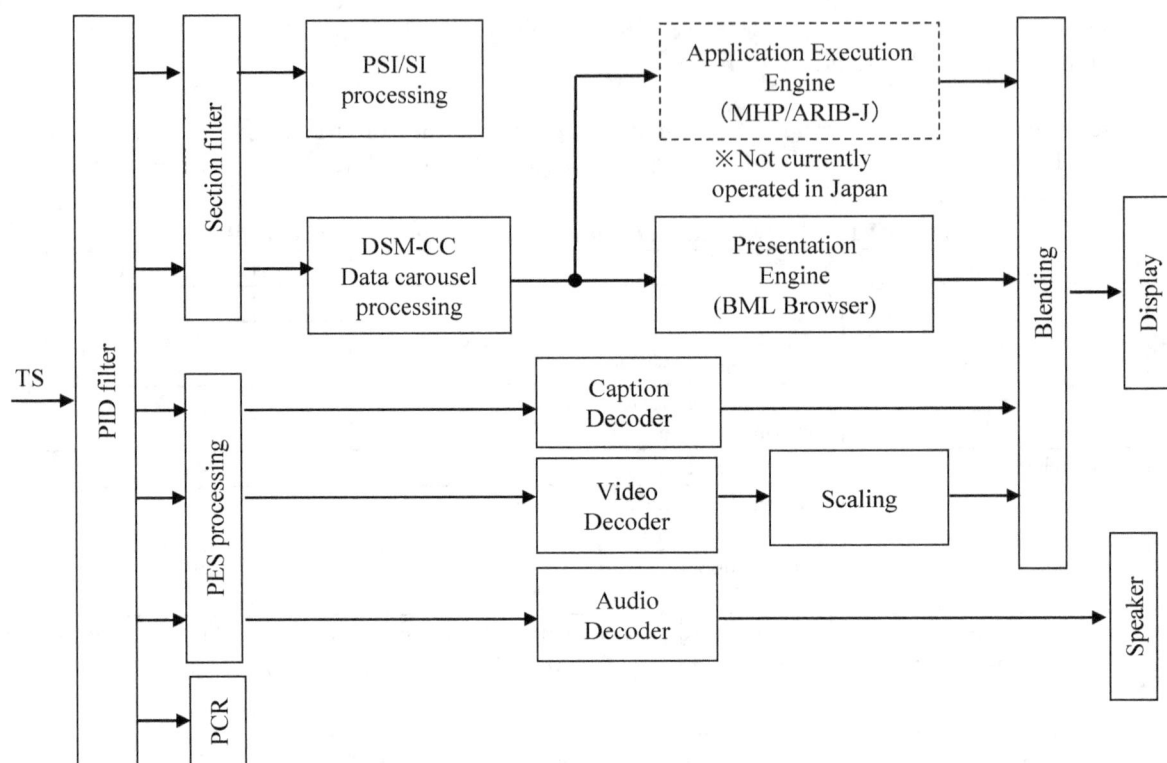

Figure VII-10　Functional block diagram of backend portion of digital receiver

Decoded broadcast TS is fed to PID filter. PID filters out specific contents from the transport stream according to the PIDs. As described before, transport stream is classified into two groups, one is PES (Packetized Elementary Stream) type, and the other is Section type. PES format is basically used as video/audio data transmission for video/audio broadcasting, on the other hand, Section format is basically used for text/character data transmission for data-casting.

In this Section, each function is introduced briefly. For details see Chapter IV,V,VI

(1) Mixing of graphic / data plane
Graphics and data as an output of BML browser is blended with motion/still picture. The blending is done according to the display model of digital broadcasting shown in Figure VII-11. The motion picture plane and the still picture plane are switched by rectangular display area. This motion/still picture plane and the text/object plane are mixed with the ratio of α and $(1-\alpha)$. Mixing ratio of α, which is output from CLUT, means a degree of blending of transparency.

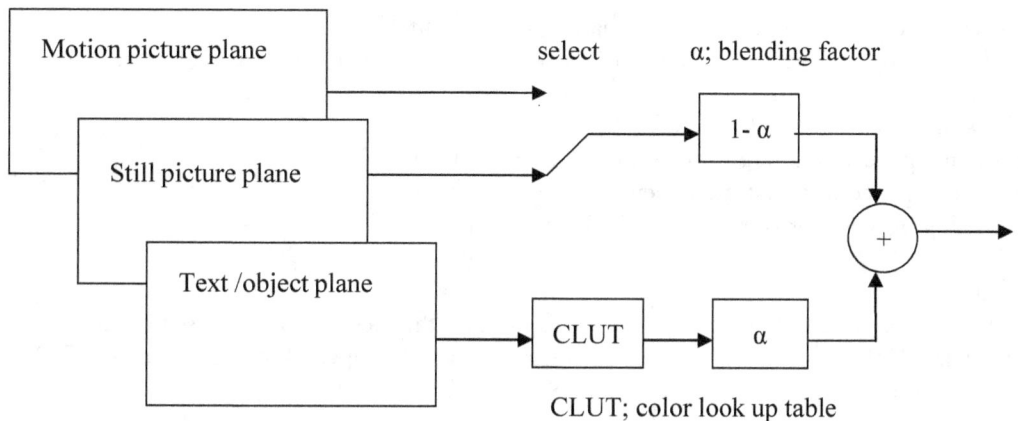

Figure VII-11 Display model of ISDB-T receivers

(2) Video decoding and video format
The function of video decoder is to decode an encoded MPEG-2 video signal, these are, MP-HL, MP-14L, MP-ML, MP-LL.

MPEG-2 is used in Japan, but in other countries that adopts ISDB-T, another coding system such as H.264 is adopted.

For video format, 4 kinds of video formats, 480i, 480p, 720p, and 1080i, are specified in Japan for SDTV and HDTV broadcasting. Note that different SDTV video formats are used in 8MHz bandwidth country. For LDTV (Low Definition TV, this quality is generally used for portable receiver) service, see Chapter VIII (One-Seg service).

For details of video format specification in Japan, see ARIB STD-B21 Section 6.1.1 and 6.1.2.

(3) Video display format
Currently two types of display format are widely used, that is 16:9 and 4:3. For the relationship between the format of video source and display format, see previous Chapter V.1.

(4) Audio decoding
ISDB-T adopts MPEG-AAC coding system for audio coding, conforming to the LC profile of MPEG2-AAC (ISO/IEC 13818-7) and ADTS (Audio Data Transport Stream) system. Furthermore, it shall conform to the following restrictions.
  (a) Sampling frequency: Corresponds to 48KHz, 44.1KHz, 32KHz, 24KHz, 22.05KHz, 16KHz
  (b) Quantized bit number: Corresponds to reproduction at 16 bits
  (c) Decodable number of channels: Corresponds to AAC stream up to 5.1 channels per ADTS.
  (d) Number of maximum multiple ADTS: Corresponds to a maximum of 8 ADTS streams within the same program.
  (e) Audio decoding functions: Decodes following audio mode
    (i) monaural
    (ii) stereo
    (iii) Multi-channel stereo (3/1, 3/2, 3/2+LFE)
      This mode means the number of audio channels to the assumed front and rear speakers. (Ex: 3/1 = 3 speakers in front + 1 at rear, 3/2 = 3 speakers in front 4 + at rear). LFE is an abbreviation of Low Frequency Enhancement, which means low frequency enhanced channel.
    (vi) Decoding process when switching the audio mode and coded parameter at the transmission side
      It shall return to normal operation without making noise within the muting time of audio parameter switching, in accordance With ARIB STD-B32.

(vii) Down mixing function from multi-channel to 2-channel stereo.
For down-mixing function, three types of down-mixing, shown below, are defined.

See ARIB STD-B21 Section 6.2.1 for details.
(a) Down mixing process from multi-channel to 2-channel stereo
(b) Down mixing process for external pseudo-surround processor
(c) Down mixing process for stereo audio field extension

(5) Data decoder
The datacasting system is described in Chapter VI already, so this Section only introduce the how datacasting decoder (especially BML browser) is located in the receiver. In Japan, Execution engine (ARIB/J) is not operated. Only presentation engine (ARIB STD-B24) is in practical use.

VII.4.3 Software of digital receiver
(1) Hierarchy of software
A digital terrestrial broadcasting receiver has many functions not only for data-processing but also management/control of receiver. Figure VII-12 shows an example of software hierarchy of Hi-Vision class terrestrial receiver

Note that Figure VII-12 is an image of software hierarchy and does not show all software. Especially, in lower middle ware layer, there are more software module which are closely related to device driver.

Of course, according to service menu of the receiver, more software should be prepared.

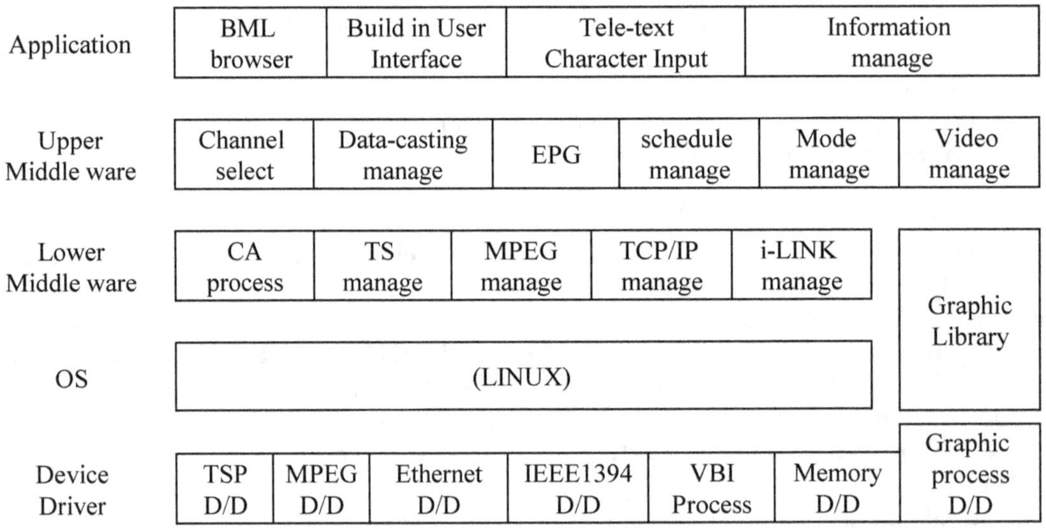

Application	BML browser	Build in User Interface	Tele-text Character Input		Information manage	
Upper Middle ware	Channel select	Data-casting manage	EPG	schedule manage	Mode manage	Video manage
Lower Middle ware	CA process	TS manage	MPEG manage	TCP/IP manage	i-LINK manage	Graphic Library
OS	(LINUX)					
Device Driver	TSP D/D	MPEG D/D	Ethernet D/D	IEEE1394 D/D	VBI Process	Memory D/D	Graphic process D/D

Figure VII-12 An example of software hierarchy of digital receiver

In general, as shown in figure, a receiver software is composed by five hierarchy, "device driver layer, "OS layer, "Lower middle ware layer", "Upper middle ware layer" and "Application layer" Followings are some key points of software

(2) Channel selection
As for an example of receiver control technique, the control flow of "program selection" is explained in this

Section.[115]

Figure VII-13 shows a basic flow of program selection. At first receiver is switched on, and then the required channel is selected. The receiver detects NIT and display Broadcaster's name ,etc. If selected channel is correct, then the receiver's operation goes to next step. If other network is selected, the receiving channel moves to another channel.

In the next step, the receiver detects the service information and display. If this information is correct, the receiver selects this stream, and then detects PAT, PMT. If these parameters are correct, then the required contents will be shown on the screen.

It is also acceptable to provide branching/short-circuit routes, etc. on the flow chart shown in Figure VII-13 as an additional receiver function.

In usual reception, the flow would not proceed into the exceptional processing. Usual reception means that the received signal has NIT and PSI/SI normally. If these PSI/SI cannot be normally detected, receiver goes into exceptional processing.

In exceptional processing, processing such as re-setting to capture the broadcast wave correctly, or special processing when the receiver is used for non-broadcasting reception (uploading of the system management ID, etc.) is conducted.

In case of "Initial scanning" of which mode is used at initial setting of receiver to get the channel information at this area, the receiver receives some channel and detects NIT, etc. and memorizes this information. Then the receiver moves to next channel and detects/memorize the NIT of this channel. By repeat this process, the receiver gets the Information for channel selection. This process is named "Initial scanning".

---

[115] Quoted from ARIB STD-B21 Figure 13-2

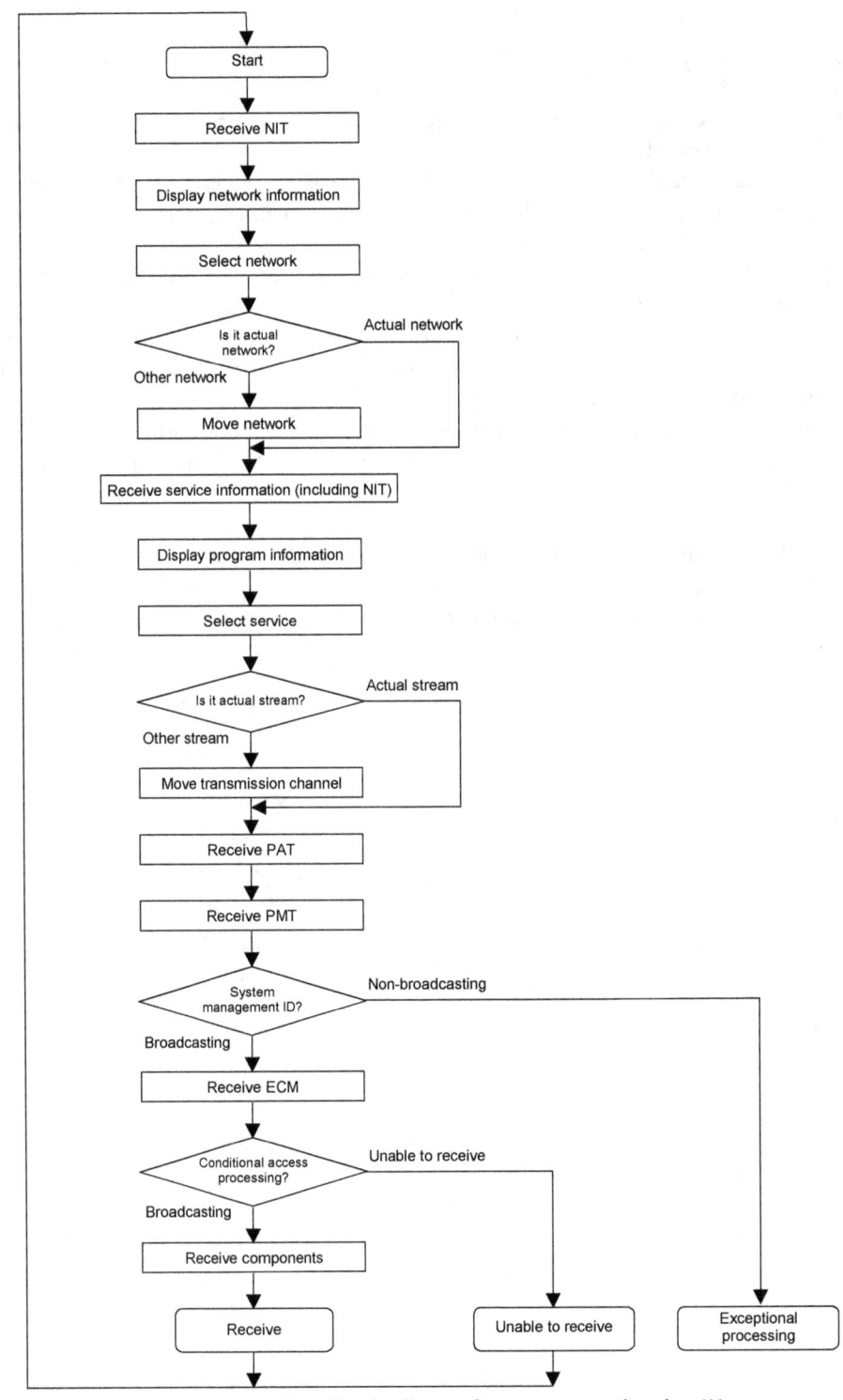

Figure VII-13 Basic flow of program selection[116]

---

[116] Quoted from ARIB STD-B21, Figure 13-2

(3) Software download

The purposes of software download in digital broadcasting are to enable (a) receiver built-in information update service, and (b) engineering service. For software down load service, following terms are defined
   (i) Notification information:
      Information used for notification such as the service ID for downloading, scheduling information thereof, and the targeted model of receiver to be updated. It is transmitted using SDTT.
   (ii) Receiver information:
      Information on the receiver set, such as maker ID, model number, group number, version number, etc. These pieces of information are stored in nonvolatile memory such as flash memory before shipping.
   (iii) Compulsory downloads:
      Downloads that must be executed.
   (iv) Discretionary downloads:
      Executable downloads displayed on the screen, and executed upon the viewer's selections.

(4) Examples of software download service

Following two services (a) and (b) are explained as examples.
   (a) Receiver built-in information update service
      Receiver built-in information update service includes the following services that are stored in memory:
      (i) Engineering service (see next)
      (ii) PNG logo service by CDT of TS in each broadcasting station of the digital terrestrial television broadcasting.
      (iii) Simple logo service transmitted by the logo transmission description of TS's STD in each broadcasting station of the digital terrestrial television broadcasting.
      (iv) Information transmission service related to necessity of updating the receiver software, method and re-packing for mobile receiver transmitted by the SDTT for strong hierarchy of TS for all the broadcasting stations in the digital terrestrial television broadcasting.
      Services of EPG, data service, video/audio service to store in memory are not included in the receiver built-in information update service.

   (b) Engineering service
      The engineering service means the following services explained below (i) and (ii) transmitted by the data carrousel. Generally, the engineering service is reported by the SDTT transmitted by the TS of the whole broadcasting station.
      (i) Function updating of the receiver software
         Function updating and addition of the receiver software. (Digital terrestrial television, BS/broadband CS digital broadcasting)
      (ii) Common data updating
         · Updating the data used commonly in the receiver.
         · Genre code table, program characteristic code table, reservation term table (digital terrestrial television, BS/broadband CS digital broadcasting)
         · Logo data (BS/broadband CS digital broadcasting)
         · Frequency list, change information (digital terrestrial television broadcasting)

(This page is left intentionally blank.)

Chapter VIII. One-Seg Broadcasting Service

In addition to showing images on conventional fixed television, ISDB-T is designed to deliver images to mobile or portable devices including cellular phones with the mobile broadcasting technology called "One-Seg" which resulted in explosive diffusion of mobile TV in Japan, where the number of shipped mobile/portable TV devices surpassed that of fixed receivers.

Thanks to the portability of the One-Seg receivers, people in countries adopting ISDB-T enjoy watching television not only in living room with fixed television, but also in private rooms, trains or offices with One-Seg receivers.

In addition for attracting more viewership in various scenes of people's lives, One-Seg is designed to bring more advantages in business implications for broadcasters.

(1) Cost saving of broadcaster's infrastructure:
As shown in Figure II-21 in Chapter II, by making use of "Segmented OFDM transmission technology", only one transmitter is required for 2 types of services (fixed reception and portable reception). In case of non-hierarchical transmission, 2 transmitters are necessary.

(2) Frequency resource saving
As shown in the Figure VIII-1, One-Seg service is possible within same frequency bandwidth. Therefore, another frequency resource is not necessary for portable reception service. In another words, there is no need to obtain additional license for frequency use. As these regulatory process takes times and efforts in most of the countries, this advantage saves times and resources for the digital mobile TV roll-out.

In contrast to the fixed television, mobile or portable television equips smaller screen whose picture quality requirement is rather not demanding. On the other hand, mobile or portable devices requires high energy saving performance with its limited battery capacity. Also, reception in outdoor or in high-speed moving vehicle or trains requires high robustness against dynamic fading with the handicap of limited antenna sensitivity that comes from limitation of space in the mobile terminals.

Table VIII-1 Difference between One-Seg and Full-Seg services in ISDB-T

		One-Seg service	Full-Seg service
requirement	Target	Mobile or portable receivers	Fixed receivers
	Primary power source of receivers	Battery	Commercial AC supply
	Screen size	Small (cell-phone size)	Large
	Mobility	High	Low
	Data rate	Several hundreds kbps	Several - dozen Mbps
implementation	Picture Format	QVGA,15f/s (in Japan)	HDTV(1080i) or SDTV (480i)
	Coding	H.264/AVC	MPEG-2 MP/HL (in Japan)
	Data broadcasting	C-profile	A-profile
	Frequency Use	1 segments	12 segments
	Modulation	QPSK	64QAM

In order to save the power consumption of portable receiver for portable reception service, receivers receives only the bandwidth of one segment (called "narrow band reception") rather than receiving whole channel bandwidth. By doing so, the sampling frequency of Portable One-Seg receiver can be reduced, so that total amount of calculation necessary in the demodulator can also be reduced. As a result, power consumption of OFDM demodulator can be reduced.

In order for robust reception against dynamic noise and weak signal level, QPSK modulation is also designed to be selectable along with 16QAM and 64QAM by adopting Segmented OFDM transmission technology.

This Chapter will explain on the RF technology for One-Seg service, followed by Video and Audio coding, Data Broadcasting, presentation of One-Seg and basic receivers.

VIII.1 Transmission Technology for One-Seg

What is One-Seg?   What are the advantages of One-Seg system?

The name "One-Seg" comes from abbreviation of "One-Segment", which means this broadcast service uses the center segment of 13 segments illustrated in Figure VIII-1 for mobile and the rest for fixed. The ISDB-T RF signal composed of 13 segments including One-Seg can be transmitted from a single transmitter. In contrast to the One-Seg services, the services for fixed TVs using the rest of the segments are called "Full-Seg" services as shown again in the Figure VIII-1.

One-Seg service is based on "Segmented OFDM transmission technology", which is explained in the section II.4.3.

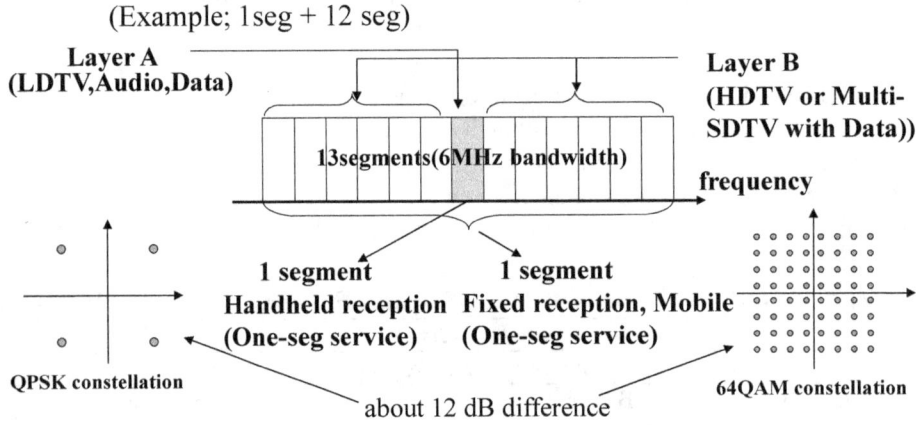

(note) same Figure as Figure II-18 of this book

Figure VIII-1   Hierarchical transmission based on segmented OFDM transmission

VIII.1.1 The relationship between transmission bandwidth and receiving bandwidth

Firstly, the relationship between transmission bandwidth and receiving bandwidth is discussed. As shown in Figure VIII-2 below, theoretically, there are four cases between transmission bandwidth and reception bandwidth in segmented OFDM transmission system.

Here "wideband" means a transmitter (or a receiver) transmits (receives) full of 13 segments; whereas narrowband means a transmitter (or a receiver) transmits (receives) only 1 segment.

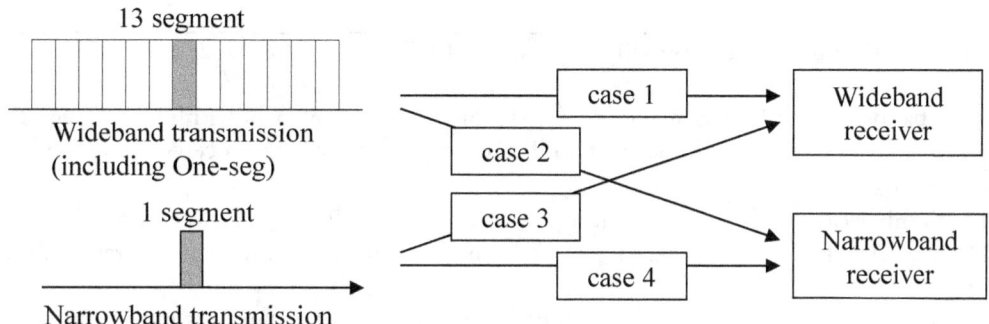

Figure VIII-2 Relations between transmission bandwidth and receiving bandwidth

While these 4 cases are theoretically possible, cases actually used in practical service are summarized in the Table VIII-2. Case 1 is ISDB-T for fixed services, and Case 2 is ISDB-T One-Seg services.

Case 4 is 1 segment digital radio in Japan called ISDB-Tsb, in which transmitter transmits 1 segment or 3 segments carrying mainly audio programs. ISDB-Tsb allows placing the system next to each other in the frequency axis called "consecutive segment transmission system" so that multiple systems can be placed in the assigned frequency slot that is generally wider than bandwidth necessary for 1 system.

Table VIII-2 Relations between transmission bandwidth and receiving bandwidth

Case	Transmission BW	Reception BW	Service type
case 1	wide	wide	TV fixed reception
case 2	wide	narrow	Partial reception
case 3	narrow	wide	None (technically possible, but not exist in service.)
case 4	narrow	narrow	digital radio

VIII.1.2   Reduction of power consumption of portable receiver and transmission rate

One-Seg service adopts unique technology, named partial reception. One of the most important features of this technology is to reduce receiver power consumption. The most important factor to reduce a power consumption is to decrease the signal processing speed in a receiver.

As shown in the Figure VIII-3, in case of partial reception (right side of the Figure), center segment of 6MHz OFDM signal is filtered by narrow band-pass filter whose pass band is as equal as 432 kHz[117].

A filtered narrow band signal is demodulated by a low sample rate FFT (Fast Fourier Transform) processor. Its sample rate is 1/8 of high sample rate of FFT which is used for full band demodulation.

---

[117] In case of 8 MHz OFDM, the One-Seg bandwidth is 576 kHz.

> Readers may have a primitive question why the 1/8 size FFT processor is used, not 1/13?
>
> The reception bandwidth of One-Seg service is 1/13 of the reception bandwidth of full segment service(=13 segments).   However, the size of FFT processor for One-Seg receiver is 1/8 of the full segment receiver.
>
> Because the Number of FFT points N should be restricted to $N=2^x$, i.e. a power of two, such as (2, 4, 8, 16, … ,1024, 2048, 4096, 8192,…), due to the algorithm of "First Fourier Transform(FFT)"[118].
>
> In case of "mode 3", 8k (8192 points) FFT processor is used for full segments reception.   The size of FFT processor for one segment reception should be no less than 1/13 and a power of two.   Because of these reasons, 1k (=1024 points) FFT processor is used for One-Seg reception.

As a result, signal process speed of demodulation block is decreased to 1/8.   Of course, in One-Seg receiver with partial reception, both demodulator circuit and backend circuit operate in low sample rate.   As described above, One-Seg partial reception is a good system from the viewpoint of power consumption.

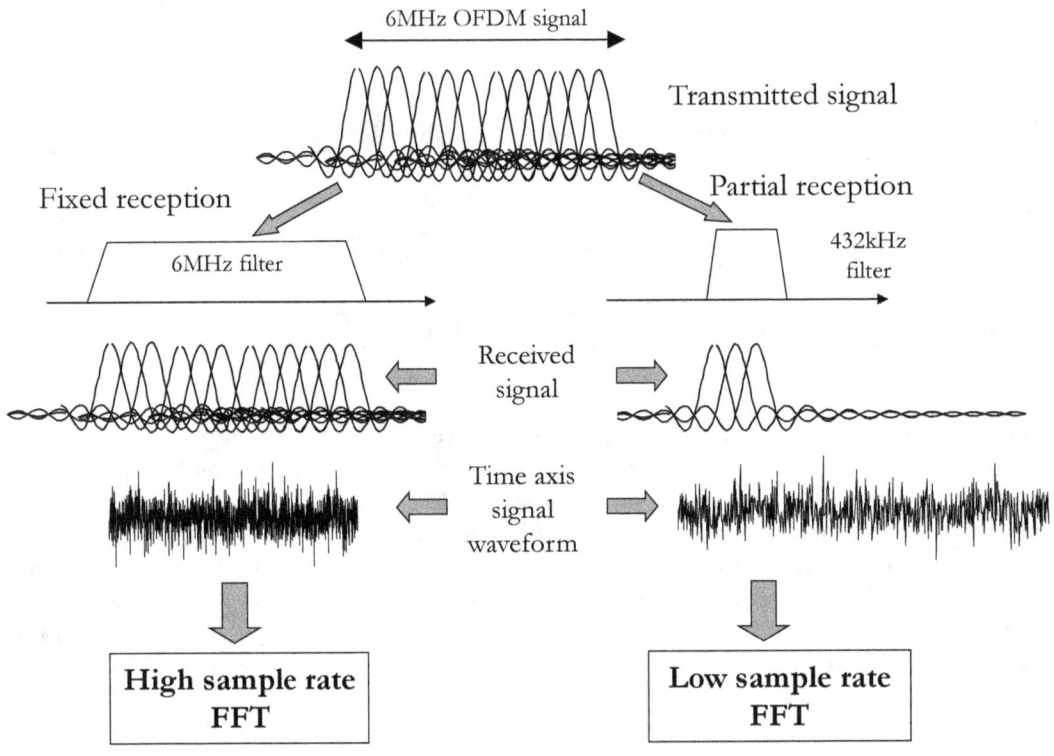

Figure VIII-3   Signal processing for wideband reception and partial reception

---

[118] Cooley, J. W., and Tukey J. W., An Algorithm for Machine Calculation of Complex Fourier Series, Math. Computation, Vol. 19, pp.297-301, April 1965

Further, we will investigate the difference of TS transmission rate of wideband/narrowband reception. As explained in the previous section, FFT sampling frequency is reduced to 1/8 in case of partial reception. According to the rate reduction, the rate of demodulated TS (transport stream) is also reduced to 1/8.

Figure VIII-4 shows the difference of TS rate for wideband/narrow band reception. "case 1" – "case 4" shown in the figure are defined in Table VIII-2 above.

In case of wideband transmission (upper portion of Figure), the receiver, which only uses for narrow band data reception, only demodulates A layer transport stream packet only. On the other hand, wideband receiver demodulates TSPs for both A and B layers.

Then the TS rate of narrowband reception is as equal as 1/8 of wideband reception.

In case of narrowband transmission defined as "case 3" and "case 4" in Table VIII-2, both transmission rate and reception rate are 1/8 of wideband transmission/reception.

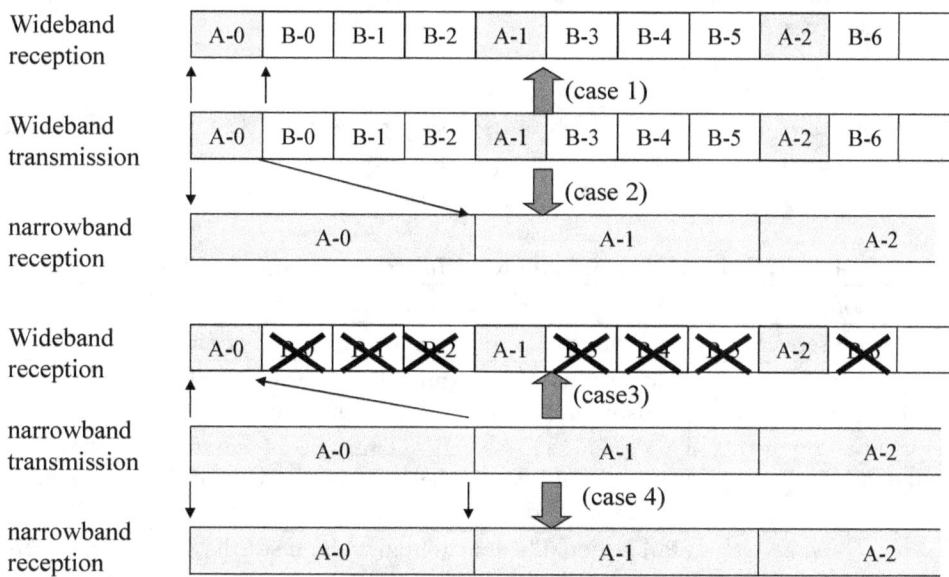

Figure VIII-4 Difference of TS rate for wideband/narrowband transmission/reception

VIII.1.3 Operational guideline for hierarchical transmission

The outline of the RF technology for One-Seg service and relation for wideband/narrowband transmission/reception is explained in the section above. To achieve a hierarchical transmission service, guidelines for PSI implementation in multiplexing are also defined in ARIB STD-B31, Attachment, Chapter 3.

(1) Multiplexing PAT, NIT, and CAT for hierarchical transmission

In case of partial reception, the Transport Stream Packet (TS) of the partial reception layer (A layer) should be decoded by itself without TSPs of another layers.

Observing the Figure VIII-4 again, in case of narrowband reception, the received TSP is in A layer only. Therefore, the control information for channel selection should be transmitted by partial reception TSP.

The transmission guideline defines the hierarchy for transmitting the control information PAT, NIT, and CAT. As shown in Table VIII-3 Hierarchical layers for transmitting PAT, NIT, and CAT is defined to be a partial reception layer.

Table VIII-3 Hierarchical layers for transmitting PAT, NIT, and CAT[119]

	Condition	Hierarchical layer for transmitting PAT, NIT, and CAT[*1]
1	Broadcasting with no partial reception	Multiplexed into the most robust layer[*2]
2	Broadcasting with partial reception	(1) Multiplexed into the layer for partial reception[*2]
		(2) Multiplexed into not only the layer for partial reception but also another layer if this layer is robuster than the layer for partial reception[*2]

*1:   CAT is required for partial reception.
*2:   If the transmission in the hierarchical layers shown above is difficult, exceptional operations are also admitted. In this case, however, detailed operational provisions shall be set separately to ensure that services in each layer will be received successfully.

(2) Multiplexing PMT
The partial reception layer also needs PMT which is necessary to detect PID. Therefore, as described in Table VIII-4, PMT should be transmitted by partial reception layer.

Table VIII-4 Hierarchical layers for transmitting PMT[120]

	Condition	Hierarchical layer for transmitting PAT
1	Partial-reception service	Transmitted by the layer for partial reception
2	When a hierarchical transmission descriptor is used within PMT[*1]	PMT should be transmitted at the robustest layer among those transmitting elementary streams (hereinafter referred to as "ESs"). However, PMT may be transmitted with the other hierarchical layer if it has robuster ranking of the layer than the layer specified above.
3	Service other than the above	PMT should be transmitted with one of the hierarchical layers transmitting ESs. It may also be transmitted with another hierarchical layer if it has robuster ranking of the layer.

*1: Services such as those in which video and other service qualities are changed in steps, in accordance with the reception status

The condition 2 of above table is applied for the case that the service qualities can be changed in steps in accordance with the reception status through the use of the hierarchical transmission descriptor

The condition 3 of above table is applied for the case that a service can be provided only when all service-multiplexed ESs are received.

For the background of condition 2 and 3, see ARIB STD-B31 Attachment 3.2.3.

---

[119] Quoted from ARIB STD-B31 Attachment Table 3-1
[120] Quoted from ARIB STD-B31 Attachment Table3-3

The maximum length of transmission period of PMT are defined as 100ms originally. As the transmission rate of partial reception is not so high, it is a good idea to reduce the transmission period of PMT. Considering this request, in ARIB TR-B14, the maximum period of PMT transmission can be extensible for partial reception.

(3) Multiplexing PCR packets at the partial-reception layer
PCR (Program Clock Reference) is essential information for synchronization of receiver/ decoder. Therefore, PCR is also necessary for partial reception.

In order to ensure that PCR packet is included into TS of partial reception layer, PCR packets for this service must be transmitted in accordance with Table VIII-5.

To reduce power consumption, the rate at which a single-segment receiver reproduces TS is likely to be lower than that for a 13-segment receiver. Therefore, the intervals at which TS packets are reproduced by the single-segment receiver do not always match those at which TS packets at the partial-reception hierarchical layer are reproduced by the 13-segment receiver, resulting in PCR jitter as shown in Figure VIII-5 may occurs.

To avoid such jitter, certain limitation for PCR transmission are placed for each mode described in Table VIII-5. These limitations ensure that the interval of PCR packets reproduced by single and 13-segment receivers are equal, with certain offsets set to adjust the difference in arrival time at the receiver demodulator output.

As an example, the position of PCR Packets with the limitation described above is illustrated in Figure VIII-6. As shown in Figure VIII-6, it is understood that the period of PCR is same both wideband receiver and narrowband receiver by this process. For the case of mode 2 and mode3, see ARIB STD-B31 Attachment Section 3.2.4.

Table VIII-5   Regulations for PCR-packet transmission at the partial-reception layer[121]

Mode	PCR-packet transmission regulations
Mode 1	For the duration of a single multiplex frame, only one PCR packet must be multiplexed per service, and the multiplexing position must remain constant for all multiplex frames (see Figure IX-7).
Mode 2	For the duration of a single multiplex frame, two PCR packets must be multiplexed per service at the same intervals .
Mode 3	For the duration of a single multiplex frame, four PCR packets must be multiplexed per service at the same intervals.

---

[121] Quoted from ARIB STD-B31 Attachment Table3-5

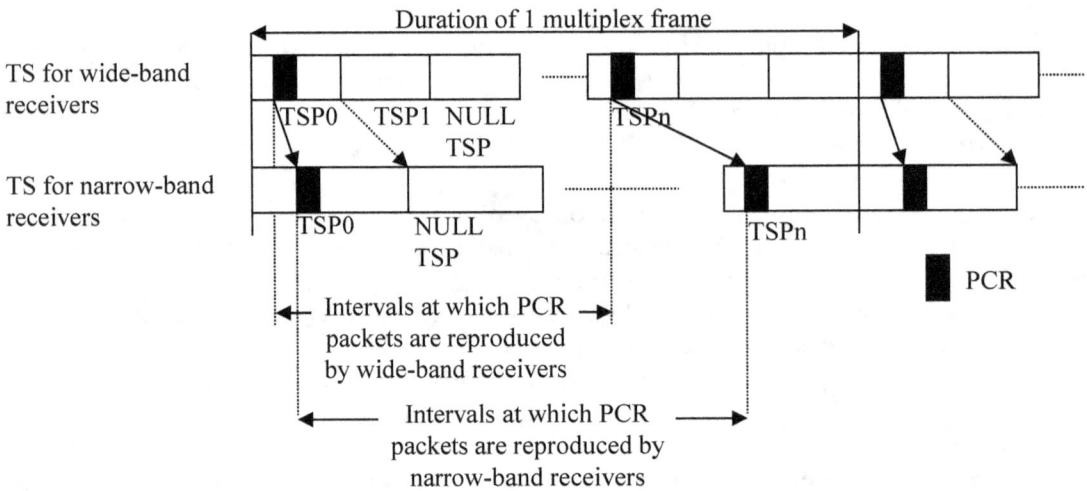

Figure VIII-5 TSs Reproduced by Wide- and Narrow-Band Receivers (No Limitations on PCR Transmission)[122]

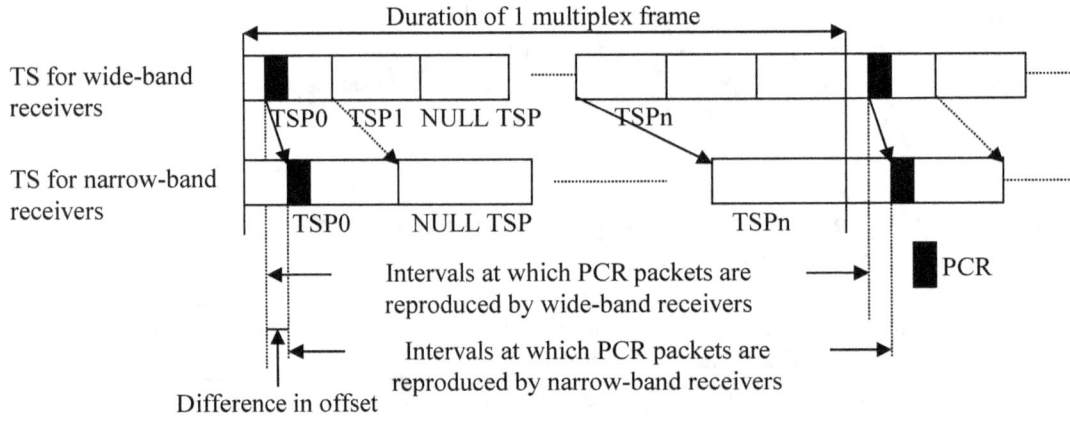

Figure VIII-6 TSs reproduced by wide- and narrow-band receivers (mode 1) (with limitations on PCR transmission)[123]

VIII.1.4 Block diagram of One-Seg receiver

In Chapter VII, a RF portion of full segment receiver block diagram is illustrated in Figure VII-7. An block diagram of RF component of One-Seg receiver is similar to a full segment receiver except the number of decodable hierarchy.

Figure VIII-7 illustrates and example of One-Segment receiver. Number of decodable hierarchy is one (number of decodable hierarchy of full segment receiver is three).

Except the number of decodable hierarchy and FFT size, other functions are almost same as full segment

---

[122] Quoted from ARIB STD-B31 Attachment Figure 3-2
[123] Quoted from ARIB STD-B31 Attachment Figure 3-3

receiver. Please note that the FFT size of One-Segment receiver is 1/8 of full segment receiver[124].

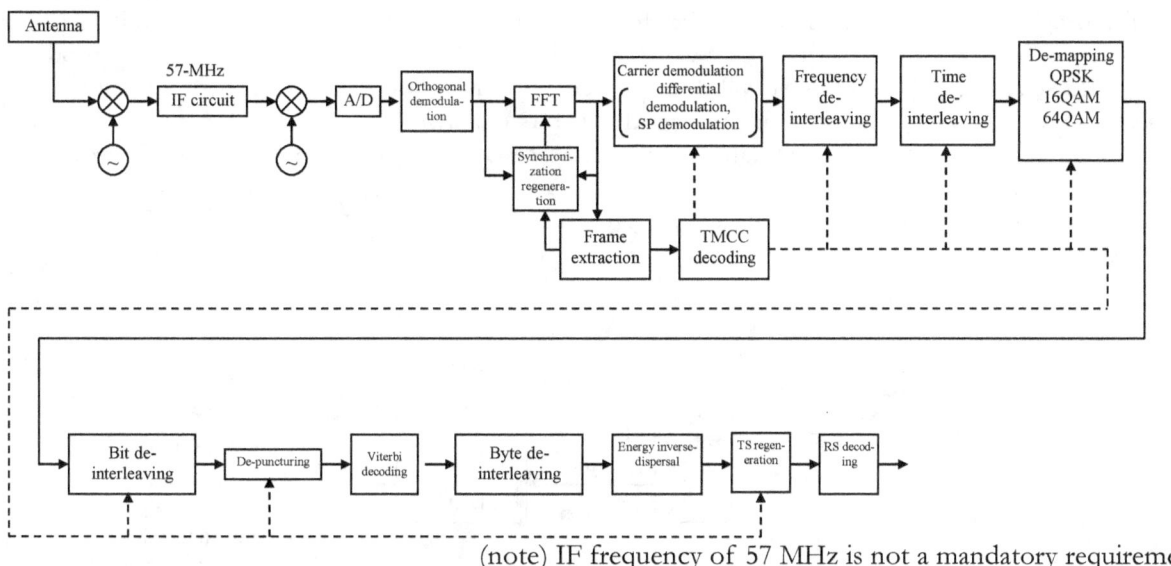

Figure VIII-7 Functional diagram of RF component of one-segment receiver[125]

VIII.1.5 Adaptive reception in mobile reception

As an example of hierarchical transmission service, the mobile receiver with adaptive reception is commercialized. Figure VIII-8 shows the transport stream (TS) of transmission side and reception side in case of 2 layers transmission.

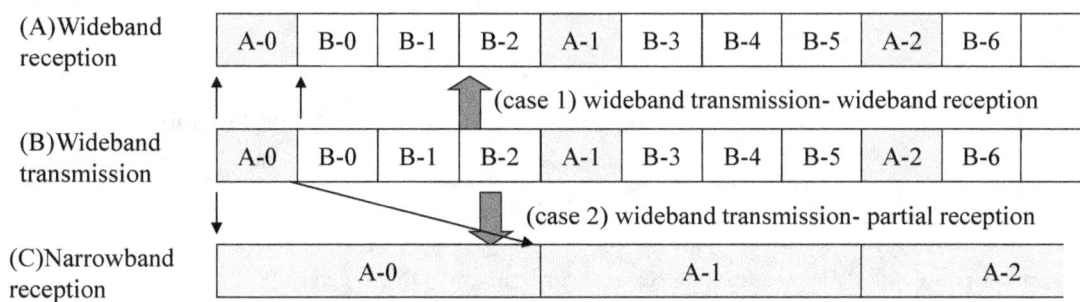

Figure VIII-8 Wideband transmission wideband/narrowband reception

Case 1 in the Figure shows wideband transmission and wideband reception. Wideband receiver is usually used for fixed receiver and mobile receiver. Case 2 in the Figure shows wideband transmission and partial reception. Partial reception receiver is usually used for portable receiver.

---

[124] See VIII.1.2 for details.
[125] Quoted from ARIB STD-B21 Chapter 5.2.6

As shown in the Figure, Transport stream of transmission side includes the TS packet of 2 layers, these are A layer and B layer. As a result, wideband receiver receives both A layer packets and B layer packet.

If the program of A layer and B layer are the same (simulcast for A and B layer), mobile receiver can display any program of A layer or B layer. So, if receiving condition of mobile receiver is not good because of low field strength, etc, the receiver will display layer A. On the other hand, receiving condition is good, in such time receiver will display layer B program that is HD quality.

By selecting A layer data or B layer data according to receiving condition, the mobile receiver can continue TV reception service.

An image of this type receiver is illustrated in Figure VIII-9

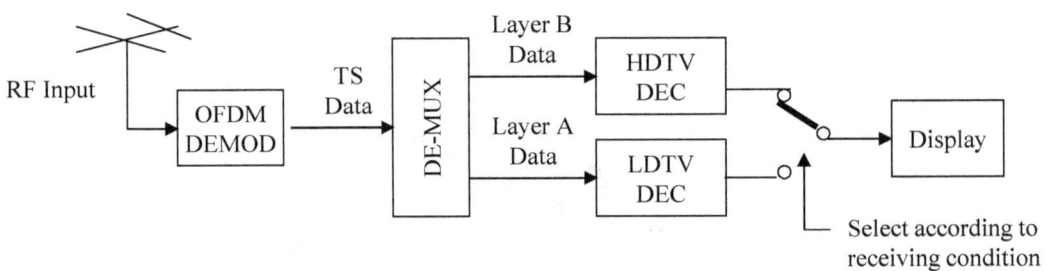

Figure VIII-9 Image of adaptive reception mobile receiver

VIII.2 Video Coding for One-Seg Service

In each service, there are different presentation models of receiver units and different operational styles of mono-media as the contents target. Therefore, ARIB TR-B14, Volume 3 defines profiles to specify the service type and detailed specifications for certain type of variations of the receiver units. TR-B14 defines following 3 profiles. Profile B is tentatively defined, but no details have been defined yet. Profile C is also referred to C-Profile

Profile A:
   Basic operation profile mainly targeting fixed receiver units (Stationary TVs, STBs, Portable TVs, etc.)
Profile B:
   (T.B.D) Basic operation profile of data broadcasting services mainly targeting transportable receiver units (car TVs, portable TVs, PDAs, etc.)
Profile C:
   Basic operation profile of data broadcasting services targeting portable receivers.

One-Seg receiver (C profile receiver) also follows international standard for its moving picture compression/coding. The MPEG standard scopes wide variety of receivers, however, incorporating various functionalities. In order for receiver design to be simpler, ISDB-T defines its specification as a subset of the international standard.

H.264/AVC is the international standard defined in ISO and ITU with variety of receivers and encoding functionalities in scope. ARIB STD-B24 includes the definition of video coding of ISDB-T as a subset of H.264. Countries adopting ISDB-T also have these standards.

Further, in order to specifically define coding method for A and C profile receivers, ARIB TR-B14 includes definition of the coding method for these profile receivers.

- H.264/AVC, ITU-T H.264, ISO/IEC 14496-10
- ARIB STD-B24, Volume 1 Data Coding, Part 2 Monomedia Coding, Chapter 4 Video coding, 4.4 H.264|MPEG-4 AVC
- ARIB STD-B24, Volume 1 Data Coding, Part 2 Monomedia Coding, Annex G Operation guidelines for H.264|MPEG-4 AVC video coding
- ARIB TR-B14, Vol. 3 Specifications for Data Broadcasting Operations, Section 4 operational provisions related to C-profile, 5 Operation of mono-media coding, 5.1 Image coding

VIII.2.1   Video coding definition in ARIB STD-B24

ARIB STD-B24 defines video coding definition for data casting, which is applicable to the One-Seg receivers. Please note One-Seg video and audio broadcasting service is defined as data casting in the ARIB standard. Here, incorporated functional parameters are categorized as "profiles" based on the definition of H.264. One is baseline profiles which is mainly for portable receivers with limited capacity, and the other is for main profile which is mainly for regular size fixed television.   Please note in actual One-Seg service only baseline profile are used and main profile are not.   In addition, video coding for fixed service (other than One-Seg) are defined separately as described in Chapter V.

Recommended operation guidelines for Baseline profile
(1) Supposed service requirements
   Bit rates: 64 kbps through 384 kbps
   Video formats: SQVGA, 525QSIF, QCIF, QVGA, 525SIF, CIF
   Frame rates[126]: 5 Hz, 10 Hz, 12 Hz, 15 Hz, 24 Hz, 30 Hz (actual number must be an integral multiple of 1000/1001) Frame skips are allowed.
   Picture display aspect ratio: 4:3, 16:9
(2) Levels
   Depending on a video coding format, a level must be selected among the applicable options: Level 1, 1.1, and 1.2.
(3) Other major operational constraint
   FMO (Flexible Macroblock Ordering), ASO (Arbitrary Slice Order), and RS (Redundant Slices) must not be operated.   Sequence Parameter Set must contain constraint_set0_flag =1 and constraint_set1_flag =1.

Recommended operation guidelines for Main profile
(1) Supposed service requirements
   Bit rate: up to 4 Mbps
   Video formats: SQVGA, 525QSIF, QCIF, QVGA, 525SIF, CIF, 525HHR
   Frame rates[127]: 5 Hz, 10 Hz, 12 Hz, 15 Hz, 24 Hz, 30 Hz (actual number must be an integral multiple of 1000/1001) Frame skips are allowed.
   Picture display aspect ratio: 4:3, 16:9
   Interlace pictures can be used.
(2) Levels
   Depending on a video coding format, a level must be selected among the applicable options: Level 1, 1.1, 1.2, 1.3, 2 and 2.1.

VIII.2.2   Video coding definition in ARIB TR-B14

Focusing on the One-Seg use, only limited set of functionalities are necessary, and redundant functionalities should be removed in order to minimize production and operation cost.   In this sense, ARIB TR-B14

---

[126] For 8MHz ISDB-T system, 25 Hz is added.
[127] For 8MHz ISDB-T system, 25 Hz is added.

defines "Operational provisions related to C-profile" to define specification for One-Seg terminals, including video coding definition for One-Seg service.

Major parameters
- H.264 | MPEG-4 AVC
- Baseline-level 1.2.
- Picture format is QVGA (4:3) (screen size : 320x240), QVGA (16:9) (screen size :320x180).

Restrictions in the bit stream
(1) Transmission cycle of the IDR access unit
  every 2 seconds
(2) Access Unit configuration
  The number and order of NAL units to configure the IDR AU and non-IDR AU are defined.
(3) Operation restrictions on syntax
  SPS (Sequence Parameter Set) restrictions
  VUI parameters restrictions
  HRD parameters restrictions
  PPS (Picture Parameter Set) restrictions
  Buffering period SEI message
  Picture timing SEI message
  Pan-scan rectangle SEI message
  Restrictions on the Buffering period SEI message and Picture timing SEI message insertion
  Slice header restrictions
  Reference picture list reordering restrictions
  Decoded reference picture marking restrictions
  Operation of PanScan in the picture display area

VIII.3 Audio Coding for One-Seg Service

ISDB-T adopts MPEG-2 Advanced Audio Coding (AAC) (ISO/IEC 13818-7) as its audio coding system. MPEG-2 AAC audio defines 3 profiles to specify the service type and detailed specifications for certain type of variations of the receiver units.

Main Profile
  The Main profile is used when memory cost is not significant, and when there is substantial processing power available. With the exception of the gain control tool, all parts of the tools may be used in order to provide the best data compression possible.

Low Complexity Profile
  The Low Complexity profile is used when RAM usage, processing power, and compression requirements are all present. In the low complexity profile, prediction, and gain control tool are not permitted and TNS order is limited.

Scalable Sampling Rate Profile
  In the case of a reduced audio bandwidth, the SSR profile will scale accordingly in complexity. In the Scalable Sampling Rate profile, the gain-control tool is required. Prediction and coupling channels are not permitted, and TNS order and bandwidth are limited. Gain control is not used in the lowest of the 4 PQF sub-bands.

VIII.3.1 Audio coding definition in TR-B14
One-Seg adopts AAC Low Complexity profile as its audio coding system. In addition to that, One-Seg also supports AAC SBR as optional audio coding system.

Spectral band replication (SBR) is a technology to improve coding efficiency at a given quality level, especially at low bit rates. Audio information in lower frequency is encoded by the usual process, whereas the information in higher frequency is encoded using spectrum information and decoded by prediction method based on the information in lower frequency. AAC+SBR is now widely used for 3G cellular phones or portable audio players.

Table VIII-6 Audio coding for digital broadcasting (for fixed reception)

Audio Coding	MPEG2-AAC (ISO/IEC 13818-7) Low Complexity Profile
Sampling Frequency	48kHz, 44.1kHz, 32kHz
Quantizer	16 bits
Audio Channels	1(mono), 2, 3(3/0), 4(3/1), 5(3/2), 5.1(3/2+Low Frequency Effects)
Bit rates	144kbps (stereo)

Table VIII-7 Audio coding for One-Seg

Audio Coding	MPEG2-AAC (ISO/IEC 13818-7, and Amendment 1)
Sampling Frequency	48kHz, 24kHz (Half Sample Rate: 24kHz only for SBR)
Audio Channels	1(mono), 2(stereo/dual)
Bit Rates	48kHz: 24kbps〜256kbps (mono) 32kbps〜256kbps(stereo) 24kHz: 24kbps〜96kbps (mono) 32kbps〜96kbps(stereo)

VIII.4 Data Broadcasting for One-Seg Service

The One-Seg service also supports Data Broadcasting or datacasting, including BML. BML, as described in the Chapter VI, bases on XHTML 1.0 and CSS, incorporating ECMAScript, DOM and some expansion for broadcasting unique functions. While BML for fixed receivers (or profile A) adopt version 3, BML for One-Seg service (or C profile) adopts version 12, with some modifications and additional functions. This Section describes the difference between BML for A profile and C profile.

1) Difference in BML Language Structure
BML for C profile is upgraded from BML for A profile in the following points:
　　Version number is 12.0
　　Character encoding is Shift-JIS
　　root element is <html> (not <bml>)
　for elements/attributes following points are different:
　　<pre>, <textarea>, <form> and <img> element were added
　　"type" attribute of <input> element can be "submit"
　　<a> and <textarea> elements can have "onclick," "onkeydown" and "onkeyup" elements

2) Difference in CSS
BML browser for C profile will implement focus movement control function. Its implementation is dependent on the manufacturer's policy. Therefore, control for focus movement, such as "nav-index"

"nav-left" "nav-right" "nav-up" "nov-down" attributes and ":focus" and ":active" pseudo class is not implemented. Color Look Up Table (CLUT) are not used, rather color values are directly indicated. "clut" "color-index" "border-color-index" attributes are depreciated. Font is limited only "maru gothic"

In addition to that, attributes to display characters in ticker style "-wap-marquee-style" and to set detailed formation parameter for <input> "-wap-input-format" were added.

3) Events

Major events for C profile receivers are listed in Table VIII-8.

Table VIII-8 Major events for C profile

Value of Type	Event
keydown / keyup (note)	Key specified in used-key-list is pushed down / released.
click (note)	An Element is selected by pushing the "Enter" key or an access key.
focus / blur	the element is in the focus / out of the focus
load	a document is loaded.
unload	The advanced notice of unloading the document
change	A change of the "value" attribute of an element is detected when the focus of the element blurs.
submit	"submit" button of the form element is pushed or the "submit" method of the form element is called.
EventMessageFired	Event message is received.
ModuleUpdated	Module update is detected.
ModuleLocked	Module is being locked.
TimerFired	Timer conFigured in beitem is fired.
MediaStopped	presentation by monomedia decoder is terminated. (audio/X-arib-mpeg2-aac only)
DataEventChanged	Detected update of data_event_id.

(note) Whether the occurrence of the key event by the "Enter" key in the input or textarea element is receiver unit dependent. (Excludes input elements with a type attribute of "submit".)

4) Extended Functions for Broadcasting for C profile

In order to implement Broadcasting unique functions, BML has extended functions. BML implement Browser pseudo object for ECMAScript to access these functions, that is the extended functions for broadcasting are provided as methods of the browser pseudo object. Major functions added only for C profile is listed in Table VIII-9.

Table VIII-9 Major extended functions for broadcasting for C profile

Function Name	Function
X_DPA_mailTo()	Sends text mail by mail applications of receivers with subject, body toAddress argument.
X_DPA_phoneTo()	Makes calls by the phone_number in argument
X_DPA_tuneWithRF()	Tunes in a channel of mobile reception by specifying the reception frequency in the physical channel number and service_id.
X_DPA_startResidenApp()	Initializes receiver's native application. Name of the application appName is given as an argument.

X_DPA_getCurPos()	Obtains receiver geographical position, such as longitude and latitude. The position value can be obtained from GPS or base station of cellular network.
X_DPA_writeSchInfo()	Writes information in the schedule notebook of the receiver features. Date, title, text, alarm is given as argument.
X_DPA_writeAddressBookInfo()	Writes information in the address book of the receiver features with name, telephone number, e-mail address as argument.

VIII.5 Presentation on One-Seg

The presentation function of C-profile basic receiver units is modeled by both virtual planes which output the decoded results of each mono-media and a display buffer that depend on the display device of the receiver units.

Similar to the A profile receiver, C profile receiver also can decode and present video, closed caption and BML contents using virtual plane concept. However, C profile receiver does not allow transparency blending of each virtual planes, rather C profile receiver defines region and designates the region for specific planes. The following is the process how C profile receiver displays the contents:
1. The receiver decodes video stream with a video decoder and outputs it to a video virtual plane (Y, Cb, Cr 4:2:0 format).
2. The receiver decodes closed caption data with a closed caption decoder and outputs it to a closed caption virtual plane.
3. The receiver decodes BML contents with BML browser and outputs it to a virtual plane for BML browser (width of 240 pixels, height of 480 pixels: each 8 bit format of RGB).
4. The receiver executes a scaling and color space conversion suitable for the display format of each receiver unit in each output from video, closed caption and BML virtual plane in order, then outputs them to the display buffer of the receiver unit. It is not necessary to overlap each virtual plane by alpha blending.

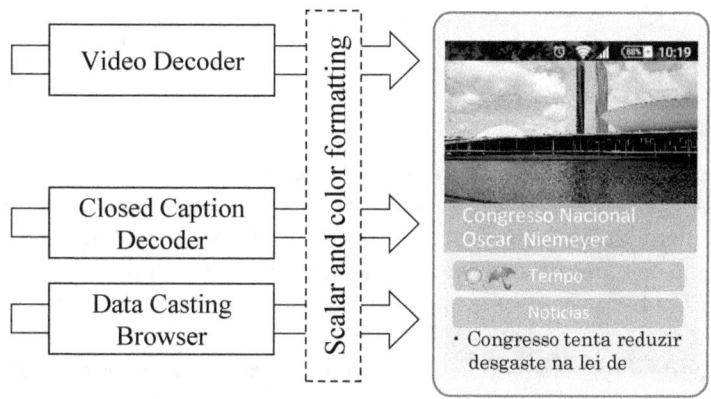

Figure VIII-10　Presentation technique of C-profile receiver units[128]

---

[128] Quoted from ARIB TR-B14, Vol. 3, Section 4, Figure 3-1

Also given the limitation of screen size and energy consumption, the size and presentable mono-media coding of each plane is defined as illustrated in Table VIII-10. Details of each encoding method will be discussed later in this Chapter.

Table VIII-10 Resolution and presentable mono-media codes of screen planes

Item		Contents of specification
Video Virtual Planes	Resolution	320x180x16, YCbCr(4:2:0), 16:9 320x240x16, YCbCr(4:2:0), 4:3
	Presentable Mono-media Codes	H.264, MPEG-4 AVC
Closed Caption Virtual Planes	Resolution	Character writing direction normal size, more than 12 characters x 4 lines, or character writing direction normal size, more than 16 characters x 3 lines. 8 bits of each RGB
	Presentable Mono-media Codes	8-bit character codes for C-profile Transmission method: Independent PES
BML Virtual Planes	Resolution	240 x 480 x 24, RGB each 8bit
	Presentable Mono-media Codes	JPEG, GIF, animation GIF Font Coding: Shift-JIS

C profile receiver also limits the number of Elementary Streams (ES) because of its limited processing capacity as follows.

Basic moving picture (H.264|MPEG-4 AVC):   maximum 1ES
Audio (MPEG-2 AAC):   maximum 2ES
Data carousel for C-profile: maximum 2ES
Event massage for C-profile: maximum 2ES (do not operate NPT)
Closed caption: maximum 1ES
PCR: maximum 1ES

VIII.6 Basic receivers

One-Seg's Service type and detailed specifications are defined as Profile C.

The digital broadcasting RF signal, fed into receiver units, are transformed into transport streams by tuners and demodulators. The demodulated transport stream is, by a transport stream decoder, de-multiplexed into video, audio and other data, and are then output as video stream by video decoding process, and audio stream by audio decoding process. By those processes, C-profile basic receiver unit plays back video and audio in a stream format.

The receiver receives and stores data within a data carousel in main memory or non-volatile memory and then processes data with CPU.

By using two-way communication lines, better interactive performance is expected than analog TV viewing. From the respective viewpoint of the above mentioned hardware process performances, it is necessary to provide the specifications for the following functions.

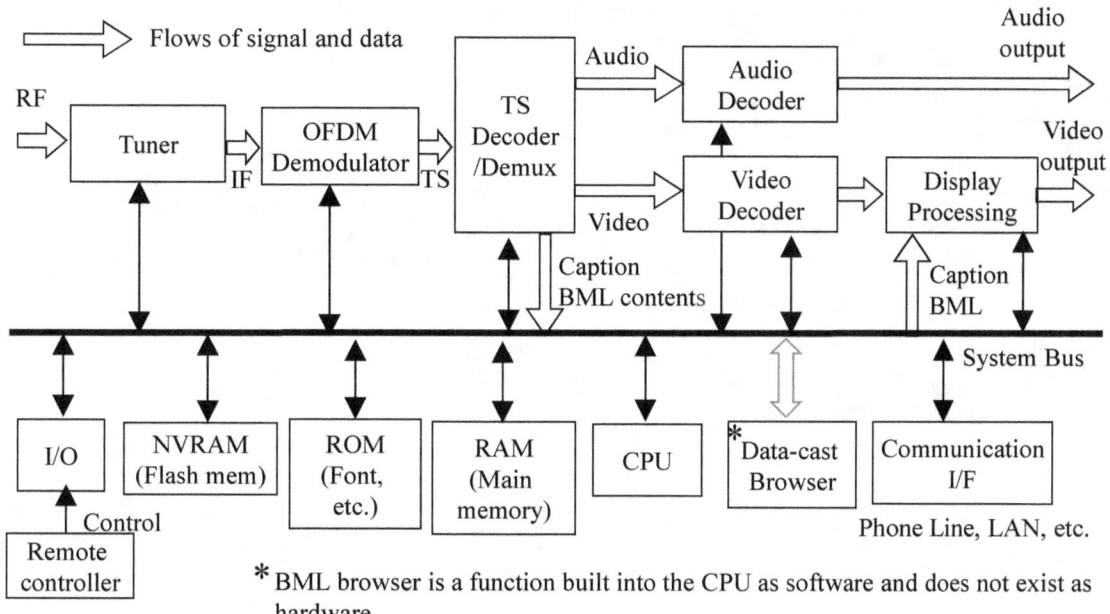

Figure VIII-11　Hardware configuration of C-profile basic receiver units[129]

---
[129] Same figure as Figure VII-2 in Chapter VII

(This page is left intentionally blank.)

Chapter IX.   Emergency Warning Broadcasting System

IX.1   Overview of EWBS

The Emergency Warning Broadcasting System (EWBS) [130] [131] enables a broadcasting station to issue a special control signal that is used to activate EWBS equipped radio/TV receivers even at stand-by mode. Once activated, the EWBS enabled receivers will wake up automatically and start receiving the emergency news program on air. This is done specifically in times of emergency.

In Japan, studies on Emergency Warning Broadcasting System began in 1980 based on the earthquake research in Tokai region. Actual EWBS service started in 1985 for analog radio and TV broadcast. In analog EWBS, a specific audio pattern is used for the receiver to automatically switch on and tune in to the emergency broadcast program.

This concept was followed and embedded in digital broadcast technology. The terrestrial digital broadcast system (ISDB-T) and digital satellite broadcast system (ISDB-S) now operates EWBS in the digital domain.

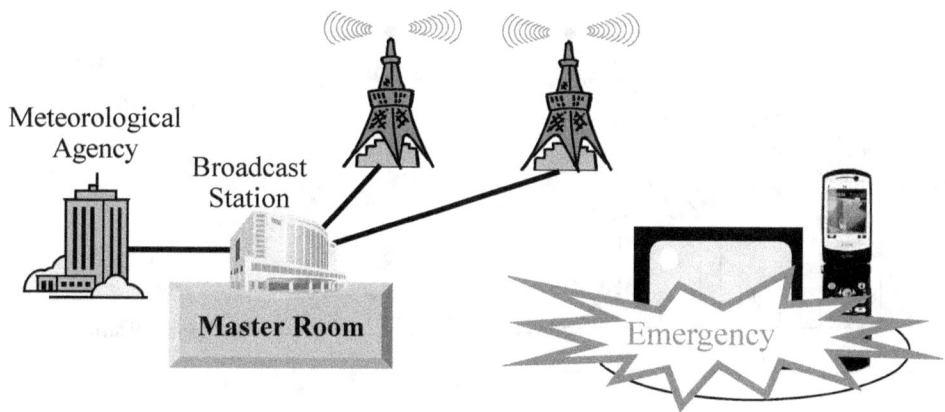

Figure IX-1   Overview of EWBS

A receiver equipped with the EWBS functionality will automatically activate even when they are on stand-by mode. The EWBS control signal circuit keeps receiving on the broadcast. When it receives EWBS signal, it immediately wakes the receiver up, switches to full operational mode and instructs the receiver to tune into the channel that transmits the emergency program.

On the broadcast station's side, the specific EWBS equipment should be connected to the multiplexer or the transmitter as an addition to the regular set-up that is conformable to the existing configuration of the broadcaster.

---

[130] "Handbook on Emergency Warning Broadcasting Systems," Kazuyoshi Shogen, ABU
[131] "UN-ESCAP - ABU Early Warning Broadcast Media Initiative," http://www.abu.org.my/abu/index.cfm?pageid=688

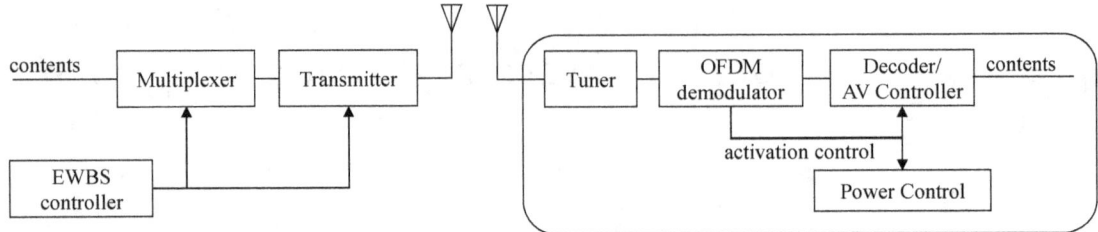

Figure IX-2 Block diagram example of EWBS

IX.2 Signaling of EWBS

There are two types of information dedicated for EWBS in the ISDB-T standard: emergency_warning_flag in the TMCC Signal and the Emergency Information Descriptor in the PMT.

IX.2.1   Emergency alarm broadcasting signal in the TMCC

First is the emergency_warning_flag in the Transmission and Multiplexing Configuration Control Signal, this is a data that is placed in a special carrier in the OFDM signal.   Details of the flag in the TMCC are defined in ARIB STD-B31.   The TMCC is used to inform the transmission parameters such as modulation, guard interval and time interleaver to receivers[132].

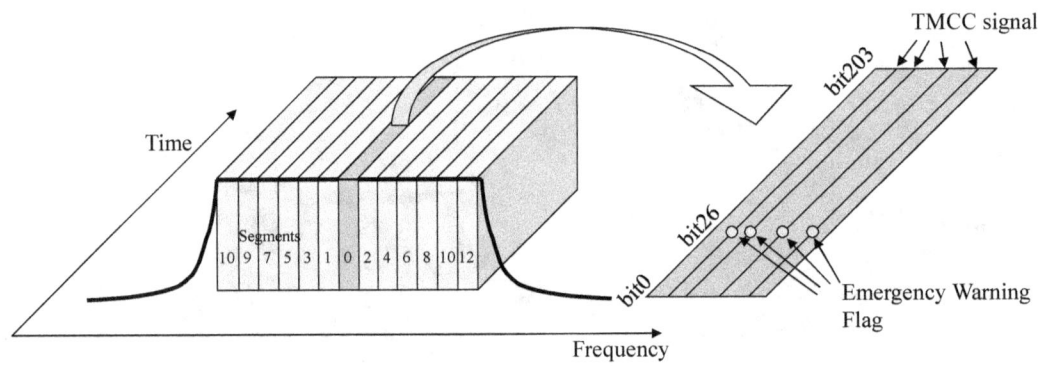

Figure IX-3   EWBS signal allocation in TMCC

One bit is assigned to "on_control_flag_for_emergency_broadcasting" in the TMCC signal.   This is used to the start-up operation of the receiver when the EWBS is activated.   The flag is "on" during the period when the emergency broadcast programs are being transmitted.

The on_control_flag_for_emergency_broadcasting in the TMCC can be passed from the multiplexer to the OFDM modulator both by:
  1) In the dummy byte of the 204 byte Transport Stream Packet, or
  2) In the ISDB-T Information Packet (IIP).
Details of both 1) and 2) are explained in Chapter III.

---

[132] See Table III-14 of this book.

IX.2.2   Emergency Information Descriptor in PMTs

ARIB STD-B10 and ARIB TR-B14 define the details of EWBS[133] [134] [135].

After receiving the activation control flag from the TMCC[136] and switches on, the receiver looks for the Emergency Information Descriptor (tag value=0xfc) in the PMT.   This descriptor describes the information and functions necessary for the emergency alarm signal.

Syntax	No. of bits	Identifier
emergency_information_descriptor(){		
descriptor_tag	8	uimsbf
descriptor_length	8	uimsbf
for (i=0;i<N;i++){		
service_id	16	uimsbf
start_end_flag	1	bslbf
signal_level	1	bslbf
reserved_future_use	6	bslbf
area_code_length	8	uimsbf
for (j=0;j<N;j++){		
area_code	12	bslbf
reserved_future_use	4	bslbf
}		
}		
}		

Figure IX-4   Syntax of the emergency information descriptor[137]

In this descriptor, service_id indicates the program in which the emergency warning broadcasting content is being transmitted.

When switch_on_control_flag in TMCC is on and the start_end_flag is "1", it means that emergency alarm signal is being transmitted.   When this bit becomes "0", it means that the emergency alarm signal has been concluded[138].

The signal_level (category of the warning) specifies the warning's legal basis: a law on earthquake response or weather related emergencies including tsunami's.

Table IX-1   The assigned value for signal_level[139]

Signal level	Description	Classification of usage
0	1st type start signal	· When broadcasting that alarm declaration is issued by the specification of article 9, clause 1 of "Large

---

[133] ARIB STD-B10, "Emergency Information Descriptor" on Part2, 6.2.24 and "Regulation on Emergency Warning Signals" Annex D
[134] ARIB TR-B14, Vol. 2, 6.11.4, "Reception of Emergency Warning System"
[135] ARIB TR-B14, Vol. 7, 7.9, "Operation of Emergency Warning System"
[136] ARIB STB-B31, 3.15.6, "TMCC Information,"
[137] Quoted from ARIB STD-B10, Part 2, Table 6-40
[138] When start_end_flag=0 and switch_on_control_flag in TMCC is on, this means EWBS test transmission.
[139] Quoted from ARIB STD-B10, Part 2, Annex D

			scale earthquake countermeasure exceptional action law" (Law No. 73 in 1978). ・ When broadcasting in accordance with the specification of article 57 of "Disaster countermeasure basic law" (Law No. 223 in 1961) (including when applying article 20 of "Large scale earthquake countermeasure exceptional action law".)
1		2nd type start signal	・ When broadcasting that tidal wave alarm has been issued by the specification of article 13 clause 1 of "Weather business law" (Law No. 165 in 1952.)

The area_code indicates where the warning is intended for broadcast. If the receiver notices that the start_end_flag=1 and it is in the area shown in area_code, it automatically selects the specific program shown in sevice_id.

Table IX-2 Area code for EWBS in Japan[140]

Local code	Description		Local code	Description	
0011 0100 1101	Local common code		1101 0100 1010	Prefecture code	Yamanashi
0101 1010 0101	Wide area code	Wide area of Kanto	1001 1101 0010		Nagano
			1010 0110 0101		Gifu
0111 0010 1010		Wide area of Chukyo	1010 0101 1010		Shizuoka
			1001 0110 0110		Aichi
1000 1101 0101		Wide area of Kinki	0010 1101 1100		Mie
			1100 1110 0100		Shiga
0110 1001 1001		Tottori, Shimane area	0101 1001 1010		Kyoto
			1100 1011 0010		Osaka
0101 0101 0011		Okayama, Kagawa area	0110 0111 0100		Hyogo
			1010 1001 0011		Nara
0001 0110 1011	Prefecture code	Hokkaido	0011 1001 0110		Wakayama
0100 0110 0111		Aomori	1101 0010 0011		Tottori
0101 1101 0100		Iwate	0011 0001 1011		Shimane
0111 0101 1000		Miyagi	0010 1011 0101		Okayama
1010 1100 0110		Akita	1011 0011 0001		Hiroshima
1110 0100 1100		Yamagata	1011 1001 1000		Yamaguchi
0001 1010 1110		Fukushima	1110 0110 0010		Tokushima
1100 0110 1001		Ibaraki	1001 1011 0100		Kagawa
1110 0011 1000		Tochigi	0001 1001 1101		Ehime
1001 1000 1011		Gunma	0010 1110 0011		Kochi
0110 0100 1011		Saitama	0110 0010 1101		Fukuoka
0001 1100 0111		Chiba	1001 0101 1001		Saga

[140] Quoted from ARIB STD-B10, Part 2, Annex D

1010 1010 1100		Tokyo	1010 0010 1011		Nagasaki
0101 0110 1100		Kanagawa	1000 1010 0111		Kumamoto
0100 1100 1110		Niigata	1100 1000 1101		Oita
0101 0011 1001		Toyama	1101 0001 1100		Miyazaki
0110 1010 0110		Ishikawa	1101 0100 0101		Kagoshima
1001 0010 1101		Fukui	0011 0111 0010		Okinawa

IX.2.3 Operation of EWBS

In order to start EWBS, the organization in charge of issuing the warning needs to coordinate with the broadcasters to be able to deliver early warning to the people in case of emergency. In Japan, considering the impact of EWBS in the society, the broadcast law limits its application[141] only in the event of:

- Tokai Earthquake Warning Declaration is issued,
- Tsunami Warning is issued, and
- The governor or mayor of the local government(s) request based on the Disaster countermeasure basic law

The flow of time-critical information between government agencies and broadcasting stations is a key factor for the effective operation of EWBS.

Upon receipt of these warnings, broadcasters will take following operation.
(When starting the EWBS)
  The Broadcaster activating the EWBS will:
1. Change the PMT value with the Emergency Information Descriptor that specifies EWS conditions (start_end_flag, the classification of Type 1 and Type 2, and area code),
2. Set the start flag for emergency warning broadcasting in the TMCC signal to "1" for transmission, and
3. Switch the contents so that viewers can recognize the program is on emergency warning broadcast.

(When concluding the EWBS)
  The Broadcaster deactivating the EWBS will:
1. Switch the contents to the normal program
2. Set the start flag for emergency warning broadcasting to "0" for transmission and
3. Delete the Emergency Information Descriptor from the PMT.

IX.3 Emergency Earthquake Warning

In October 2007, the Japan Metrological Agency started providing Earthquake Early Warning Information for the general public. Early earthquake warning is a technology to issue public earthquake warnings before earthquake tremor arrives with application of sensor network and communication networks[142]. As electronic signal travels much faster than the earthquake wave, we can detect and issue warnings before earthquake wave arrives.

---

[141] "Enforcement Regulations of the Broadcast Law (in Japanese only)", Article 17-27, http://law.e-gov.go.jp/htmldata/S25/S25F30901000010.html
[142] "Earthquake Early Warning," Japan Metrological Agency, http://www.jma.go.jp/jma/en/Activities/eew.html

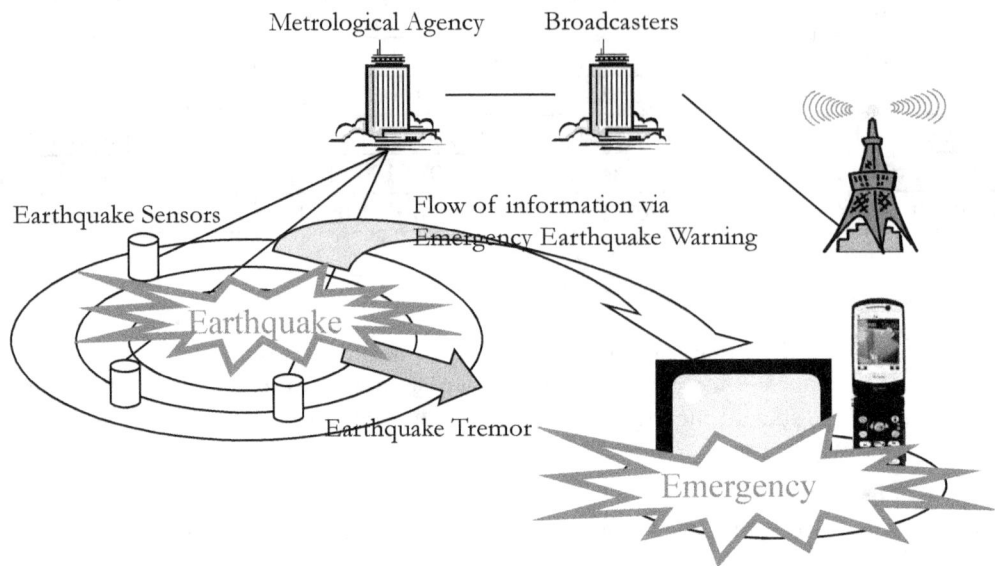

Figure IX-5   Concept of emergency earthquake warning

Tremors of an earthquake generally extend out from the seismic focus in a shape like a concentric circle. According to past observations, there are two main types of seismic waves: P-waves, or initial tremors, and S-waves, or main tremors. P-waves are the first to travel outward, followed by S-waves, which causes stronger tremors. Most earthquake-induced damage are attributed to these S-waves.

Seismographs deployed nationwide detects a P-wave tremor, continuously gathers tremor data and sends them to the Metrological Agency. By analyzing the data, a P-type tremor can be detected followed by the determination of the seismic focus and the magnitude of the earthquake immediately.

When a tremor (P-wave) is detected, the earthquake early warning notification will be sent to the government agencies, Nippon Telegraph and Telephone East Cooperation, Nippon Telegraph and Telephone West Cooperation and Japan Broadcasting Corporation (NHK) based on the Meteorological Service Act[143]. The notice is issued automatically and in electronic means. Information can be provided before the beginning of strong tremors occur (those farther away from the seismic focus will have more time to prepare for the quake). Upon receipt of the Emergency Earthquake Warning, NHK will broadcast the information to the public.

In the earlier implementations of ISDB-T, Emergency Earthquake Warning is transmitted and superimposed in the video and audio signal of regular broadcast contents. This is based on the assumption that, when emergency occurs, news program will interrupt the regular programs on air. Equipment in the master control room generally has functions that allow automatic information insertion upon receipt of the warning.

Also, datacasting can be used. One form is Superimposition in the form of an image, character and sound indicating earthquake information such as location and magnitude. Originally, EWBS with superimposition was an option. In 2015, the ISDB-T Harmonization Document was updated[144] incorporating use of superimpose in EWBS being a priority.

Another form is BML. When warning is issued, the BML of earthquake information will be transmitted

---

[143] "Meteorological Service Act (in Japanese only)" Article 15, http://law.e-gov.go.jp/htmldata/S27/S27HO165.html
[144] ISDB-T Harmonization Document, Part 3: Emergency Warning Broadcast System (EWBS), Nov. 2015

along with an event message to instruct the receiver to display the BML information. In this operation, various patterns of BML data are prepared and stored so that, when a warning is issued, BML can be transmitted immediately.

Since there is only a limited amount of time for the detection, transmission and reception/display the earthquake warning, efforts are being done to further improve the display of Emergency Earthquake Warning. When digital broadcast technology is used for the delivery of the warning, it actually introduces a longer delay than analog broadcast mainly because of the digital compression technology. In order to shorten the latency, a new implementation is standardized[145], in which the Auxiliary Channel in the OFDM transmission carries information about the warning. Here, updating the standard in the modification of the bit allocation and format setting of the AC data, OFDM framing and TS format has been anticipated and determined beforehand.

---

[145] ARIB STD-B31, 3.16, "Auxiliary Channel"

(This page is left intentionally blank.)

Chapter X. Deployment of Digital Terrestrial Transmission Network

This Chapter mainly explains the differences between analog and digital transmission system, SFN system construction, and the degradation factors for digital transmission network system. These are the basic technical issues that need to be addressed prior to the actual digital transmission network design.

X.1 The Digital Transmission Network

In this section, it explains the differences between the analog and digital transmission network including the key points for digital transmission network design.

X.1.1 Differences between analog and digital transmission network

Most broadcast engineers involved in the analog transmission have relevant experience in the design, manufacturing, installation and maintenances of transmitter/transmission network systems and related hardware. Therefore, it would be easier to establish an understanding of a digital transmission system by highlighting the differences between the two transmission systems.

Figure X-1 shows both analog and digital broadcast system.

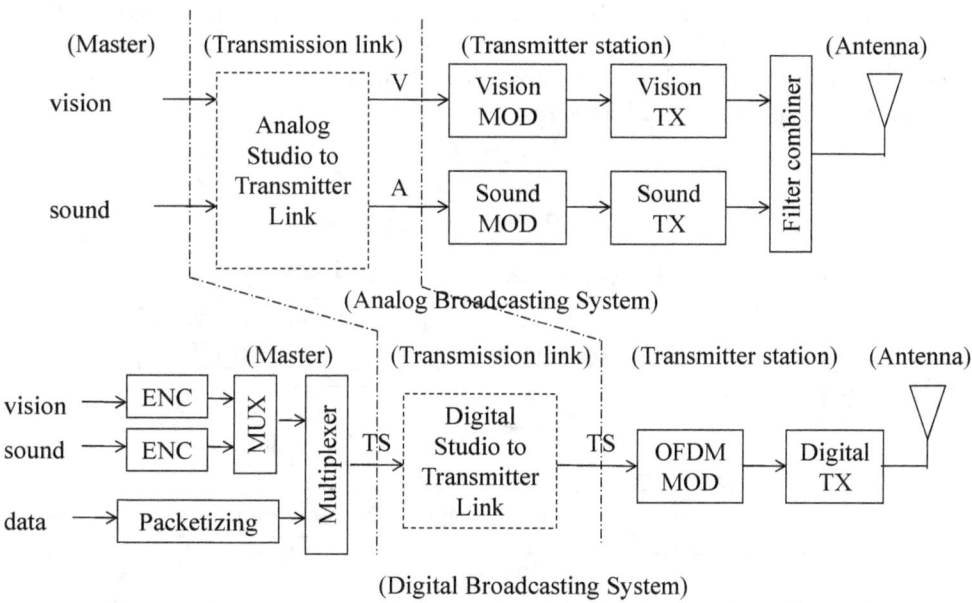

Figure X-1 Composition of broadcast system (comparison of analog/digital system)

In the Figure, analog transmitter system has two transmitters, a vision and a sound transmitter. (For low power analog transmitter system, sometimes a common amplifier for vision and sound RF signal is used.) On the other hand, digital transmitter has only one modulator and one power amplifier block.

While the composition of digital transmitter system is rather simple as described above, each hardware component that makes up a digital transmitter system has more sophisticated functions as compared with the components of an analog transmitter system.

The similarity that exists between an analog and digital transmitter is the function of the power amplifier which is basically for RF emission. In terms of the type of signal to be used, a digital transmitter is requested a more stringent design and better linearity of power amplifiers.

For the parameters for evaluating transmitter system, please see Section X.3.4.

X.1.2   Required signal to noise ratio

In analog TV broadcast system, Vestigial Sideband (VSB) Amplitude modulation is used. In general, the signal to noise ratio (S/N ratio) of the demodulated video signal is proportional to the carrier to noise ratio (C/N ratio) of an RF signal. The curve (a) of Figure X-2 shows the theoretical relationship between the field strength and Quality of Service of analog TV. For example, five grade scale are used to evaluate the received signal quality.

In digital systems, Quality of Service (QoS) is not proportional to the input signal strength. The curve (b) shows a typical relationship of the decoded video quality and RF signal strength. If an RF signal strength is greater than the certain threshold level called cliff point level indicated in the figure, the decoded video quality is rated as excellent in the specified area. When the RF signal level becomes lower than the level of cliff point however, serious disturbance in the decoded audio/video will occur, such as large block noise, freeze frames and ultimately picture black out. This characteristic is known as the "cliff effect".

It is important for the design of the digital systems to maintain RF signal levels greater than that of cliff point in a targeted service area. The signal level of cliff point varies depending on the transmission parameters chosen.

The theoretical values of the required C/N for ISDB-T can be found in the ARIB STD-B31[146]. This standard serves as a design reference. Please take note that the standard is based on the model receiver's capabilities of the time when it was published. Current capabilities of the model receivers have significantly improved and will exhibit a superior performance result.

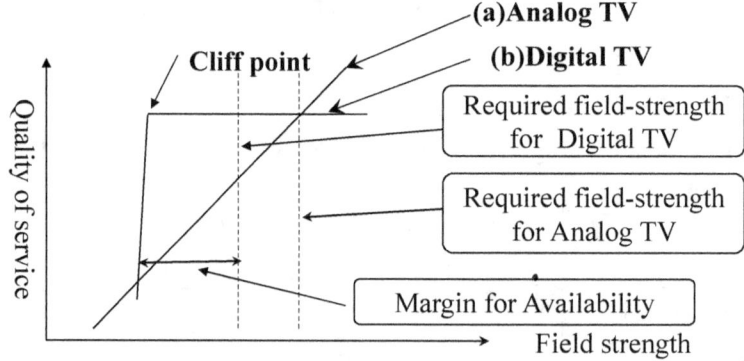

Figure X-2   Input/output characteristics of an analog/digital TV system

---

[146] ARIB STD-B31, Appendix 3; Considerations in the Link Budget for ISDB-T, Reference 3, Table A.3.3-2

This Section has tackled only the effect of thermal noise as a degradation factor. Unfortunately, other factors may and will affect the "Margins for availability". The details of these degradation factors are described in the section X.3.

X.1.3 Transmission network configuration

In analog systems, if two transmitters transmit the same TV content in the same frequency, the receiver picks-up a combined signal from two sources with different time of arrival in the receiver. As a result, a phenomenon called "ghosting" occurs which generates serious degradation in the picture quality. Therefore, it is difficult to assign the same frequency for multiple transmitters for analog TV broadcast without generating substantial picture quality impairments even if the TV contents being transmitted are identical.

On the other hand, as described in the section II.3.3, SFN operation is possible by making use of OFDM transmission technology with Guard Interval. SFN technology enables the efficient utilization of the limited frequency resources and provides an opportunity to efficiently manage the network coverage.

X.2 SFN and Signal Transmission in a Network

With OFDM technology, the broadcast signal can cover beyond the coverage of a single transmitter with a combination of multiple transmitters and relay stations operating in either a single or multiple broadcast frequencies. This combination of multiple transmitters and relay stations is called a "Transmission network."

X.2.1 Types of transmission network

There are two types of transmission networks based on the assigned frequencies of multiple transmitters that constitute a "transmission network".

Single Frequency Network (SFN) uses the same frequency for all of the transmitters in the service area (right side of Figure X-3); the other is Multi Frequency Network (MFN) in which each transmitter in the network utilizes different frequencies to cover the defined service area (left side of Figure X-3). The transmission network of an analog TV broadcast is a classic example of an MFN. For MFN, it is not necessary to be cautious about synchronized operation for multiple transmitter sites because they operate independently.

In the case of SFN, as described in the section II.3.3, the transmission timing (synchronization) of each transmitter site should be managed carefully within specified range, as all transmitters in this network operates inter-dependently. Further details are provided in the next section.

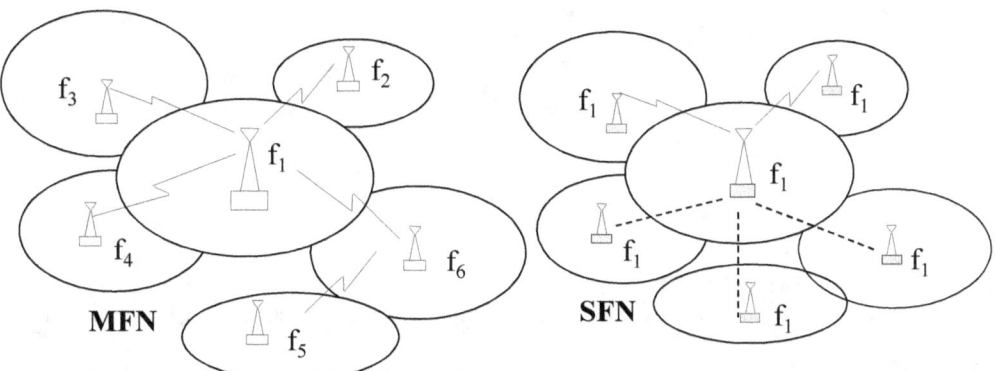

Figure X-3 Examples of network configuration

X.2.2   Synchronization for SFN
For SFN operation, it is necessary to satisfy the following conditions[147]:

(1) Transmission frequency:
To prevent interference between carriers within the SFN service area, the "frequency error" in the transmission frequency of a transmitter that forms the SFN must be within a specified range.   In Japan, the frequency error is defined within 1Hz.

Please note that the term "frequency error" is defined in the IEC standard[148].   Here, it is "the difference between the assigned frequency and the characteristic frequency as identified and measured".   In addition, "frequency error" includes "frequency drift" and "frequency setting error".

(2) IFFT sampling frequency:
For the transmission network in which multiple modulators are used, all IFFT sample-clock frequencies must be identical.

If one of the frequencies differs from the rest, then the degradation effect applies to all the OFDM symbol period within the network.   A symbol shift beyond the guard interval length and the OFDM signals results to interference between symbols.

Please note also that the frequency of each of the uppermost and lowermost carriers of the frequency band must not drift more than the specified range.

(3) OFDM signals:
When multiple OFDM modulators are used, the output OFDM signal waveforms must be identical with all the SFN stations.   Note that it is preferable to utilize a common transmission timing for the ease of delay control in order to make the difference in delay time within the service to be shorter than the guard interval.   All contents of the modulated signal from the multiple transmitter stations must also be identical.

X.2.3   Types of interface point in transmission network
Before proceeding to the details of the transmission network, it is necessary to define the interface points and signal format within the transmission network[149].   The following Figure X-4 shows the interface point and signal format.   In the Figure, 3 interface (I/F) points are defined, I/F(1), I/F(2) and I/F(3).

---

[147] ARIB STD-B31, Attachment, Section 4.1
[148] IEC 62273-1, Method of measurement for radio transmitter – performance characteristics for terrestrial digital television transmitters
[149] Uehara, M. Proceedings of the IEEE, Vol.94, Issue 1, Jan. 2006, Application of MPEG-2 Systems to Terrestrial ISDB(ISDB-T)

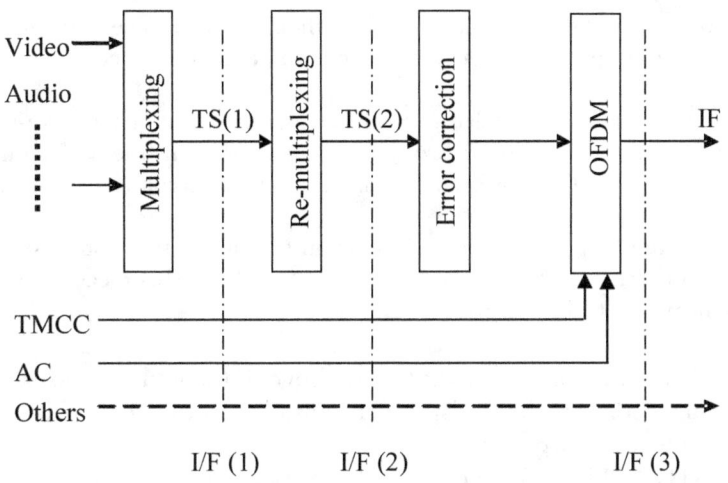

Figure X-4　Interface points in a transmission network[150]

The signal formats for each interface points are defined as follows:

TS(1):　The point in which the TS signal is so-called MPEG-TS and does not have the multiplexed frame structure of ISDB-T.　This format is referred to as "usual TS."

TS(2):　The point in which the TS signal has the multiplexed frame structure of ISDB-T, referred to as "broadcast TS".　The signal structure of BTS is explained in Chapter III.2.2 of this book.

IF:　OFDM modulated signal (section III.4)

A re-multiplexer shown in Figure X-4 should be unique in the same network to satisfy the condition (3) indicated in Section X.2.2.

In general, both service multiplexing function and re-multiplexing functions are implemented in a single hardware, called a "Multiplexer", which is located in the broadcast studio.　See Chapter XI.2 as an example of the hardware.

Please note that the re-multiplexer is defined in the "transmission systems" (see section III.2.2), but it is not necessarily mean that the re-multiplexer has to be a part of an OFDM modulator.　It is up to the manufacturer how to implement the re-multiplexer.　The multiplexed frame structure is arranged at the re-multiplexer which prepares the final format of the contents.　Therefore, there should be only one output of the re-multiplexer in a transmission network.

Because of the above stated explanations, I/F(1) is not used as an interface of SFN, rather, I/F(2) and I/F(3) are mostly used.

X.2.4　Types of synchronization scheme for SFN

Three types of synchronization schemes for SFN are introduced in the ISDB-T standard[151], as follows:

Complete synchronization:
　　The clock signal used in a certain transmitter station in the network is delivered to other transmitter stations

---

[150] Quoted from ARIB STD-B31 Attachment, Section 5.1
[151] Quoted from ARIB STD-B31 Attachment, Section 5.2

and used as a reference. Other transmitter stations should synchronize with this clock in some way. In some cases it is necessary to prepare a network that distribute the reference clock signal separately.

Slave synchronization:
The clock and frame timing of a modulator in each transmitting stations is slave-synchronized to the clock of the multiplexer or re-multiplexer in the broadcast studio. For this method, a bit clock signal of STL can be used for synchronization. In this system, a delay adjustment circuit is necessary.

Reference synchronization;
The clock and frame timing of a studio and all transmitter stations that compose the transmission network are synchronized to a common reference signal, such as a GPS. This measure needs a common reference receiver in each transmitter station.

Please note that complete synchronization explained above is theoretically possible, but it is not used in practice because providing dedicated links for synchronization proves to be costly. For consideration purposes, this type method may be used only at the least priority.

In addition, ARIB STD-B31 [152] also describes another synchronization method, "Synchronization conversion (Quasi-synchronization)". The idea of this method is to transmit TS(1) as illustrated in Figure X-4 from a studio to transmitter stations. A transmitter station independently generates the synchronization signal and transmits by their own sync signal. The synchronization difference between the studio and the transmission station is adjusted by inserting a null packet at the transmitting station side. In this method, the synchronization of each transmitting signals cannot be achieved, therefore, SFN cannot be constructed. Because of this reason, this method is not used in practice.

X.2.5 Examples of SFN construction

One of the key point in designing an SFN is deciding the signal interface format and synchronization method especially if you have to deal with the various types of relays in a particular service area. Below are the four types of SFN construction methods used.

Table X-1 Types of SFN construction methods

Case	Interface point	Synchronization/Relay network
Case 1	I/F (2) "Broadcast TS", a digital bit stream in the format of MPEG-2 Transport Stream with the additional control signal and bit rate adjustment is transmitted with the synchronization information.	Slave synchronization
Case 2		Reference synchronization
Case 3	I/F (3) The modulated ISDB-T OFDM signal from a transmitting station is picked up by the next transmitting station and then re-transmitted. Depending on the frequency to be used for the relay, there are two categories: IF or Broadcast wave relay.	IF relay network
Case 4		Broadcast wave relay network

---

[152] ARIB STD-B31 Attachment "Operation Guidelines for Digital Terrestrial Television Broadcast", Section 5-2

Case 1: I/F(2) interface format – slave synchronization:
The conceptual block diagram of a network configuration utilizing the TS(2) interface format – slave synchronization is shown in Figure X-5.

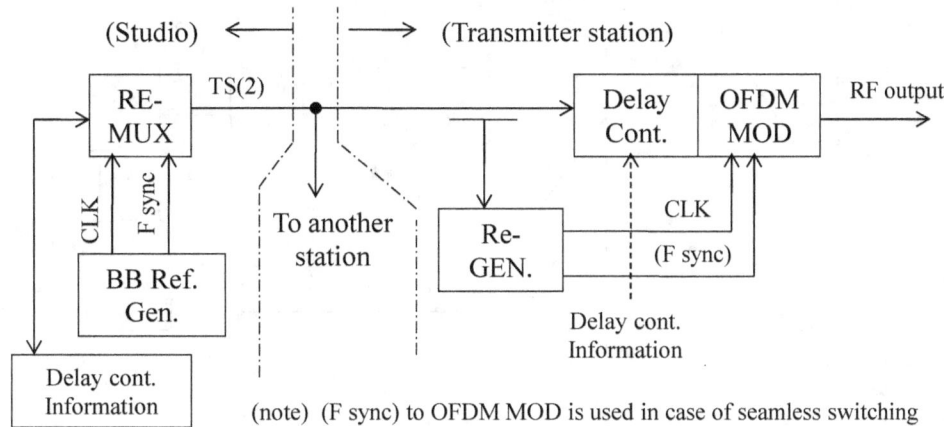

Figure X-5   Block diagram of an SFN (I/F(2) - Slave synchronization)

In this case, the signal format between studio to transmitter station is TS(2) (broadcast TS). Identical signal needs to be delivered to all transmitters in the SFN. A clock is re-generated from TS(2) signal and provided to the OFDM modulator, so that re-multiplexer found in the studio and OFDM modulators in the SFN can be synchronized. Usually, a transmission link is composed of primary and backup. If time delay exists between a primary and a backup link, an interruption may occur when switching each other. To avoid this interruption, the time delay between the primary and the backup link should be compensated by comparing a frame synchronization (or F sync) signal each other. This switching method is named "seamless switching".

Regarding the reference frequency used for an RF signal, if the output frequency error is required to be precise and within the specified range, an independent reference oscillator needs to be highly stabile such as crystal oscillator, or rubidium oscillator. GPS signal source is also an option for Base Band (BB) Reference Generator (to provide clock reference and OFDM frame timing).

Regarding the delay management, "Delay control device" (in the Figure, illustrated as "Delay Cont.") placed in front of modulator controls delay time, by compensating the delay difference between each transmitter output timing and assigned delay time.

The assigned delay time value is estimated based on the delay time of each devices and transmission link. In case that the same type of OFDM modulator and Transmitter are used in each transmitter station, the delay time of these devices is same. Therefore, only the delay time difference between each STL (Studio to Transmitter Link) should be considered for SFN operation, otherwise, you need to examine the interface points that has been used to be able to determine the amount of delay it contributes to the system.

Case 2: I/F(2) interface point – reference synchronization:
Same as Case 1, the signal format between studio to transmitter is TS(2) (broadcast TS). The difference is

that in case 1 the synchronization signal was transmitted from the studio via the relay link, whereas case 2 use common reference signal for synchronization for studio and transmitter stations[153].

The conceptual diagram of network configuration is shown in Figure X-6.

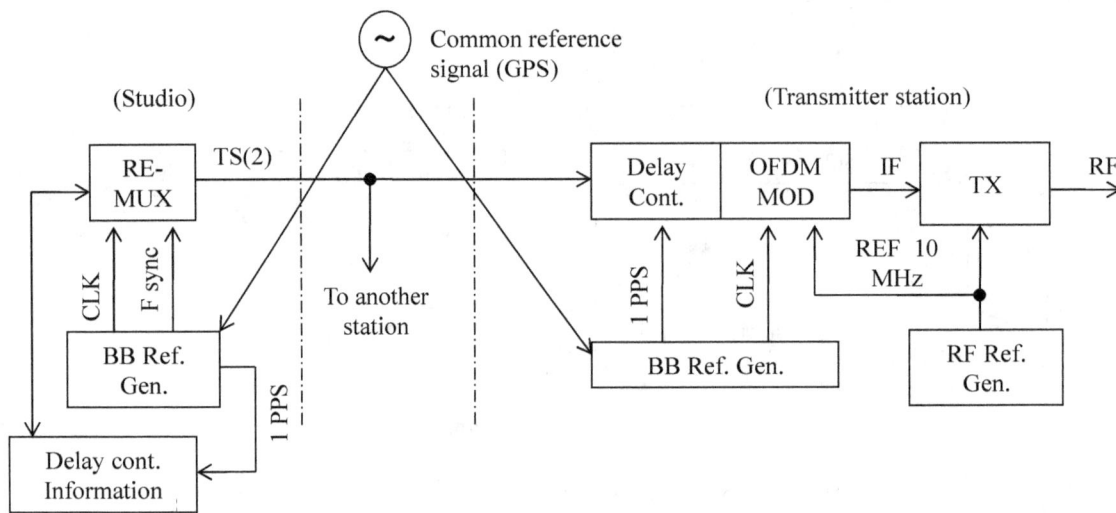

Figure X-6   Block diagram of an SFN (I/F(2) INF- Reference synchronization)

The signal in the form of TS(2) are delivered from the studio to all transmitter stations in the transmission network. "Baseband reference generator" (BB Ref. Gen) equipped in both the studio and transmitter station are slave synchronized to a common reference source.   GPS are often used as this common reference source.

Regarding the delay management, as shown in the Figure, a studio and transmitter station(s) have a common reference time based on GPS.   The OFDM frame transmission timing (at the output of re-multiplexer) is measured based on this common reference time, and this timing information for the transmission is multiplexed as an additional information (see section X.2.6 later) into the TS signal and then sent to the various transmitter stations.

At a transmitter station, additional information in NSI for assigned transmission timing based on the common reference signal (such as GPS) described above is decoded.   Then it will be compared with the measured time difference of current transmission timing.   With that, the delay control devices illustrated in the figure controls the delay time so that the actual transmission timing matches the assigned transmission timing.

For the additional information and network timing control procedure, see Section X.2.6.

Case 3: I/F(3) interface point – IF relay network
In this case, a modulated OFDM signal is sent from studio to the relay stations.   The frequency interface used for relays is called an Intermediate Frequency (IF) which is different from the frequency being used for the actual broadcast.   Microwave is usually used for IF relay network.

The conceptual block diagram of a network configuration with IF interface format is shown in Figure X-7.

---

[153] ETSI TS101 191, V1.4.1, June 2004, Digital Video Broadcasting (DVB);DVB mega- frame for Single Frequency Network (SFN) synchronization

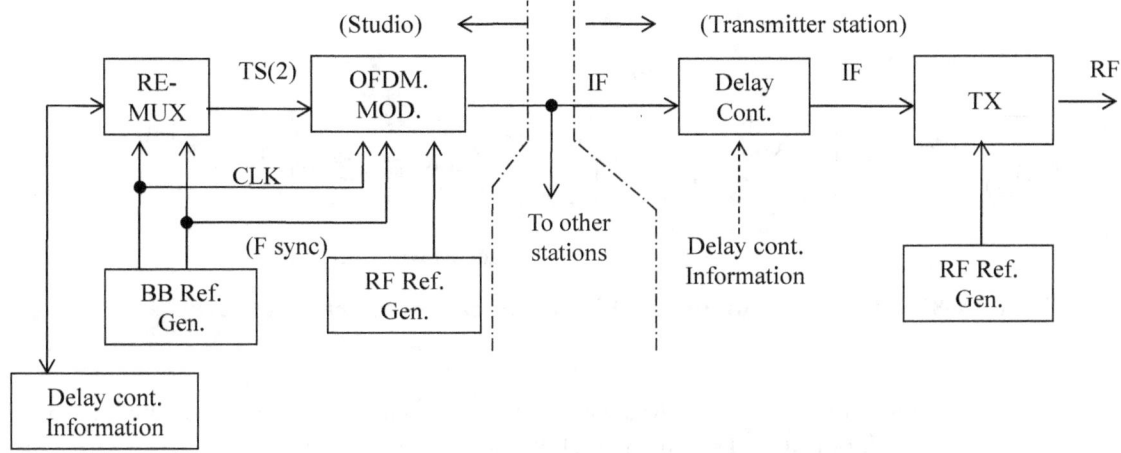

Figure X-7　Block diagram of an SFN (IF Interface)

This transmission network has only one OFDM modulator and this modulator is typically installed in the studio or in the main transmitter station. Therefore, it is not necessary to synchronize the IFFT sampling clock between the studio and the transmitter station. For the delay management, delay control device on each station adds fixed delay time independently.

Each assigned delay time value that compensates the delay difference between each transmitter output time is set in "Delay control device" (in the Figure, it is illustrated as "Delay Cont.") equipped in each transmitter station.

Using IF, the broadcast frequency of a relay station need not to be the same with that of the main transmitter station. In this case, independent oscillators in each stations are requested so that the their "frequency error" and resulting reception performance degradation caused by the frequency difference of transmitters in a SFN are within specified range. In case of Japan, the range is within 1Hz. This applies to Case 1 and Case 2 as well.

Case 4: I/F(3) interface point - broadcast wave relay network
In this case, the modulated OFDM signal is sent to a relay station. The difference from Case 3 is that the frequency for the link is the same frequency being used for on-air broadcast.

Broadcast wave relay network is representative of an analog transmission network configuration and is also popular in digital MFN broadcast in Japan as its hardware configuration is simple and can be manufactured economically.

The conceptual block diagram of this network type is shown in Figure X-8.

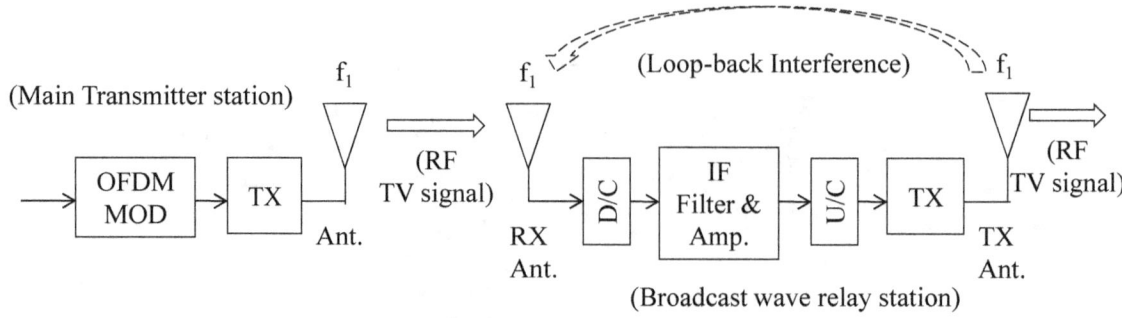

Figure X-8   Block diagram of an SFN (Broadcast wave relay network)

The signal interface format is a modulated OFDM signal, which is the same as the one described in Case 3. For a network interface, OFDM modulated signal format (IF interface format) is used.

This concept is applicable for both SFN and MFN. If the relay station(s) transmits in the same frequency (f_1), it is for SFN, whereas if the relay station transmits in a different frequency from f_1, then it is for MFN.

As shown in the Figure, in case of an SFN, when the transmit and receive frequency of the same transmitter station(s) are the same (indicated f_1 in Figure), a coupling loop interference from the transmit antenna to the receiving antenna may occur. Coupling loop interference is a phenomenon in which if the coupling loop interference level is beyond the threshold level, mutual oscillation occurs in the worst case. Please note that even though the coupling loop level is not high (less than the threshold), the coupling loop interference component may still contribute some degradation in the transmission signal. To reduce this degradation, the coupling loop canceller is proposed (see Chapter X.3.4). Because of above reason, the SFN with broadcast wave relay network is not popular; MFN is mostly used.

X.2.6   Additional information multiplexed on a broadcast TS for network control

The additional information for a network control is multiplexed in the "Broadcast TS" (defined as TS(2) in X.2.3), which is used for the transmission network management which includes a control information such as delay in SFN. This control information is used not only for network synchronization control of SFN but also serves as a way to measure the delay time at each transmitter station for configuring delay control device in Case 1 and Case 3 described above.

This Section will explain multiplexing method, types of additional information and network synchronization process, etc.

(1) Additional information and multiplexed positions

The types of transmission control information which is useful for the transmitter, the transmission network control and the multiplexed positions are described in Table X-2. Two types of multiplexed configuration are available: (i) Multiplexed to the dummy byte as part of each TSP (Transport Stream Packet) and (ii) Multiplexed as invalid hierarchical TSP (such as IIP, ISDB-T Information Packet). Please note that the multiplexed positions of each information are specified each by each in Table X-2.

Table X-2 Additional information and multiplexed position[154]

No.	Transmission item	Description	Multiplex position (note 1)	
			Dummy byte	Invalid hierarchy
1	TMCC identifier	00: BS digital 10: Terrestrial digital TV 11: Terrestrial digital audio	X	-
2	Buffer reset flag	Synchronization device buffer reset control	X	-
3	Switch on control flag for emergency broadcast	Designating the duration of the emergency broadcast	X	X
4	TMCC initialization timing head packet flag	Designating the top packet from which the TMCC will be modified	X	-
5	Frame head packet flag	Designating the header packet of the "multiplexed frame"	X	-
6	Frame synchronization identifier (w0,w1)	Designating the duration of an even or odd number frames	X	X
7	Layer indicator (hierarchy information of each TSP)	Hierarchy discriminator of A, B, C, NULL Designation of the TSP that carries IIP or that carries AC data	X	-
8	Transmission parameter switching index		X	X
9	TSP count down index	The top packet of a multiplexed frame is 0, then increments the number in the order of the packet.	X	-
10	TMCC (including mode and GI)	TMCC and modulation device control information	-	X
11	Broadcast network control information	Control information such as delay in SFN	-	Optional
12	AC data (note 2)	Information transmitted by AC	Optional	Optional
13	Service provider's organized data	Data multiplexed to BTS independently by service providers		Optional

(note 1) for overlapping items such as the dummy byte and invalid hierarchy, it should be multiplexed so as not to contradict each other.
(note 2) Detailed information for AC data is described in ARIB STD-B31[155]

(2) ISDB-T information and multiplexed position
Additional information that is multiplexed as "dummy byte" is named "ISDB-T information."[156] As shown in Figure X-9, Broadcast TSP is composed of 204 byte, 188 byte for information, a rest 16 byte area is for parity data. The original parity data area (16 byte) is divided into two 8 byte areas, one is for "Dummy byte area" in which ISDB-T information is multiplexed and the other is maintained for parity data area.

The parity data is generated as the shortened Reed-Solomon code (204,196) which is generated from an

---

[154] Quoted from ARIB SRD-B31 Attachment, Section 5.5.1 Table 5-5
[155] ARIB STD-B31, Attachment, "Operation Guidelines for Digital Terrestrial Television Broadcast", Chapter 6
[156] ARIB STD-B31, Appendix, v 1.6, "Operation Guidelines for Digital Terrestrial Television Broadcast", Appendix, table 5-6 and table 5-7

original Reed-Solomon code (255,247) by adding 00HEX of the 51 byte in front of the input data byte, and deleting the first 51 byte after calculating the RS code.

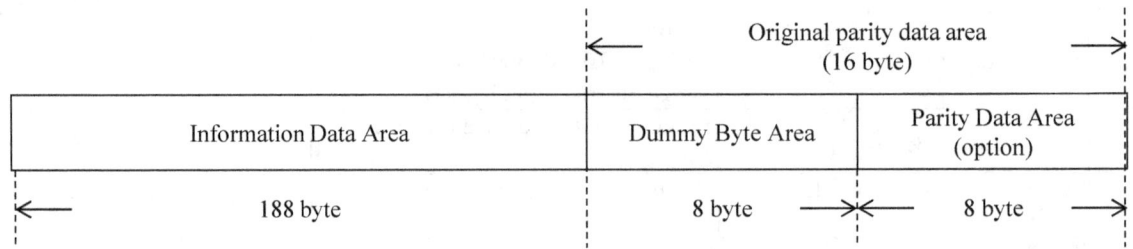

Figure X-9 Multiplexed position of dummy byte in TSP[157]

(3) Details of IIP (ISDB-T Information Packet) and multiplexed position

Table X-2 explained that two configurations are prepared for the additional information for the multiplexing area.   One is "dummy byte" area, and the other is "Invalid hierarchy packet".

"Invalid hierarchy packet" is defined as the "Null packet" which is inserted into the "Broadcast TS" to adjust the transmission speed[158].   This Null packet carries no information data, hence, this null packet is also called "invalid hierarchy packet".   Null packet is also identified as "0000" of the layer indicator which is described in Table X-2.

One of these null packets in every multiplexed frame is used for the ISDB-T information container.   This packet is named ISDB-T Information Packet (IIP).   IIP is one of Null packets, but carries IIP information. Its layer indicator is defined as "1000."

IIP specifies the general information for the signal and the network system management.   On the other hand, Network Synchronization Information (NSI) conveys the data needed for SFN operation.   NSI is explained in the next section.   The syntax of IIP is shown in Table X-3.

Table X-3 Syntax of IIP (ISDB-T Information Packet) Description[159]

Data Structure	Number of Bits	Bit String Notation
ISDB-T_information_packet(){		
TSP_header{		
sync_byte	8	bslbf
transport_error_indicator	1	bslbf
payload_unit_start_indicator	1	bslbf
transport_priority	1	bslbf
PID	13	uimsbf
transport_scrambling_control	2	bslbf
adaptation_field_control	2	bslbf
continuity_counter	4	uimsbf
}		
payload{		

---

[157] Quoted from ARIB SRD-B31 Attachment, Section 5.5.2 figure 5-10
[158] See Section III.2.2 of this book
[159] Original table from ARIB STD-B31 Attachment, table 5-8

Syntax	Bits	Format
IIP_packet_pointer	16	uimsbf
modulation_control_configuration_information( )	160	bslbf
IIP_branch_number	8	uimsbf
last_IIP_branch_number	8	uimsbf
network_synchronization_information_length	8	uimsbf
network_synchronization_information( )		

for(i=0;i<(159-network_synchronization_information_length);i++){
    stuffing_byte(0xFF)
}
}
}
	8	bslbf

(note 1) the syntax of modulation_control_configuration_information is not shown in this table.[160]
(note 2) for the details of network_synchronization_information, see Table X-4 in this Chapter.

The IIP packet[161] is defined to be one TSP per multiplex frame. The payload of one IIP packet is 184 bytes. For the transmission information exceeding 184 bytes, it should be divided and transmitted using multiple multiplex frames.

When the IIP is made up of multiple TSP's, the TSP inserted in the multiplex frame is called the sub IIP packet.

(4) Network_synchronization_information (NSI)

As described in the last Section, network_synchronization_information are used for network synchronization especially for SFN operation.

Table X-4 shows what information is inside the "network_synchronization_information". For example, the synchronization_time_stamp (STS) indicates a delay time from 1 pps signal of GPS and maximum_delay indicates a maximum delay time between studio output and each transmitter output are useful for SFN network operation. Details will be discussed in the following section.

Table X-4 Description of the network_synchronization_information[162]

Syntax	Description
Synchronization_id	0x00 : SFN_synchronization_information is added 0x01~0xFE : For future extension 0xFF : SFN_synchronization_information is not included.
SFN_synchronization_information	Synchronization control information including delay time control in SFN network.
synchronization_time_stamp	Time difference from the reference time. Indicated in 10MHz periodic unit (on the 100ns time scale). Indicates the delay time of the head of the multiplex frame (start time) in which the next tmcc_synchronization_word is '0' against the latest 1pps signal gained from the time reference such as GPS at the delivery output (Ex: output to STL) of the

---

[160] ARIB STD-B31, Attachment, table 5-11
[161] For the structure of IIP packet, see ARIB STD-B31, Attachment, figure 5-11
[162] ARIB STD-B31, Attachment, table 5-13

		line to the broadcast station.
maximum_delay		Maximum delay time. The time interval between the delivery output (Ex: output to STL) to the broadcast station at the studio and the broadcast wave emission from the transmission antenna of each broadcast station in the SFN. Indicated in 10MHz periodic unit (on the 100 ns time scale). This value should be set to less than 1 second [within the range between 0 (0x000000) and 9999999 (0x98967F)].
equipment_loop_length		Indicates the total length of equipment_loop. Indicated in byte units.
equipment_control_information		Information to control the offset of delay time or fixed delay time individually for each broadcast station.
equipment_id		Designates each broadcast station to control by the equipment_control_information.
renewal_flag		When renewing the values of static_delay_flag, time_offset_polarity, and time_offset, this field in equipment_control_information of the targeted equipment_id will be renewed. When renewing the value of maximum_delay, this field in all equipment_control_information syntaxes (all equipment loops) will be renewed. This field toggles between '1' and '0' for renewal.
static_delay_flag		For the delay control of the SFN, the delay time may be adjusted from the reference time such as a GPS in one case. Typical and static delay time may be allocated to the broadcast station not using a reference time in another case. The static delay flag should be '1' when the latter control is employed. In this case, the only effective control information is the time_offset which is used for delay control.
reserved_future_use		Reserved bit for future extension. The value should be '1'.
time_offset_polarity		Indicates polarity of the following time_offset. '0' should be designated for a positive value and '1' for a negative value. When static_delay_flag is '1', '0' should be always designated for the time offset polarity.

For the details of the bit composition of the network_synchronization_ information see ARIB STD-B31[163].

(5) An example of delay time measurement and adjustment using NSI

In this section, using a simple network model, the delay time measurement and adjustment process through the use of NSI is explained as follows:

1) Transmission timing measurement in SFN

Figure X-10 shows a very simple network example of I/F(2) interface point – reference synchronization (Case 2). In this example, the following pre-requisites are assumed, (a) the delay time between the studio output and the output of TX-1 and TX-2 are different ($t_{STL1} < t_{STL2}$), (b) delay time from transmitter station TX-1 and TX-2 to reception point are same ($t_{PATH1} = t_{PATH2}$).

In this example, it is assumed that both t_{STL1} and t_{STL2} include a delay caused by the signal processing in the transmitter station.

---

[163] ARIB STD-B31Attachment, table 5-12

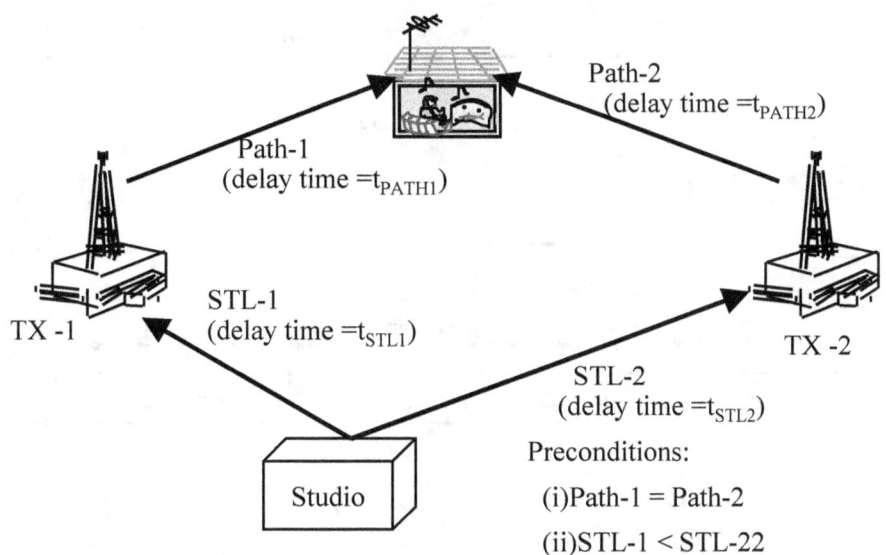

Figure X-10   An example of an SFN model

Next, it explains the signal process to construct the NSI data from a multiplexed TS(2) in the studio.

Figure X-11 shows the relationship between TS(2) (Broadcast TS) and 1 pulse per second (pps) signal, which is generated in a GPS receiver as illustrated in Figure X-6. In this case, a 1 pps signal is utilized as a time reference for both studio and transmitter stations.

Initially, it is important to measure the time difference between a 1 pps timing and the start timing of a multiplex frame, defined in Chapter III.4, of the re-multiplexer (located in the studio). The measurement result is defined as Synchronization Time Stamp (STS), which is shown in Figure X-11.

The delay time is also measured with respect to the 10MHz clock signal from a GPS. STS on the other hand is multiplexed into the IIP and sent to the transmitter site, the STS is extracted and used as the time reference in establishing the delay time in the transmitter site(s).

Another important parameter is the "maximum_delay_flag" defined as the maximum permissible delay time of the output timing of any transmitter in the network. The value of "maximum_delay_flag" is defined so that the output timing of the transmitter in an SFN can be synchronized. This parameter is also multiplexed into the IIP.

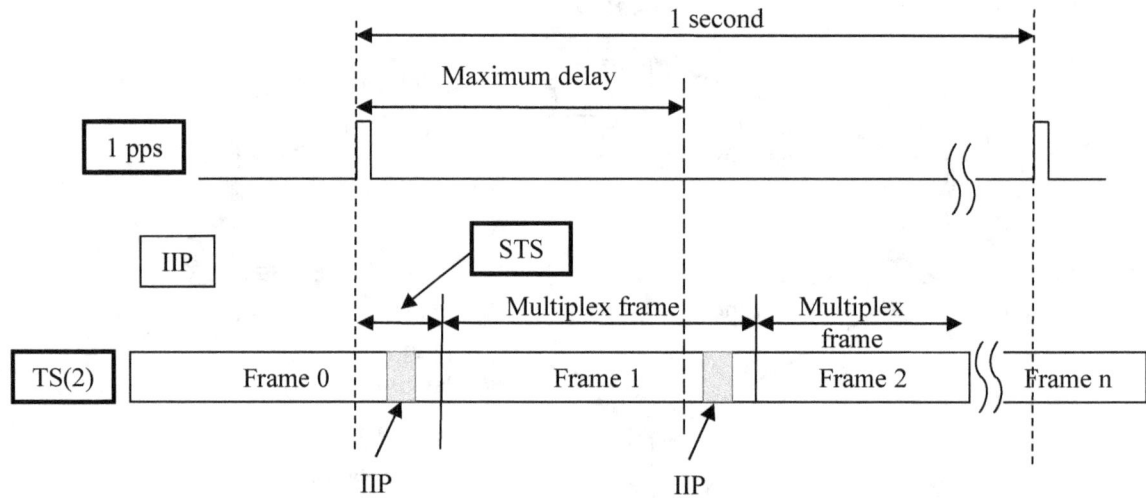

Figure X-11 Relationship between 1 pps and STS

Figure X-12 shows the relationship of delay time including Studio to Transmitter Link (STL) and the transmitter stations. The TS(2) signal is sent from studio to the transmitter stations via an STL which is then modulated and amplified in the transmitter station. At the output timing of the transmitter station, the delay time of the STL, modulator and transmitter (t_{STL}) is added to TS(2) transmission timing at the studio output. (see Figure X-10 which provides a simpler illustrative explanation).

The "Measured value" shown in Figure X-12, which is defined as the time difference between 1 pps and multi-frame header at transmitter output, is given as the sum of STS and t_{STL}.

In other words, the total delay time of the STL and transmitter station can be calculated by following equation.

Measured value − STS = delay time of STL and transmitter

STS can be extracted at each transmitter site by decoding the embedded IIP in TS(2) from studio. By this measurement process, each transmitter station can determine its own transmission timing (the output timing of multi frame header).

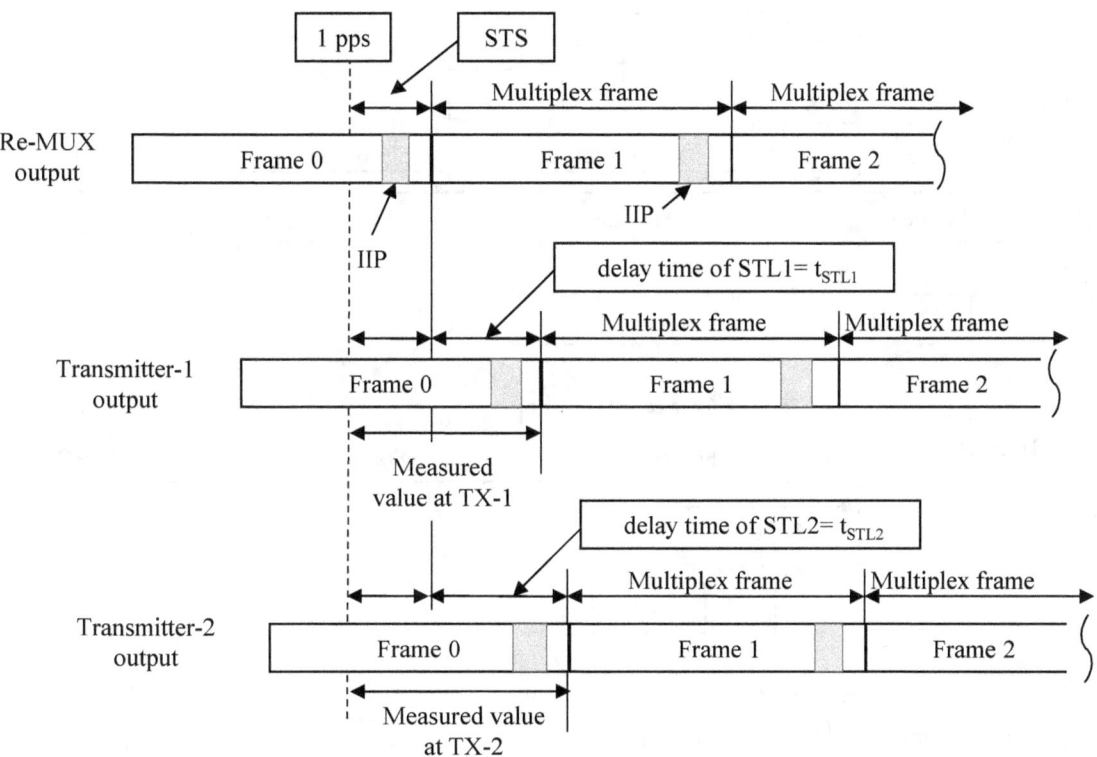

Figure X-12 Relationship between the measured value and the delay time of the STL

2) Transmission timing adjustment in SFN
An example of the procedure for adjusting the transmission timing is illustrated and explained in Figure X-13.

The example shown in the figure is in the simplest case, all the transmitter output timing should be adjusted at "maximum_delay" (T_{MAX}). As the signal is relayed in number of stages from main station, satellite stations to gap fillers, delays are accumulated in each stage. T_{MAX} is the longest total accumulated delay time in the farthest station in the SFN. The delay timing in SFN is generally set to the maximum delay, because emission timing of each station can be delayed by inserting delay time but total accumulated delay time cannot be shortened.

Each transmitter site adjusts the output timing to T_{MAX} by adding the appropriate delay time which is calculated by following equation:

Added delay = $T_{MAX} - t_{STL} = T_{MAX} -$ (measured value $-$ STS)

The delay time added for each station is indicated as $Td_{(TX-n)}$ in the figure.

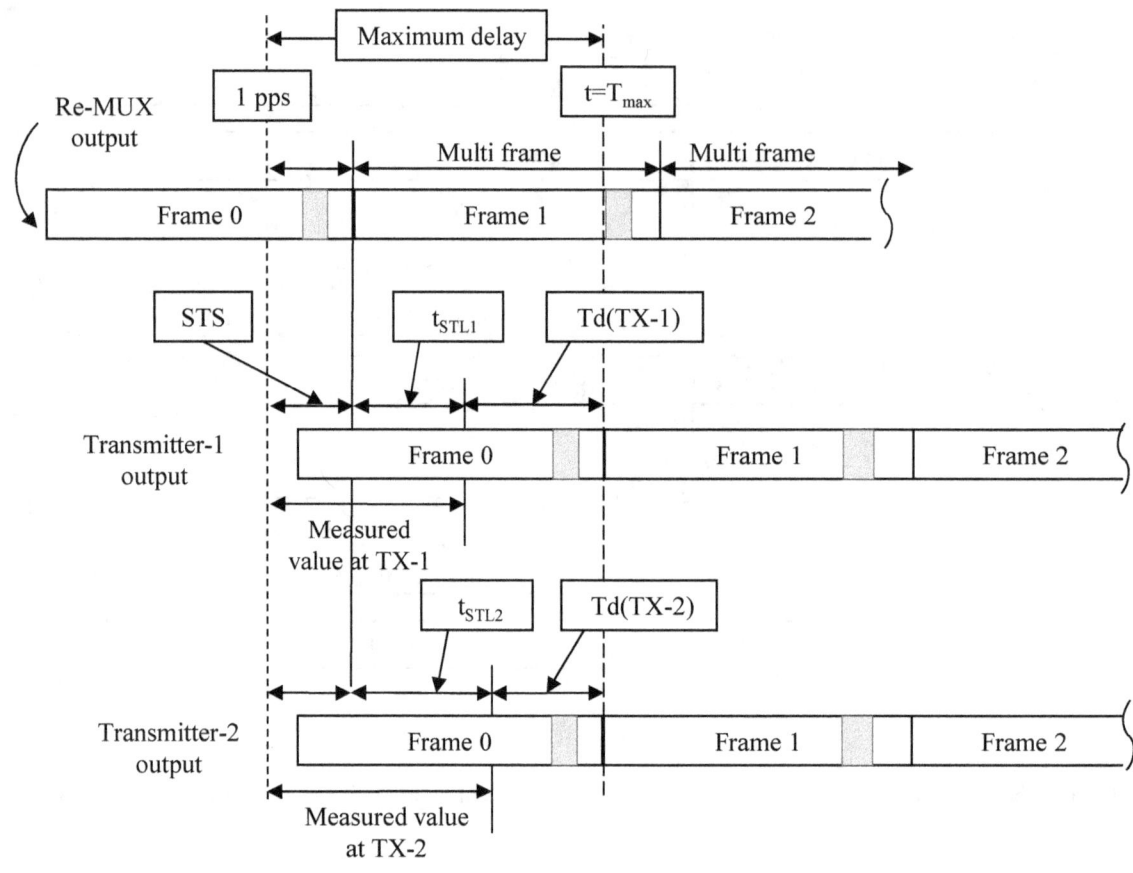

Figure X-13 An example of the output timing adjustment procedure (adjusted to maximum_delay)

As explained in the above equation, if all transmitter output timing is adjusted to its maximum_delay, all transmitter output timing can be synchronized to the 1 pps reference.

If the output timing for each station operates independently, the parameters in the NSI (Table X-3) such as the "equipment_control_information", consisting of the "equipment_id", "time_offset", can be used. By setting parameters in NSI, the transmitter output timing for each station can be controlled independently.

The above explanation applies for the timing adjustment of Case 2 SFN (reference synchronization type). However, NSI and delay time measurement is only useful for measuring and verifying the actual transmission time in for Case 1 SFN (slave synchronization type) and Case 3 SFN (IF relay type).

Although the delay time for each transmitter station is assigned for Case 1 SFN and Case 3 SFN, in the event of transmitter station maintenance and STL route change, the determined delay time of each transmitter station should be and can be altered. In this case, it becomes necessary to repeat the transmission time measurement.

In case of Case 3 SFN (IF relay network), the signal format is "OFDM modulated signal" where no dummy byte and IIP exist. In this case, the information for network control should be carried by other communication network or by AC in OFDM modulated signal.

X.3 Degradation Factors in a Transmission Network and the Methods of Evaluation

In this section, it explains the model of the transmission network from the signal quality, degradation factors, evaluation methods, etc.

X.3.1 Network model
(1) Overview of a network model
Figure X-14 shows an example of a model transmission network in SFN, including relay stations whose interface format is in based on TS(2), IF and broadcast wave[164].

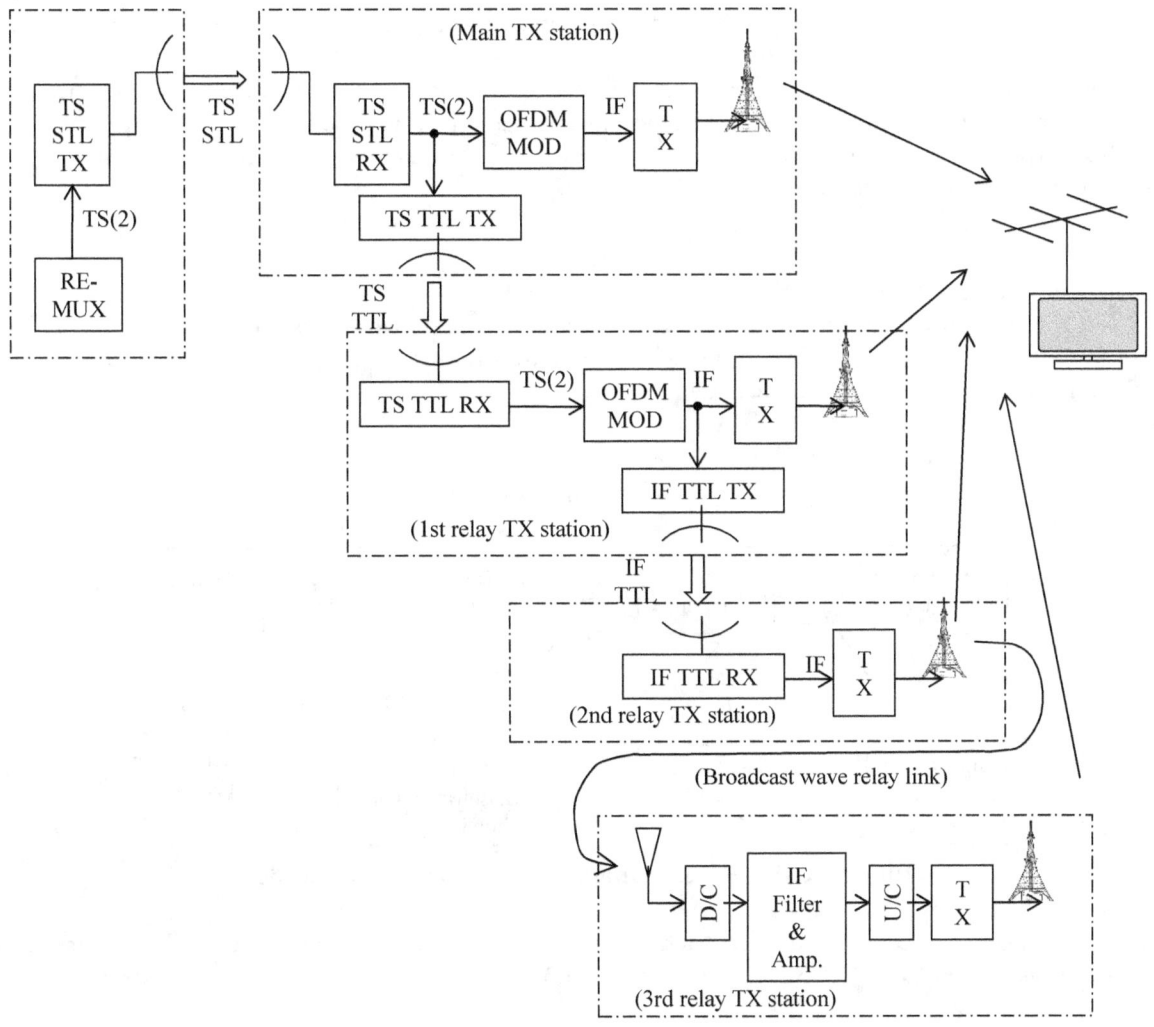

Figure X-14 An example of a Terrestrial Broadcast network model (SFN)

---

[164] JEITA handbook, Methods of measurement for digital terrestrial transmission network

In the above case, TS STL (Studio to Transmitter Link) transmission system is used for the 1st link between the studio to the main Transmitter station. For the 2nd link between the main TX station to 1st relay station is TS TTL (Transmitter to Transmitter Link) system.

TTL is in general a microwave signal link from a transmitter to another transmitter. This type of signal link is used in case that a relay station is located too far not to send a signal directly from Studio, then the signal is sent via the main transmitter station. In the case of Figure X-14, TTL is assumed to be used as for a signal relay link.

The 3rd link between the 1st relay station and 2nd relay station uses IF TTL.

For 4th link between 3rd relay station and 4th relay station, this type of link, called a "broadcast signal wave link", is often used when neither a microwave frequency nor fiber link are available.

For the network design, it is important to check the signal quality which may have been degraded by some factors introduced in the various stages of the network. Most importantly for digital systems, as explained in Chapter X.1.2 Quality of Service (QoS) is not proportional to input signal strength. The relevant factor of QoS in digital network is the C/N margin with respect to the "cliff point".

In calculating the C/N margin as described above, it is essential to establish or determine the various types of degradation factors that exist in the transmission network and how it affects the C/N.

For this purpose, the typical link budget model and degradation factors in a transmission network will be explained in the succeeding section.

(2) Degradation factors in the network model

Figure X-15 shows an example of a model transmission network that consists of an IF transmission for studio-transmitter link and broadcast wave relay for a transmitter-transmitter link. In addition, various degradation factors in the links are indicated[165].

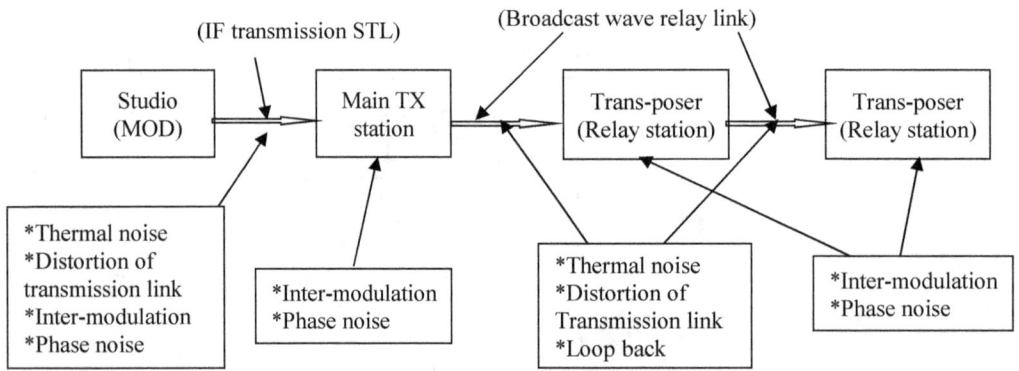

Figure X-15　An example of transmission link model

The degradations are cumulative from each stage of the relay stations and eventually affects the total "equivalent noise degradation" (END, see Chapter XII.3.7) at receiver side[166] [167]. Therefore, the network

---

[165] ARIB STD-B31 Appendix 3; Considerations in the Link Budget for ISDB-T
[166] IEC 62273-1, Method of measurement for radio transmitter – performance characteristics for terrestrial digital television transmitters
[167] JEITA handbook, Methods of measurement for digital terrestrial transmitters

design should carefully consider these factors from the start.

The details of these factors are explained in Chapter XII.3..

In case of TS transmission system, the error introduced at the TS transmission link is corrected at the input portion of transmitter station. Therefore, the degradation due to the TS transmission link does not affect the output signal quality of the transmission.

(3) Accumulation of degradations

The equivalent noise degradation for each stage is cumulative. For this reason, the equivalent C/N of the final stage should be carefully inspected in order to establish the number of transmitter stages and required C/N for each stage.

As an example, the relationship between the number of stage and estimated equivalent C/N are shown in Figure X-16. Here, several examples with changes of an equivalent C/N for the main station are illustrated in the figure. In this example, the equivalent C/N of each relay station is assumed as 50dB for phase noise, and 48dB for intermodulation. Other degradation factors, such as multi-path interference, co-channel interference, etc, are ignored.

An equivalent C/N of the transmitter output exhibits gradual degradation due to the cumulative effect of the various stages.

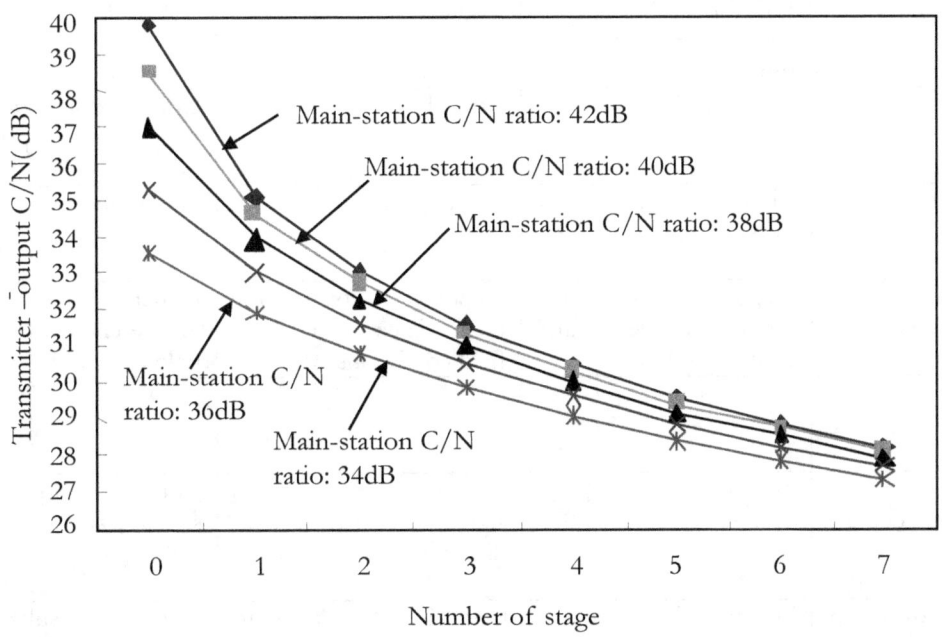

Figure X-16 Relationship between the number of stages and total C/N[168]

X.3.2 Individual degradation factors

This Section provides a general explanation of the degradation factors with respect to the measurement methods, details are described in Chapter XII.

---

[168] Quoted from ARIB STD-B31 Appendix 3; Considerations in the Link Budget for ISDB-T, Figure A3.4-2

(1) Degradation factors and their effects
As shown in Table X-5, there are various types of degradation factors in both the propagation links and transmission equipment. In addition, the degree and circumstances of degradation varies from system to system.

Table X-5 shows major degradation factors and effects[169].

Table X-5 Degradation factors and effects

Degradation factor	Digital broadcast		Analog broadcast
	Multi-carrier(note 1)	Single-carrier(note 1)	
1. Propagation link			
1.1 Multipath (static)	Not critical (due to the OFDM signal + guard interval, ISI is not affected) (note 2)	Critical (Inter Symbol Interference)	Critical (ghosting on picture)
1.2 fading (dynamic multipath)	Degraded, but with the introduction of time interleaving in ISDB-T, the degradation is reduced (note 3)	Critical (Adaptive filtering may not improve the effects of degradation of time varied ISI.	Critical (ghosting, etc.)
1.3 Urban noise (Impulse type noise)	same as 1.2	Critical (impulse noise degrades BER characteristics)	Critical (dotted/snowy interference present on picture)
1.4 Coupling Loop interference (note 4)	Critical only for SFN-(Coupling Loop interference may occur)	Not applicable	Not applicable
1.4 Co-channel & Adjacent channel interference	Depends on the interference level and field conditions	Depends on the interference level and field conditions	Depend on the interference level and field conditions
2. Equipment performance degradation			
2.1 Distortion of amplitude- frequency response (note 5)	Not critical (note 5)	Critical (due to signal/waveform distortion)	Critical (results due to signal/waveform distortion)
2.2 Distortion of group delay- frequency response	Not critical (note 5)	Critical (results to signal/waveform distortion and creates ISI))	Critical (results to signal/waveform distortion)
2.3 Intermodulation	Critical	Critical (results to signal/waveform distortion)	Critical (Differential Gain, Differential Phase)
2.4 Phase noise	Not Critical due to the	Not critical	Not critical

[169] JEITA handbook, Methods of measurement for digital terrestrial transmission network

| | minimal C/N requirement of the TMCC | | |

(note 1) Multi-carrier system: ISDB-T and DVB-T, Single carrier system: ATSC
(note 2) see Chapter II.3
(note 3) see Chapter III.3.7
(note 4) this phenomena applies only to "broadcast wave relay station in SFN." See Chapter X.2.5.
(note 5) for the influence of amplitude-frequency response distortion, see Chapter X.3.4(1).

As described in the table above, the degradation factors and their influences varies with the available transmission systems. For further explanation of these factors, the mechanism of the degradation are explained in the following sections.

X.3.3   Signal degradation in the transmission link

This Section explains the phenomena and influence of degradation in the transmission signal. The degradation introduced by the transmission equipment is explained separately in section X.3.4.

(1) Static multipath
There are 2 degradation factors caused by multipath interference.

One is ISI, the other is the distortion of the frequency response characteristics. Regarding ISI, as explained in section II.3.3, with the introduction of the guard interval, OFDM transmission system is naturally robust against multipath (see Figure II-15 of this book). In the case of single carrier system, due to the shortness of the symbol duration, (not longer than 0.2 us), the multipath component introduces serious interference to the signal.

The distortion on the frequency response characteristics is directly attributable to the multipath interference as illustrated in Figure X-17.

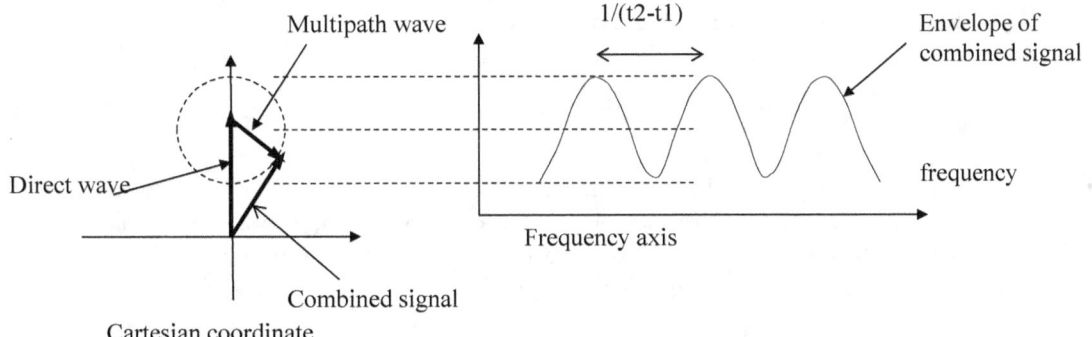

Figure X-17 Amplitude-frequency distortion due to multipath

Effects of amplitude-frequency distortion will be explained in section X.3.4(1).

(2) Fading (dynamic multipath)
Figure X-18 illustrates the mechanism of "fading" effects.

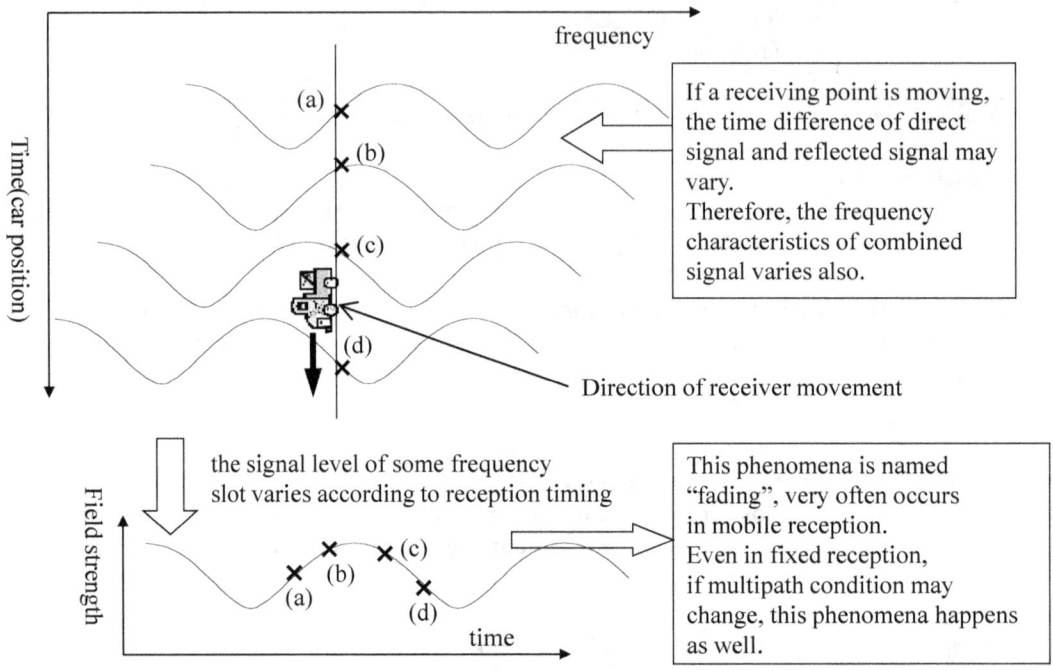

Figure X-18 An image of the mechanism of "fading" effects

As illustrated, the reception point of moving receivers at various time are indicated from (a) to (d). The arrow shows the direction to where receiver is moving. The amplitude frequency characteristics of each reception point is different resulting to a time varying signal level for certain frequency slots of OFDM signal.

Fading occurs when the time difference between the direct path and multipath changes dynamically. This phenomenon occurs more often for mobile reception, but, in the case of fixed reception, slow variations of such phenomena may occur, for example, over the sea propagation, urban area propagation (caused by temporally occurring mobile objects such as an airplane passing overhead), etc.

In order to cope with these phenomena, the following measures are proposed and used in the ISDB-T system:

Time interleaving:
For high speed variations of received signal level, time interleaving technique is very effective. ISDB-T utilizes this technique to enable effective reception performance most especially in mobile/ portable receiving condition. Take note that this technique is not as effective especially against slow fading. Therefore, for transmission relay links, diversity reception technique is proposed.

Diversity reception:
Diversity reception is a technique that combines two or more received signal and improve the reception performance. This technique was developed in analog communication and is still an effective technique for digital communication system in handling the effects of both fast and slow fading. This technique is widely used in mobile reception systems (receiver in car) and "broadcast wave relay station".

(3) Urban noise
Urban noise factors such as car engine noise, electric switching noise due to electric motor/generator

generates an impulse noise spread all over the spectrum.

Time interleave technique is effective for urban noise.

(4) Coupling Loop Interference

In case of broadcast wave relay station for SFN in which the receiving frequency for the transmitter input and the output frequency for broadcast is the same (indicated f_1 in FigureX-8), coupling loop interference may occur from the transmitter antenna to the receiving antenna.

If the Coupling Loop Interference level is over the threshold level, self oscillation may happen in the worst case. Even though the Coupling Loop level is not high, the Coupling Loop Interference component may affect the transmitting signal.

To counteract this type interference, the following are proposed:
(i) Reduce the coupling coefficient between the transmitting antenna and receiving antenna: i.e. extend the distance between both antennas, control the antenna direction, etc
(ii) use Coupling Loop Interference Canceller: see Chapter X.4.3

X.3.4 Degradations caused by the transmitting equipment
(1) Distortion of amplitude- frequency response of a transmitter
This type of distortion creates a waveform signal distortion. Figure X-19 illustrates the mechanism of this distortion. The left side of this Figure shows a case of "no distortion", while the right side shows a case of "linear distortion".

The lower side of this Figure shows a vector diagram for each case. As shown in the right side (in case of "linear distortion"), the upper and lower sideband vectors are not equal due to the distortion of amplitude-frequency characteristics.

In case of "no distortion", both sideband vector are same level and located at opposite side. Therefore, the synthesized vector varies only amplitude compared to the carrier vector, and the phase of the synthesized vector remains unchanged. On the other hand, in case of "linear distortion", sideband vector amplitudes are not the same and not located at opposite side, so, the synthesized vector has phase variation form the carrier vector and the amplitude is different from the vector of "no distortion". This distortion caused by the amplitude-frequency characteristics results to and envelope distortion of RF signal in multicarrier systems. In case of a single carrier digital transmission system, the envelope distortion of the modulated RF signal causes a degradation of the BER characteristics.

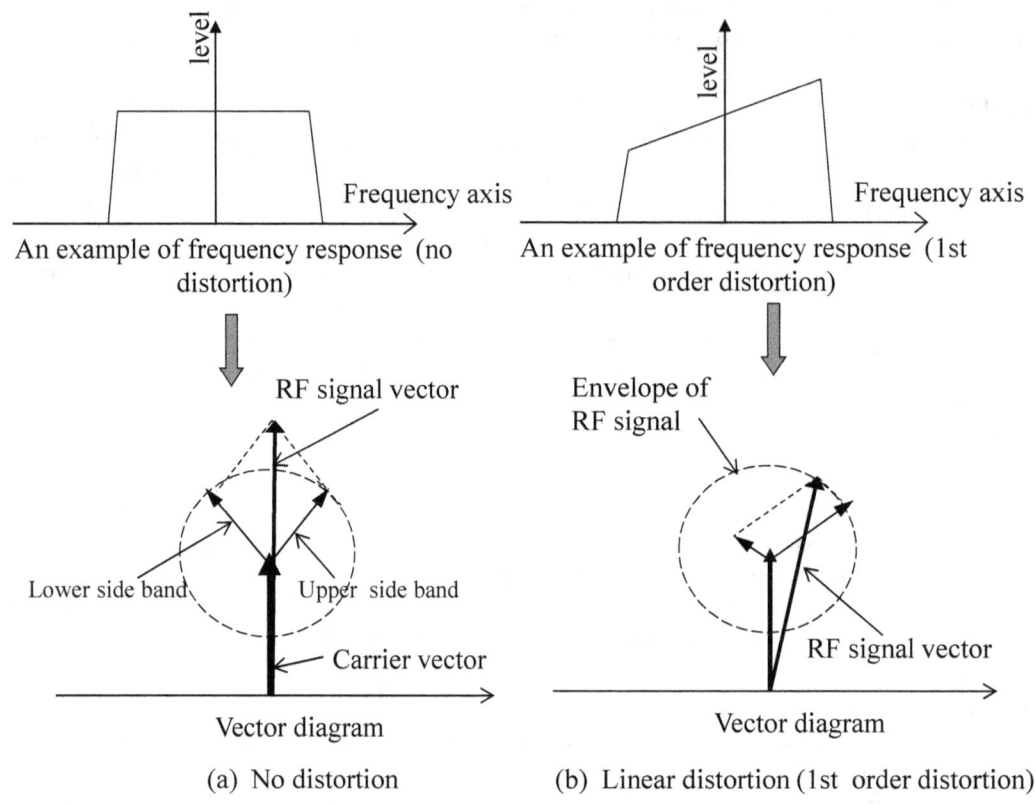

Figure X-19　Relationship of frequency response distortion and envelope distortion

In the case of a multi-carrier system shown in Figure X-20, the signal transmission bandwidth of each carrier is about 1/6000 (ratio of 1kHz and 6MHz in case of 6MHz ISDB-T system) for mode 3.

For 7MHz and 8MHz ISDB-T system, the ratio between each carrier bandwidth and a total bandwidth is same as the ratio of 6MHz ISDB-T system.

Therefore, the imbalance of both sideband levels are reduced to 1/6000 (imbalance in a 6MHz band is equal to 10%, in the case of individual carriers, the imbalance of 1 kHz band is about 0.0017%).

This means that the signal degradation in a multi-carrier system due to these factors is almost negligible compared to a single carrier system.

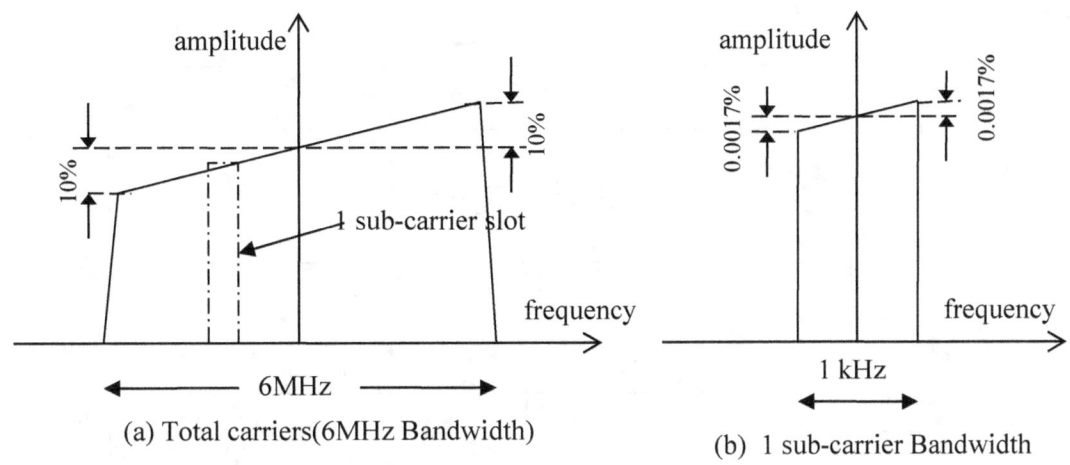

Figure X-20 An image of a wideband and narrowband frequency characteristics

(2) Intermodulation
In Figure X-21 (a), an OFDM signal is composed of multiple-carriers. The non-linear distortion of the amplifier results to a higher order of inter-modulation, most specifically, the third order inter-modulation products falls into the adjacent frequency slots (see Figure X-21(b))

An OFDM signal has a lot of sub-carriers, therefore, a lot of third order inter-modulation products are generated and falls into adjacent frequency slots of an OFDM sub-carrier. These unwanted products degrade the signal quality of an OFDM signal.

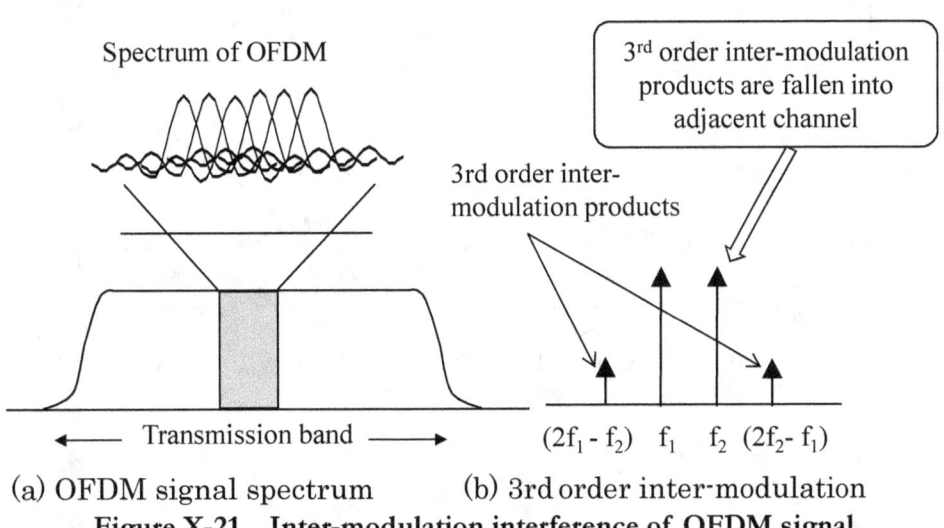

(a) OFDM signal spectrum   (b) 3rd order inter-modulation
Figure X-21 Inter-modulation interference of OFDM signal

Figure X-22 shows an example of a BER (bit error rate) of an OFDM signal. In this Figure, three BER characteristics are illustrated, linear, low and high distortion. In the case of a high distortion system, the BER characteristics are severely degraded. The OFDM transmission system is sensitive against non-linear distortion.

For the improvement techniques, see next Section.

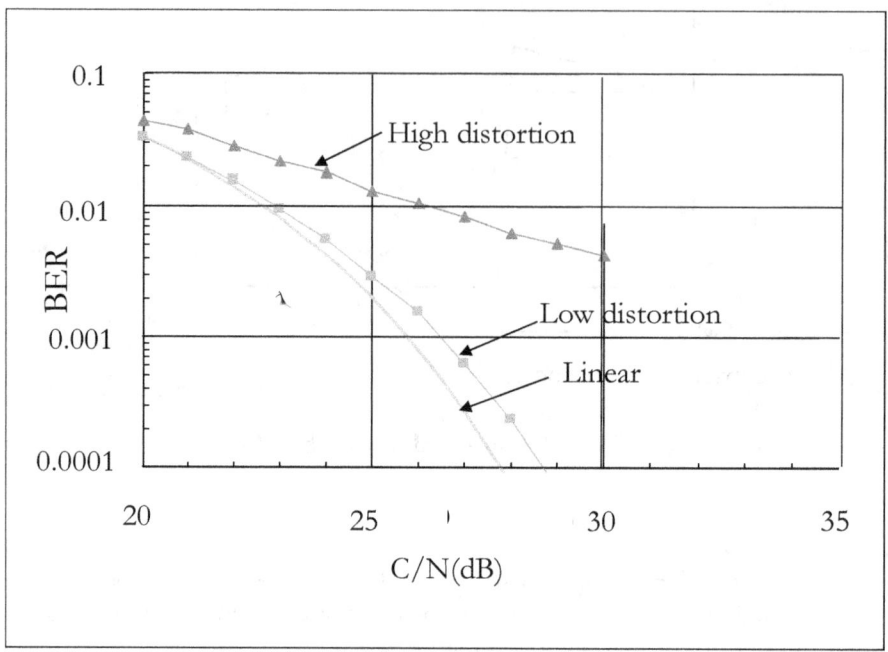

Figure X-22 Examples of BER characteristics of an OFDM signal

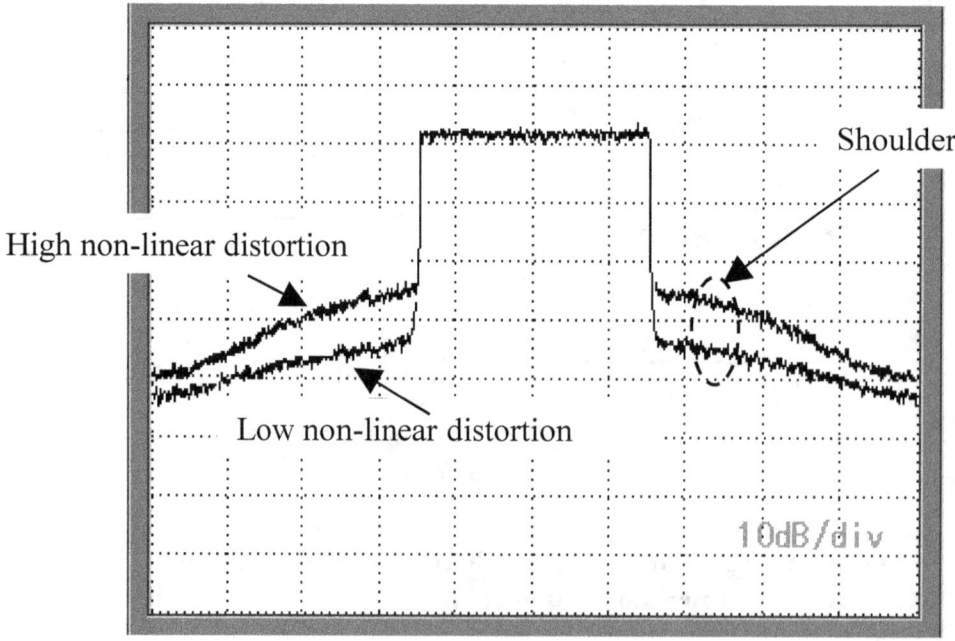

Figure X-23 OFDM signal spectrum

If an inter-modulation exist, as illustrated in the Figure X-23, the upper and lower out-band spectrum of an OFDM signal increases (inter-modulation products are generated not only at the in-band frequency portion but also on the out-band frequency portion).

This phenomenon is called "shoulder", this is one of the important measurement items for OFDM transmission system.

Please note that it is necessary to measure the RF spectrum at the output of an amplifier before the filter so that measurements are not to be influenced by the filter characteristics.

(3) Phase noise

Phase noise is due to the instability of the local oscillator(s) that are used for the up and down frequency conversion. The phase noise is one of the critical degradation factors in a narrowband transmission system. Please note that OFDM transmission system is based on a "narrowband transmission system", due to the sheer number of narrow band carriers.

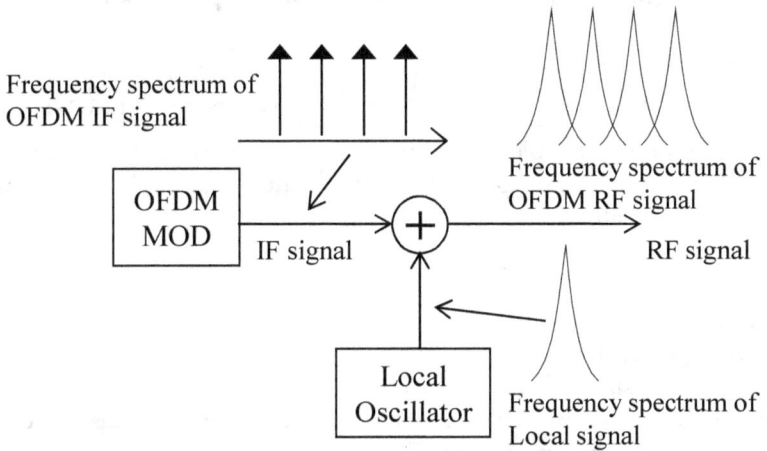

Figure X-24　Phase noise and mechanism of degradation

Phase noise results in two types of degradation, first is CPE (common phase error) and second is ICI (inter carrier interference).

The degradation mechanism is illustrated in Figure X-25. The Phase Noise within the range of each sub-carrier band (in-band) affects its own sub-carrier. This is called CPE. The in-band phase noise components lead to a phase shift of signal constellation (circular motion). This phase shift decreases a C/N margin, as a result, C/N degradation occurs.

On the other hand, the out-band components of Phase noise (ICI) affects the adjacent sub-carriers. These components behave similarly as thermal noise. As a result, C/N degradation occurs.

To reduce the degradation introduced by phase noise, it is necessary to use a high stability oscillator.

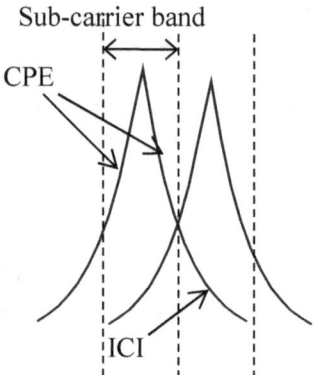

Figure X-25   Degradation caused by phase noise

X.4   Examples of Signal Improvement Technologies in a Transmission Network

This Section will introduce some examples of improvement techniques. Inter-modulation distortion of power amplifier(s) introduces serious damage to the RF output signal. The following two techniques are introduced to counter inter-modulation.

X.4.1   Feedback pre-distortion compensation type amplifier

Figure X-26 shows an image of a "feedback pre-distortion compensation" technique. The amplifier output signal and the modulator output signal are compared so that the distortion component can be identified and isolated. Then, from this data, the input RF signal of power amplifier is pre-distorted by the "compensator" to suppress the harmonic components generated in the next power amplifier stage. Through this, inter-modulation distortion is minimized.

Please note that the degree of inter-modulation distortion depends on the output level of amplifier. It is better to ask for the recommendation of the manufacturer to determine the acceptable amount of inter-modulation distortion level.

This technique is mainly used for high power transmitter.

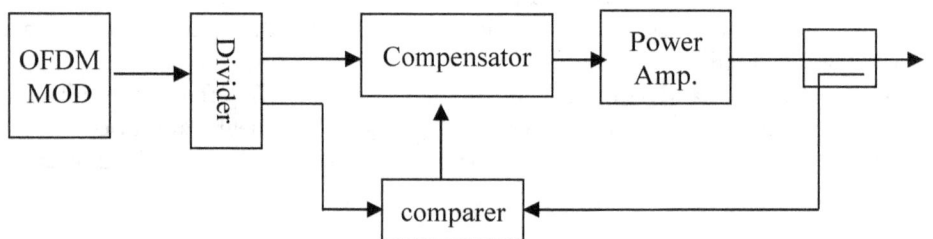

Figure X-26   An example of a feedback pre-distortion compensation type amplifier

X.4.2   Feed forward type amplifier

This technique is used as an added measure to compensate for the inter-modulation distortion. Figure X-27 illustrates an example of a general transmitter diagram. The OFDM modulator output is fed to the "main amplifier" and to the "difference component amplifier". The difference component amplifier extracts the difference of the main amplifier input and main amplifier output, followed by the distortion component being amplified. The amplifier characteristics, such as dynamic range for difference component amplifier, is desired to be similar to that of main amplifier.

Both main amplifier output and distortion amplifier output are inversely combined leading to the effective compensation of the distortion component.

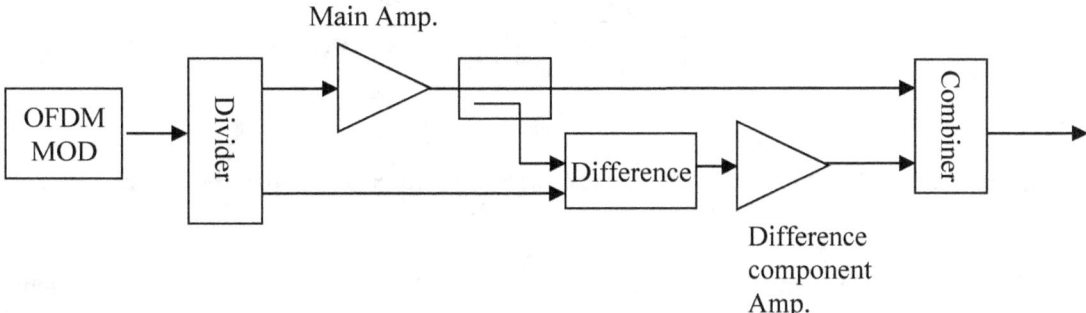

Figure X-27 An example of a feed forward type amplifier

Regarding power efficiency, feedback type amplifiers are more efficient than the feed forward type amplifier and suitable for high power amplifiers. The major difference is in that the feed forward type can handle wider bandwidth and exhibits lower inter-modulation distortion compared to feed back type amplifier. Due to these reasons, "feed forward type amplifier" is more applicable for middle power and multi-channel (wideband) transmitters.

X.4.3   Coupling loop interference canceller

Broadcast wave relay network sometimes suffer Coupling Loop Interference. The input of the system suffers interference from its own output as input and output use the same frequency. In this case, coupling loop interference canceller is effective to solve the problem.

In the case of a broadcast wave relay network, the output of the relay itself may inadvertently be fed into the relay's input, and if the parameters are not appropriately set, it may result in a severe degradation on the signal quality.

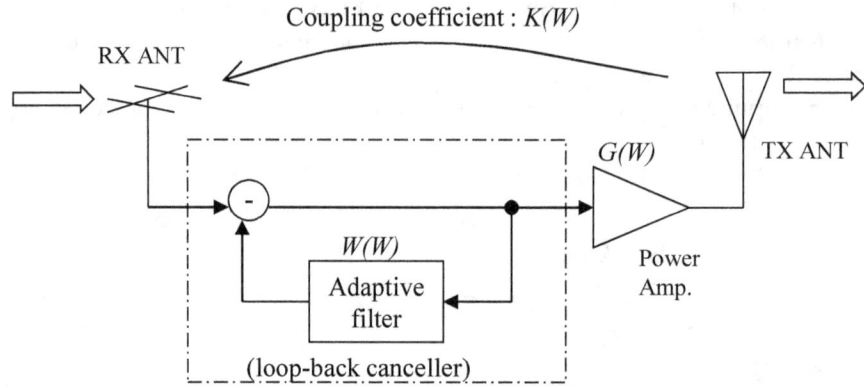

Figure X-28 Conceptual block diagram of a coupling loop interference canceller

Figure X-28 illustrates the loop-back canceller mechanism. An output of the relay's TX antenna is fed back to the receiving antenna in some part. As a result, a replica of signal is being looped-back perpetually. This Coupling Loop component introduces degradation similar to a multipath interference. In the worst case condition, this loop-backed component leads to self oscillation.

The following Figure shows the transfer function of a "broadcast wave relay station". A transfer function $T(W)$ is defined by following formula:

$$O(W) = I(W) * T(W) = I(W) * [G(W) + \{G(W) * K(W)\} - W(W)]$$

Here, $O(W)$: output signal,
$I(W)$: input signal,
$T(W)$: transfer function of the relay station,
$G(W)$: transfer function of an amplifier,
$K(W)$: transfer function of a transmitting and receiving antenna,
$W(W)$: transfer function of an adaptive filter

By using this formula, the technique called "Coupling Loop Interference canceller" can be introduced. The equation is represented as a block diagram shown in the Figure X-29.

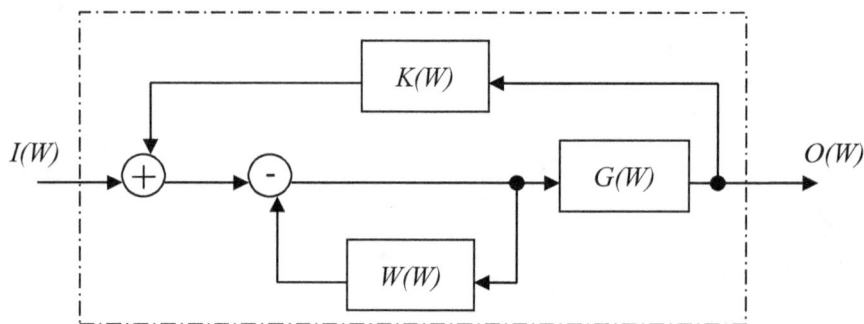

Figure X-29 Transfer function of a "broadcast wave relay station"

The loop-back interference component is described as $G(W)*K(W)$. Therefore, if $W(W)$ is equal to $G(W)*K(W)$, the loop-back interference is cancelled. This is the principle of the Coupling Loop Interference canceller. The key point of this technique is how to establish $W(W)$ including, phase of RF signal, delay time and signal level In other words, the function of this system is to measure the Coupling Loop Interference, then adjust the transfer function of an adaptive filter($W(W)$) to minimize the level of interference.

Note that in case that $K(W)*G(W) > 1$, the system becomes unstable and ultimately results into a self oscillation. Therefore, it is important to decrease the coupling coefficient $K(W)$ down to a manageable level of operation

In addition, in the case of an MFN, the transmission frequency is different from the receiver frequency, therefore, the Coupling Loop component can be rejected by input filter. In case of an MFN, $K(W) = 0$, where no loop-back degradation exist.

(This page is left intentionally blank.)

Chapter XI. Broadcaster's Infrastructure for Digital Terrestrial Broadcast

A Broadcaster's infrastructure is a unified system consisting of key subsystems whose details have been discussed so far. This Chapter will now explain how these subsystems are interconnected in a broadcasting network, in order to get a broad overview of broadcaster's infrastructure and how the broadcaster's infrastructure migrates from analog to digital.

XI.1 Construction of Broadcaster's Facility and Migration to Digital broadcasting

In this section, let's see how broadcaster's facility migrates from analog to digital. The facility of TV broadcasters can be classified into three major categories below:
  (i) TV program production, recording,
  (ii) editing of broadcast programs (generally known as "Master control room"), and
  (iii) transmission facility.
As for a communication link:
  (iv) A program contribution link from Outdoor Broadcasting (OB) Van, etc and a link for program exchange with other TV stations
  (v) a STL (Studio to Transmitter Link) for TV program signal, and a transmission network link such as microwave link, fiber link and satellite link for TV program distribution.
This book will focus on (ii), (iii), (v) for its explanation on the broadcaster's infrastructure design.

XI.1.1 Transition to digital system

There is a trend among broadcasters on their migration to digital where they start their transition from digitalizing or replacing cameras and studio facilities before their start of digital broadcast. This is partly because digital recording and editing is easier in comparing to that of analog and provides better signal quality with no degradation in recording and reproducing. Also, switching the broadcast signal from analog to digital has substantial social impacts. It requires time and careful planning. Therefore digitalizing studio tend to proceed faster than that of transmission facility.

Figure XI-1 illustrates an example of a certain broadcaster how they migrate from analog to digital. Figure XI-1 (A) shows a facility for analog TV service. From here, as discussed above, a studio equipment was digitalized to improve the video/audio quality and to ease handling video/audio source, as shown in Figure XI-1 (B). In this step, the broadcasted signal was still in analog. In order to interface between digital studio and analog STL, D/A (Digital to Analog) converter was inserted to convert digital contents into analog.

Finally, it moved to digital broadcasting shown in the Figure XI-1 (C). At this step, Encoder/Multiplexer (ENC/MUX) is added, the STL and transmitter system was replaced to digital systems.

However in practice, during the period of the transition from analog broadcast to digital broadcast, both analog and digital broadcast co-exist. This period is generally called simulcast service period where they broadcast the same program in analog and digital simultaneously. In this case, as shown in Figure XI-1 (D), digital output from studio system will be fed to the analog and digital transmission system in parallel. Figure

XI-1(D) shows a case in which the analog STL and the digital STL are installed separately. If there is a case where the analog transmitter and the digital transmitter are installed in the same station, two transmission links are not necessary. In this case, the A/D converter should be placed in the transmitter station.

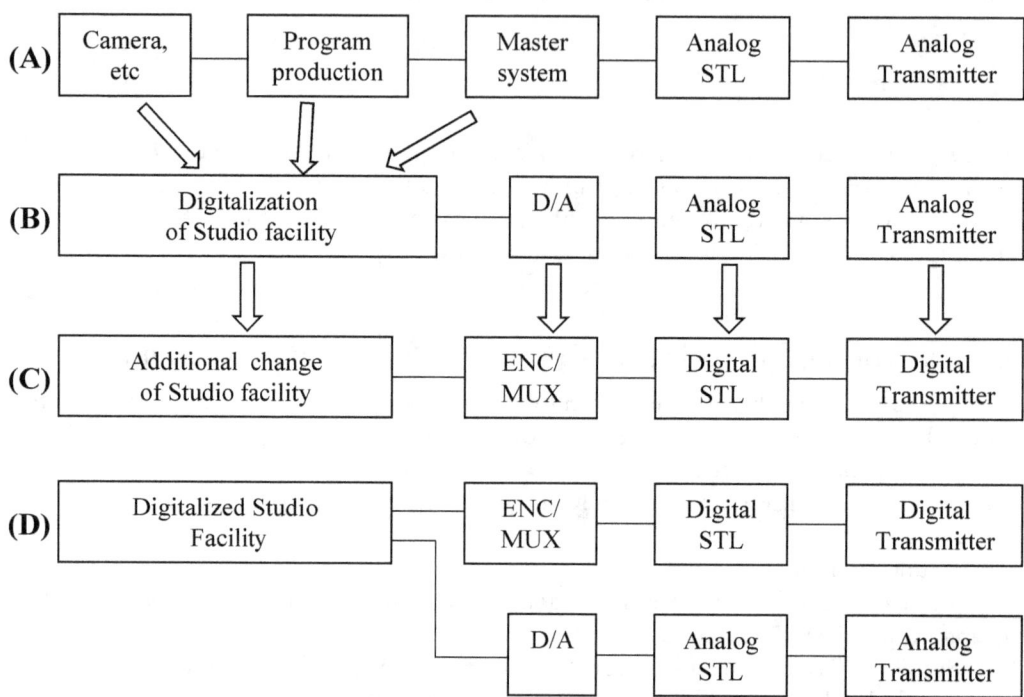

Figure XI-1 An example of a transition of broadcaster's facility

It is also possible for a broadcaster to prepare a digital transmitter first before digitalizing the studio system. Figure XI-2 shows an example of the transition of infrastructure in this case. As shown in Figure XI-2 (F), source from analog cameras and editing system are converted into digital with A/D converter and fed into the digital transmission system. Figure XI-2 (G) shows a final structure after analog service is stopped.

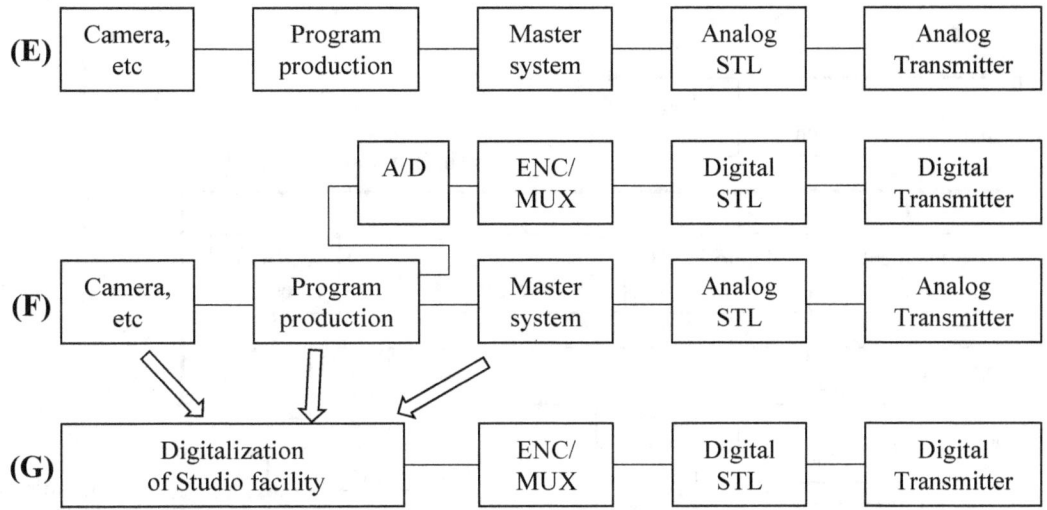

Figure XI-2　An example of the transition of broadcaster's facility (another case)

Regarding a transition to a digital broadcasting, the following issues are important, and should be studied carefully for designing a facility for digital broadcasting.

(1) Service model in digital broadcasting
When the broadcast is in analog, the service model was straightforward, SD 1 Channel with no data broadcasting. In case of the digital broadcast, there are many possible alternatives and combinations, for examples, HDTVs, SDTVs, data broadcast configuration, mobile and portable reception service, etc. Accordingly, it is important to decide the service model, a service plan and schedule to digitalize the broadcaster's facility based on the broadcaster's service requirements.

(2) Transmission networks
As shown in the Chapter X, the designing methodology of transmission network for digital broadcast is quite different from the one for analog broadcast.
The most important issue is to designate service area, and to design the transmission network configuration how to cover the area with how many transmitters and frequencies. This issue includes selecting SFN or MFN. According to a transmission network configuration especially in SFN, many parameters for network design are different from that of analog, such as, the position of relay transmitter station, distribution network system, network synchronization, etc.

(3) The simulcast service with an analog broadcasting
Undertaking the digitization of broadcast, it is necessary to prepare a simulcast period of analog broadcasting. In order to continue analog broadcasting, the broadcaster's facility should cover analog broadcast along with the digital broadcast during simulcast period.

XI.2　Master Studio Subsystem

In Figure XI-3, an example of digital studio system configuration is illustrated. As shown in this Figure, TV contents from sub-studio, program exchange or program server are fed to Matrix Switcher (Sw'er). Then the TV program to be on-air are selected and fed to ENC/MUX subsystem.

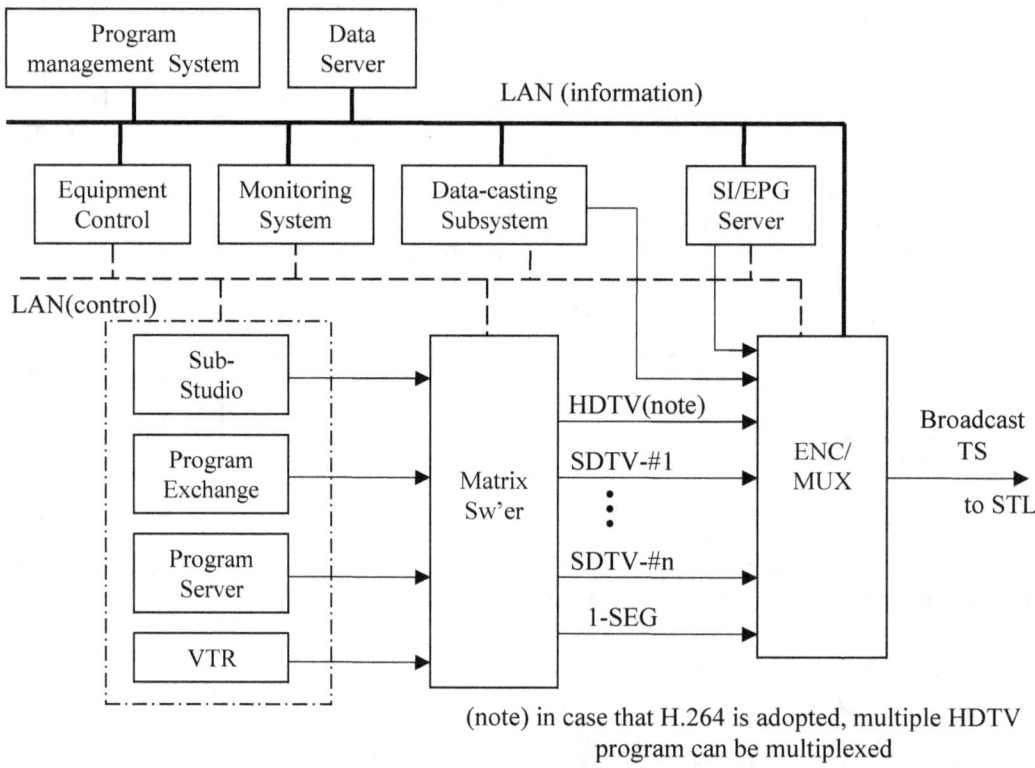

Figure XI-3 **An example of digital studio system construction**

The followings are key points for the digital studio systems design.

(i) ENC/MUX system should be introduced: the example shown in Figure XI-3 is on a system to utilize an uncompressed digital video signal. In this case, Encoder will be placed in the output of the Matrix Switcher. ENC/MUX subsystem will be explained later.
(ii) The program structure and video/audio encoding parameter set are arranged in PSI/SI. This PSI/SI information is multiplexed into broadcast TS and sent to a receiver for the receiver control. The PSI/SI data flow will be explained in section XI.2.1 later.
(iii) It has a datacasting subsystem.

ENC/MUX subsystem is the core facility of a digital master studio. Figure XI-4 shows an example of ENC/MUX subsystem configuration.

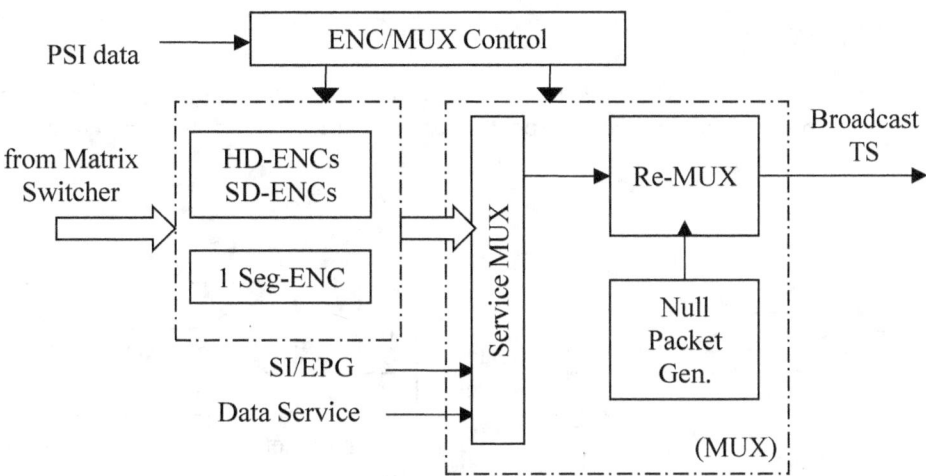

Figure XI-4   An example of ENC/MUX subsystem configuration

XI.2.1   Encoder

The SD/HD digital signals that are supplied from Matrix switcher are compressed and coded by video/audio encoders. Compression coding parameter sets of these video/audio encoders are provided from a control system of digital studio in the form of parameter sets. The video/audio coding parameter sets, such as, number of program or video/audio quality should be decided and managed considering not only technical aspect but also a service menu and business aspect of a station.

Usually, an encoder has both video encoder and audio encoder in it, and provides a TS signal in which the video and audio encoded stream are multiplexed.

In addition, many encoders correspond to both SDTV and HDTV specifications, so it may be possible for one encoder to be used for HDTV or SDTV operation according to the program change. An interruption may occur during switch of program from HD to SD or the other way round, therefore it is desirable to prepare HD and SD encoder separately.

XI.2.2   Multiplexer

As illustrated in Figure XI-4, a multiplexer for ISDB-T consists of "Service Multiplexer (Service MUX)" part and "Re-Multiplexer (Re-MUX)" part.

The service MUX multiplexes compressed video/audio data, data-casting data and SI/EPG data into one Transport Stream according to the program information. At the input portion of a service MUX, a PID filter, which selects the necessary TV program, is equipped.

On the other hand, Re-MUX is a component with a function to adjust the output TS rate to a given bit rate by inserting Null-packets, and to re-map the input TS based on Multi-frame pattern described in Chapter III. For 6MHz ISDB-T system, the Broadcast TS has a fixed bit rate of 32.50…Mbps. For 8MHz system, it is 43.34… Mbps.

XI.2.3   PSI/SI

PSI/SI from "Program management system" contains several kinds of information, such as, the number of TV programs in a broadcast service, encoding parameters of the program or Electronic Program Guide (EPG), etc. Figure XI-5 below shows an example of the data flow of PSI/SI. This information is provided to data server, which will pass it to a data-casting subsystem, SI/EPG server and ENC/MUX subsystem as the control data.

SI/EPG server generates a SI/EPG data from the given PSI/SI information, and sends to Multiplexer. This SI/EPG data is multiplexed into a transport stream as information to a receiver for TV program contents.

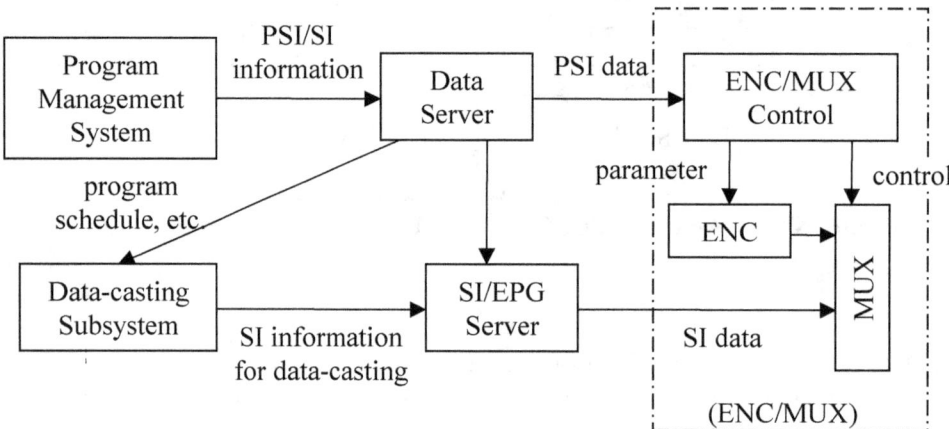

Figure XI-5   An example of PSI/SI data flow

XI.2.4   Data-casting subsystem

Data-casting subsystem mainly consists of two categories: data broadcasting and captions.

Data broadcasting in ISDB-T is transmitted in the data carousels called DSM-CC section format. The data broadcasting subsystem keeps on transmitting DSM-CC data in DII and DDB packets[170] and associated PMT packets in every certain period.

Caption and super impose is also categorized as data-casting subsystem. Caption and super impose is sent in the independent PES format. Presentation Time Stamp (PTS) can indicate timing when to show them on the display.

In order for datacasting fully functional in practice, not only sending out but also composing the code, proofing and switching contents are important as shown in Figure XI-6. Data-casting systems generally include authoring tools that eases composing the code and providing preview screen for coding. Also program server, taking input of program information, switches the contents to transmit such as news, traffic information or weather forecasts.

---

[170] See section VI.5.2 for the details of DSM-CC

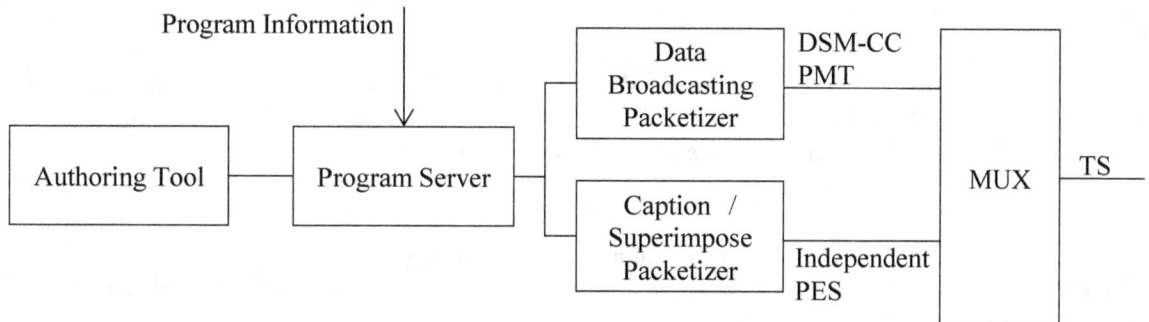

Figure XI-6　A configuration example of data-casting subsystem

XI.3　Transmitter and Transmission Network

Let's see the transmitter and transmission network with the focus on the difference between digital and analog systems.　Figure XI-7 shows an example of a digital and an analog transmitter station.

As illustrated in the Figure, a digital TV transmitter is simpler than that of analog broadcast in terms of RF signal handling, as analog TV transmitter has two streams of process for audio and video modulator and power amplifier separately and combiner, whereas the digital TV transmitter is a straight forward path.

In case of a broadcast wave relay transmitter station and a gap filler transmitter station for analog broadcasting, the combined TV signal (including vision carrier and aural carrier) is amplified by single amplifier for low power station.　It is possible to say that low power transmitter for both analog and digital has similar configuration.

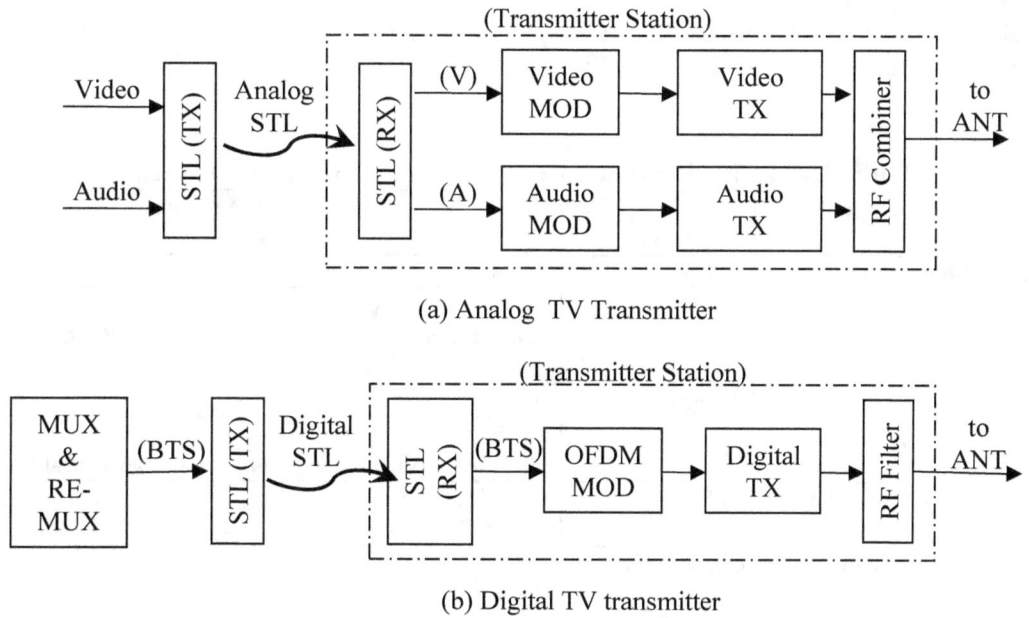

Figure XI-7　Construction of transmitter station[171]

---

[171] High power analog TV Transmitters often have filters at video and audio TX output separately and the RF combiner combines

The measurement items and the evaluation methods are quite different between the analog and digital TV transmitter because of the differences in the characteristics of the signal. The measurement items and measurement methods will be discussed in detail in Chapter XII.

XI.3.1 Important points on digital TV transmission networks design
As mentioned in Chapter X, note that the following two points should be taken into account for a digital transmission network design.

(1) SFN synchronization techniques and network link design
For SFN system design (see Section X.2.2), it is necessary to consider (a) network synchronization method (see Section X.2.4), (b) network link (see Section X.2.3 and X.2.5) and (c) the transmission of network synchronization and transmission control information.

(2) The signal quality degradation in multi-stage relay network
As mentioned in section X.3, in case of IF relay network and broadcast wave relay network, the signal quality degradation cumulates in the number of relay stages. Therefore, it is necessary to choose the number of relay stage and/or the relay transmission system to satisfy the required C/N ratio at the output of final stage.
As it will be described in Chapter XII further, the evaluation of the signal quality in the transmission network use MER, END (or ENF) or delay profile.

XI.3.2 Network relay link
(1) Classification of the network relay link
Table XI-1 shows a classification of relay link of transmission network, both studio to transmitter link (called STL) and transmitter to transmitter link (called TTL).

Table XI-1 Classification of relay link for transmitter network

Relay link type	Interface point (note)	Transmission media
TS transmission relay link	I/F(2)	Micro-wave
		Fiber (high-speed digital line) (Digital network, such as ATM, STM, IP are usable)
IF transmission relay link	I/F(3)	Micro-wave (OFDM)
		Fiber (RF) (OFDM modulated signal by dark fiber is usable)
Broadcast-wave relay link	I/F(3)	Broadcast wave

(note) Interface points are defined in Chapter IX.2.3

---

these signals into one RF output. This is because of power efficiency. Without these filters, video TX output flows into audio TX output (or vise-versa), thus combiner efficiency may get worse. Also, these filters suppress the out of band emission. Regarding digital TXs, filters are placed to reduce outbound emissions.

Figure XI-8 also illustrate the hardware images of these relay link. Please note that Figure X-5 through X-8 in previous Chapter are illustrated based on functions that consists of infrastructure, whereas Figures in this Chapter focuses on the equipment that consists of the infrastructure. Although drawing are different, the Figures illustrates basically the same concept.

(a) of this Figure shows an example of micro-wave STL for TS transmission. The TS(2) type baseband signal is digitally modulated by "64QAM MOD" and up converted to SHF (Super High Frequency) band. The up converted signal is amplified by "SHF PA (Power Amplifier)", then transmitted to STL(RX).

(b) of this Figure shows an example of micro-wave STL for IF transmission. The OFDM modulated signal is up converted to SHF band and is amplified by "SHF PA".

(c) of this Figure shows an example of broadcast wave relay link. Unlike the previous two cases, this system does not use STL.

The broadcast signal from former stage transmitter station is utilized as the input signal of latter stage transmitter station. As described in previous Chapter X.2.5 Case 4, in case of SFN, the loop-back interference from transmitter antenna to receiving antenna may be a serious problem, therefore, it is important to check the network design stage that the loopback interference is critical or not.

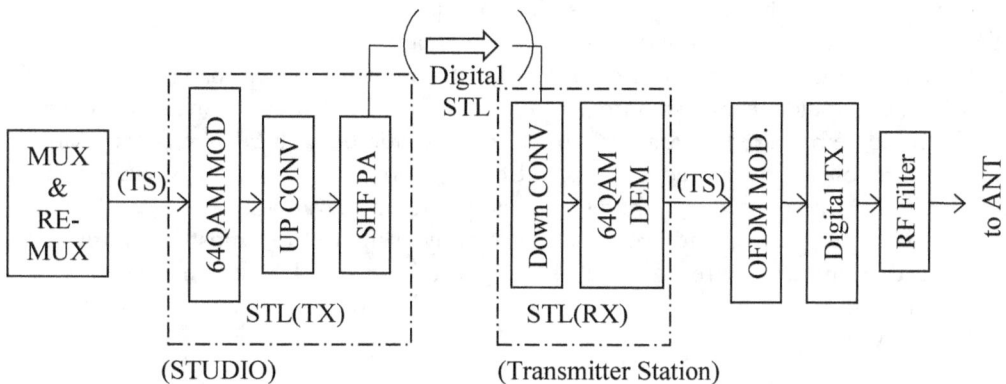

(a) TS transmission micro-wave link

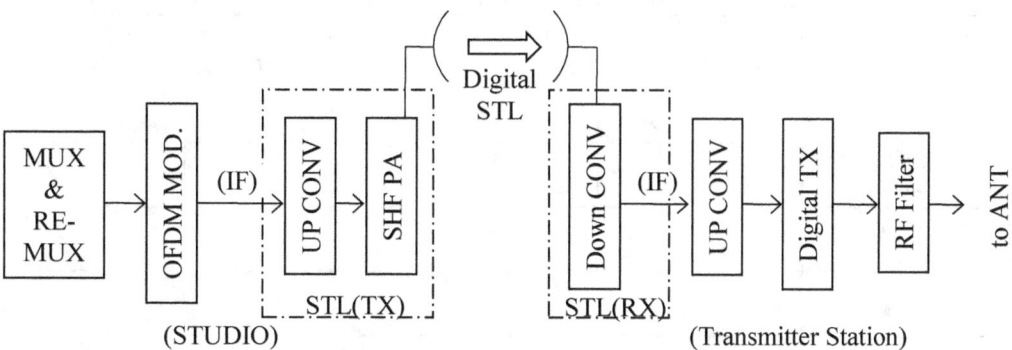

(b) IF transmission micro-wave link

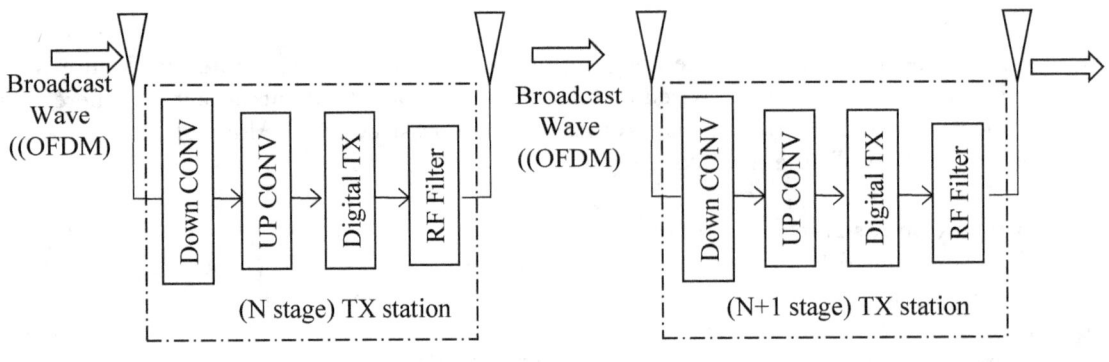

(c) Broadcast wave relay link

Figure XI-8   Classification of network relay link type

The comparison of these relay links is shown in Table XI-2. From the viewpoint of "signal quality", TS transmission system is the best. Because, as explained in Chapter X.3.1, the signal degradation of network is not accumulated in TS transmission system, whereas in other two systems, the signal degradation is accumulated in every stages of the relays in a transmission network.

For SFN timing adjustment, TS transmission system is also better than other systems, because, as explained in Chapter X.2.6 before, it is possible to utilize NSI in IIP for network timing management. For other two systems, this NSI data should be moved into AC data area in OFDM modulated signal or transmission network for NSI should be prepared separately. In either case, additional hardware is necessary.

From the viewpoint of saving a frequency resource, the broadcast wave relay system does not need additional frequencies for micro wave link, therefore, this system has advantage.

Lastly, from the cost perspective, the broadcast wave relay system is the most economical, because this system does not need an infrastructure of STL. However, in case of SFN, it is necessary to be careful for network design so that the relayed signal are not harmed by loopback interference from input and output signal of broadcast wave relay link.

Table XI-2   Comparison table for relay link systems

Relay link	Infrastructure & maintenance costs	Signal quality	SFN timing adjustment	Saving of microwave frequency resource
TS transmission- Micro wave/ Fiber	3	1	1	2
IF transmission- Micro wave	2	2	1	2
Broadcast- wave relay station	1	3	2 (note)	1

(note) in case of broadcast wave relay station, the adjustable range of transmission timing is limited.

XI.3.3   Technical topics for digital TV transmission network

In this section, some technical topics that relate to a distribution of transmission network management information are explained.

(1) "204 byte?" or "188 byte?" (with TS(2) interface)
The "additional information"(see Chapter X.2.6), which includes essential information for network timing management, such as STS, Maximum Delay, etc, is necessary for the SFN network control. Therefore, "additional information" should be multiplexed on broadcast signal to relay stations via relay link.

As described in Chapter X.2.6, the "additional information" is multiplexed on a part of dummy byte and on a part of invalid packet. Therefore, it is necessary to send not only 188 byte area of TSP including TS header, but also the dummy byte and invalid packets in which "additional information" is multiplexed for TV transmission network management.

In Japan, many transmission networks use SFN, therefore, the additional information is necessary. Because of this reason, the "204 byte transmission" system is mostly used. For the signal format of "204 byte transmission", see previous Chapter.IX.2.6.

(2) NSI data distribution by AC data area via IF transmission link
For transmission network construction, as shown in Chapter X.2.6 and Figure X-10, IF relay link and broadcast wave relay link as well as TS relay link are utilized in some networks. For the management of the transmission timing of each station, additional information in NSI is an necessary information. Regarding a distribution of NSI, 2 measures are possible, one is multiplexing on AC in OFDM signal, the other is to prepare another distribution network.

The network control information is picked at each transmitter station from received OFDM signal, and used for the management of each transmitter station.

This network control information is useful not only for a dynamic control of SFN transmission timing, but also for a time reference of a transmission timing measurement.

(3) TS(2) distribution via satellite and fiber communication link
As described in (1), to distribute "204 byte transmission" data though the existing digital communication links, such as a fiber network[172] (ATM/STM[173]) or a satellite digital communication link, it is necessary to take care of several issues shown below when designing the network.

(i) Packet size of ATM
As illustrated in Figure II-3, the structure of TSP (Transport Stream Packet) was designed to have an affinity with the cell of ATM. The information area of TSP (188 bytes) are divided into four data blocks, and each blocks size is 47 byte, just same size as the information area of ATM cell.

Therefore, for "204 byte transmission" it is necessary to re-arrange the Transport stream. In order to place 204 byte TS packets into 47 byte ATM cells, some re-map process of TSPs are necessary so that all data including both broadcast contents data which is already inserted into 188 byte area of Transport Stream and "additional information" described above should be properly carried. This re-map process depends on the product design and implementation.

(ii) Data rate reduction in communication link
As explained in section III.2.1(7), the Broadcast TS clock rate has fixed rate of 4fs (for 6MHz ISDB-T, 32.50...MHz, for 8MHz ISDB-T, 43.34...MHz), thus the transmission speed of "broadcast TS" (BTS) is fixed to constant rate (for 6MHz ISDB-T, bit rate is as equal as 32.50...Mbps, for 8MHz ISDB-T, 43.34...Mbps) in any combination of OFDM transmission parameters. This bit rate is higher than actual information bit rate.

---

[172] A digital micro-wave link is also available.
[173] Recently, an IP network becomes available for TS(2) transmission link.

Refer section III.5 for the calculation process of actual information bit rate

In order to create a fixed rate BTS, null packets are inserted to adjust the bit rate. That is, if BTS is transmitted by digital communication link, these null packets are redundant and the wider bandwidth is required than necessary. To solve this issue, SFN adaptor for ISDB-T, that removes null packets in TX side and reconstructs original BTS in RX, is proposed and developed by plural manufacturers. Figure XI-8 is an example of transmission link with SFN adaptor.

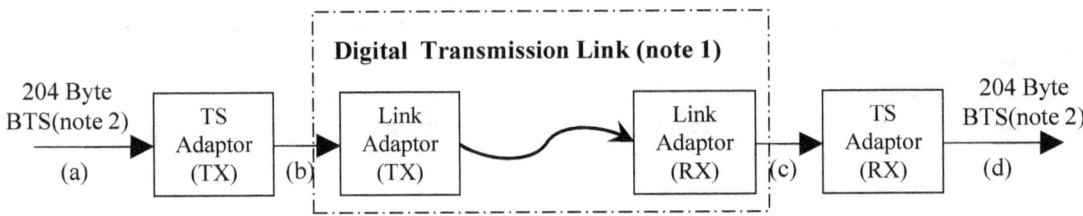

(note 1) Any kind of digital transmission link, such as fiber, satellite, radio communication, etc are usable

(note 2) 204 byte broadcast TS (BTS) format

Figure XI-9   Block diagram of digital link with TS adaptor

The SFN adaptor (TX) has two functions, one is, in order to reduce a transmission capacity, to delete null packets which are not to be sent in the link; the other is to re-map necessary TSP including dummy bytes on ATM cell. In contrast, SFN adaptor (RX) has two functions, converting the data format from cell type to TS type, and re-inserting the null packets to adjust the bit rate of BTS. The bit rate of points (b) and (c) illustrated in the figure above is lower than the bit rate of BTS. By utilizing the SFN adaptor, both interfacing to communication link and to reduce bit rate can be achieved.

Chapter XII. Measurement and Evaluation of Digital Broadcasting

In general, the measurement items for analog broadcasting system/hardware includes (i) measurement of video/audio signal quality and (ii) measurement of (broadcast) radio wave signal. On the other hand, the measurement items for digital broadcasting system/hardware in the field does not include direct measurement of video/ audio signal quality, but direct measurement of a digital baseband signal and a digitally modulated RF signal. This is because of the nature of digital signal. As shown in Figure XII-1, so long as the signal is encoded and multiplexed properly into digital bit stream, the measurement can focus if the digital bit stream are carried without losing information.

Digital broadcasting system basically consists of "baseband digital signal block" and "digital RF signal block". In general, Transport Stream (TS) format is used for the signal format in "baseband digital signal block" of digital broadcasting. In this block, the main purpose of signal measurement is to measure/evaluate the impairment caused by encoding/decoding. In RF signal block, the main purpose of signal measurement is to measure/evaluate the RF signal impairment and limitation of reception.

The measurement/evaluation of digitalized video/audio signal and the analysis of baseband digital signal are written is many text related MPEG technology, therefore this book focuses on digital terrestrial broadcasting RF signal measurement.

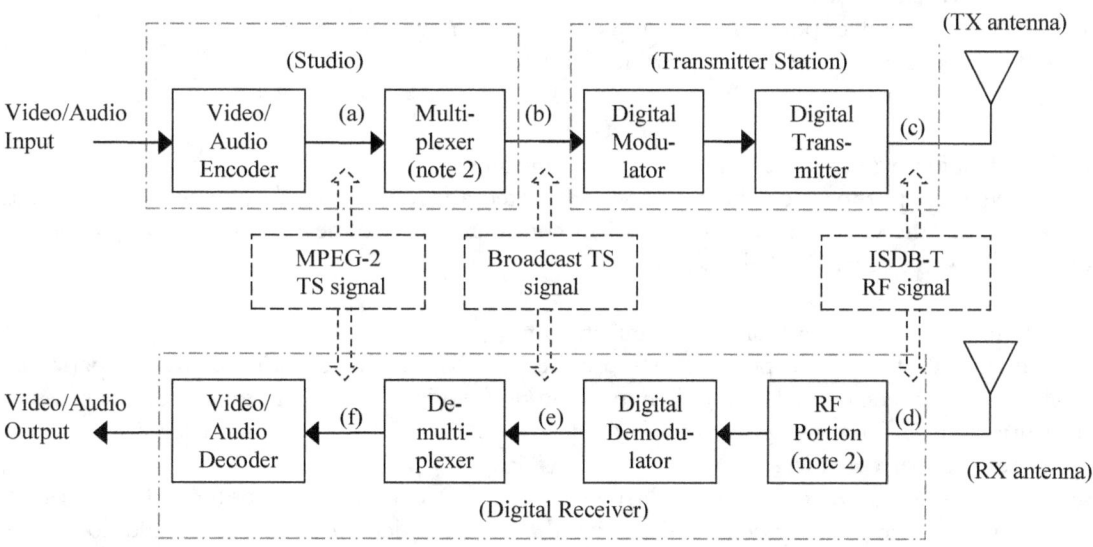

(note 1) signal format of each point: (a)(f):MPEG-2 TS, (b)(e): Broadcast TS, (c)(d):Segmented OFDM(ISDB-T) signal
(note 2) Multiplexer output signal format of ISDB-T is defined as "Broadcast TS"

Figure XII-1 Functional diagram and signal format for digital broadcasting

XII.1 Overview of Measurement for Digital Broadcasting System

Let's see why we need such a measurement and how we can do this. There are several key points of measurements in digital terrestrial broadcasting. This section firstly discusses on the overview of the purpose and configuration of the measurement.

XII.1.1 Purpose of measurement for digital broadcasting signal

The main purposes of the measurement for digital terrestrial broadcasting can be listed as follows:

(1) Measurement/evaluation of the performances of broadcast system
In order to evaluate the performance of a digital broadcasting system, and verification of the limitation of performances of the broadcasting system, it is common to measure the minimum required signal level, protection ratio, etc. These measured data are utilized as the basis of channel planning in implementing the broadcasting system. In addition, degradation by several kinds of interference data are also necessary.

(2) Measurement/evaluation of signal quality
Measurement data is utilized for evaluation of the received signal quality at certain point in the coverage area. Especially, this data is important for the transmission network construction by radio-wave relay technique.

(3) Measurement/evaluation of digital transmitter and transmission site
Used for acceptance test of a digital transmitter at factory, and verification test at transmitter site to verify if the transmitter transmits signal as designed and within the limit set by regulations.

(4) Measurement/evaluation of digital receiver
Used for evaluation of receiver performances at factory test, for example, to see how sensitive a receiver is or to what extent a receiver works properly under interference.

XII.1.2 Classification based on measurement configuration

In order to achieve the purposes of the measurement stated above, let's see how to setup the measurement environment. From the viewpoint of measurement configuration, the measurement of digital broadcasting are classified into 2 cases described below:

(1) Measurement of the reception limits under interferences
The measurement data for the purpose of (1) described in XII.1.1 above are used for comparison test on transmission parameter sets, basic data for channel planning (required field strength, protection ratio, etc).

The measurement data for the purpose of (4) described in XII.1.1 above are used for measurement/ evaluation of receiver performances.

The measurement configuration is shown in Figure XII-2 (a) below. It is configured to measure if the signal source and receiver properly send and receive the signal under the presence of additional noise and interference signal.

For measurement items, see Table XII-1 in Section XII.2.

(2) Measurement for the degradation of digital broadcasting signal
The measurement data for the purpose of (2) and (3) described in XII.1.1 above are used for measurement/ evaluation of the quality of received signal at each relay station. And the measurement data for the purpose of (4) described in XII.1.1 above is used for measurement/ evaluation of transmitter & transmission network

performances. The measurement configuration is shown in Figure XII-2(b). For the measurement items, see Table XII-3 in Section XII.2.

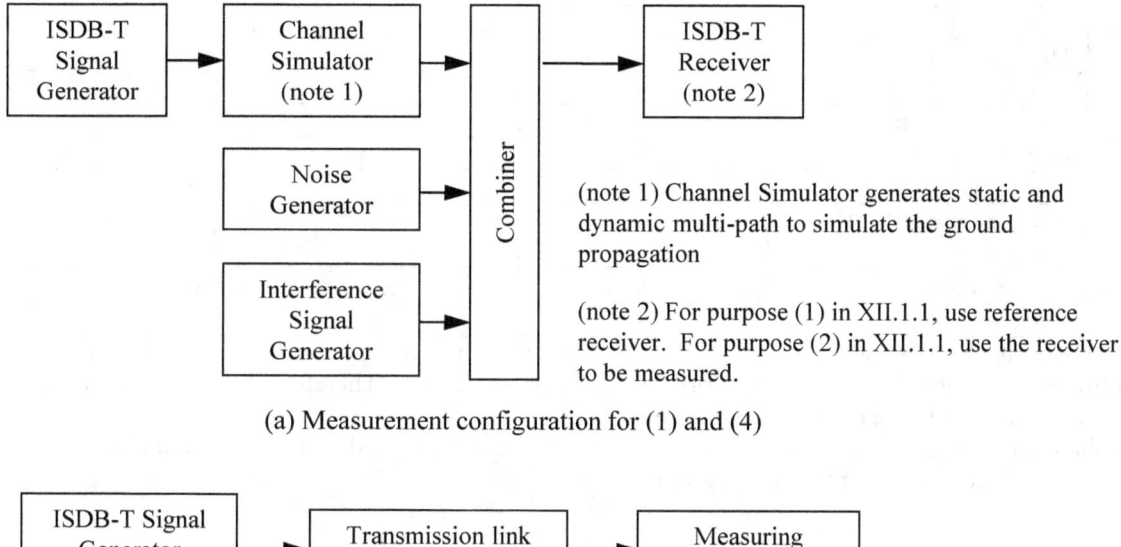

(a) Measurement configuration for (1) and (4)

(note 3) For purpose (3) in XII.1.1, replace this with transmitter or transmission system

(note 4) For purpose (3) in XII.1.1, this function should be removed

(b) Measurement configuration for (2) and (3)

Figure XII-2   Overview of measurement diagram

XII.2   Overview of Measurement/Evaluation Items for System and Receiver

This Section explains on the overview of measurement/evaluation purposes classified into (1) and (4) in Section XII.1.1 above.   (XII.3 will explain on purpose (2) and (3) later.)

Section XII.2.1 explains for measurement / evaluation items, followed by some examples of measurement configuration in Section XII.2.2.

XII.2.1   Measurement/evaluation items for system and receiver

Table XII-1 below shows the measurement items for system and receiver performance evaluation. Measurement items for system evaluation in this table are proposed in the Report ITU-R BT.2035 Chapter 2[174] for evaluation of system performance.   Measurement items for receiver performance proposed in this table are general items.

---

[174] REPORT ITU-R BT.2035-2: Guidelines and techniques for the evaluation of digital terrestrial television broadcasting systems including assessment of their coverage areas

Table XII-1 Measurement/evaluation items for system and receiver

No	Measurement item	System evaluation	Receiver performance	Note
1	Receiver input level	X	X	
2	Min. required C/N	X	X	
3	Frequency capture range		X	
4	Static multi-path	X	X	
5	Dynamic multi-path	X	X	
6	Co-channel interference	X	X	(note 1)
7	Adjacent channel interference	X	X	(note 1)
8	Impulse noise	X	△	(note 2)
9	Phase noise	X	(note 3)	

(note 1) In general, three test categories are proposed: analog to digital, digital to analog, digital to digital
(note 2) Impulse noise is critical in mobile/ portable reception condition. Therefore, this item is important for the evaluation of mobile/portable receivers.
(note 3) phase noise of local oscillator in receiver affects to "required C/N", therefore, this item is not usually tested for receiver performance evaluation.

The performances for Required C/N, survivability against multi-path, co-channel/adjacent channel interference are very important for channel planning. In general channel planning is investigated considering these data.

On the other hand, in designing/development stage of digital receiver and LSI, it is important to evaluate the receiver/LSI by testing for these items

XII.2.2 Overview of each measurement items for system/receiver evaluation
(1) Required C/N
Required C/N is defined as the minimum carrier to noise ratio at which receiver operates normally.[175] In general, normal reception is defined as the condition in which the demodulated digital data is almost error free. In case of concatenated error coding system (Convolutional coding + RS coding), normal reception means BER is under 2×10^{-4} after Viterbi decoding. BER is one of an index of performance described in Section XII.3.5 later.

Figure XII-3 below illustrates an example of measurement block diagram for "required C/N." Signal from RF signal generator is degraded by adding the noise from Noise generator and fed to the receiver. Thermal noise pattern is generally preferred for this measurement. Noise level and input level are adjusted by Attenuators (ATT) in the line. ATT 1 is used to adjust carrier to noise ratio, and ATT 2 is used to adjust receiver input level.

Starting the noise level from minimum, and increasing the noise level gradually, the BER of receiver will start to increase as well. When BER rose to certain level, the ratio of carrier level to noise level is required C/N.

---

[175] In ATSC system, this point is called TOV (Threshold of Visibility)

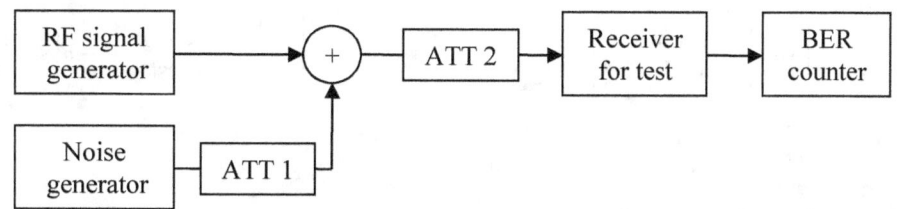

Figure XII-3 An example of measurement diagram for "Required C/N"

In case of measurement of Required C/N, to avoid the influence of receiver internal noise, the receiver input level should be set to enough level (around -50dBm).

Note that required C/N may vary depending on the transmission parameter set. Therefore, it is necessary to decide the transmission parameter set before measurement.

(2) Receiver input level
Generally, same test system for measuring required C/N is used for measuring receiver input level. For the measurement of receiver input level, noise generator is disconnected, and the input level is adjusted by ATT 2. The limit of reception is also checked by BER counter. As the input signal level decreases, the BER will increase. When BER rose to a certain level, the signal level at this point is minimum receiver input level.

(3) Frequency capture range
Frequency capture range is defined as a frequency range in which a receiver can receive an input signal. For this test item, the measurement diagram of Figure XII-3 is also used. In order to measure frequency capture range, the output frequency of signal generator will be changed, and the limit of receiver capture range will be measured.

(4) Measurement for receiving performance under static/dynamic multipath condition[176]
A fading simulator makes it possible to perform measurement of receiver performance under static/dynamic multipath condition. A fading simulator located at the output of signal generator simulates a fading. Figure XII-4 (a) shows an example of receiving performances under multi-path condition.

Figure XII-4(b) shows a functional diagram of fading simulator. The input signal is divided into path 1 to path n, and the level and delay time (including phase) of each path is independently controlled based on the multi-path pattern to be simulated.

If the level/delay of each path are fixed in time axis, this pattern is called "Static multi-path". This pattern is usually used when simulating "fixed reception" (usually roof top antenna reception). If the level/delay of each path is varied in time axis, this pattern is called "Dynamic multi-path." This pattern usually simulates "mobile/portable" reception.

In case of indoor reception, receiving signal are rather affected by moving target such as human in house and car running outside the street. Therefore, simulations on indoor reception often use the dynamic multi-path pattern.

---

[176] REPORT ITU-R BT.2035-2: Guidelines and techniques for the evaluation of digital terrestrial television broadcasting systems including assessment of their coverage areas

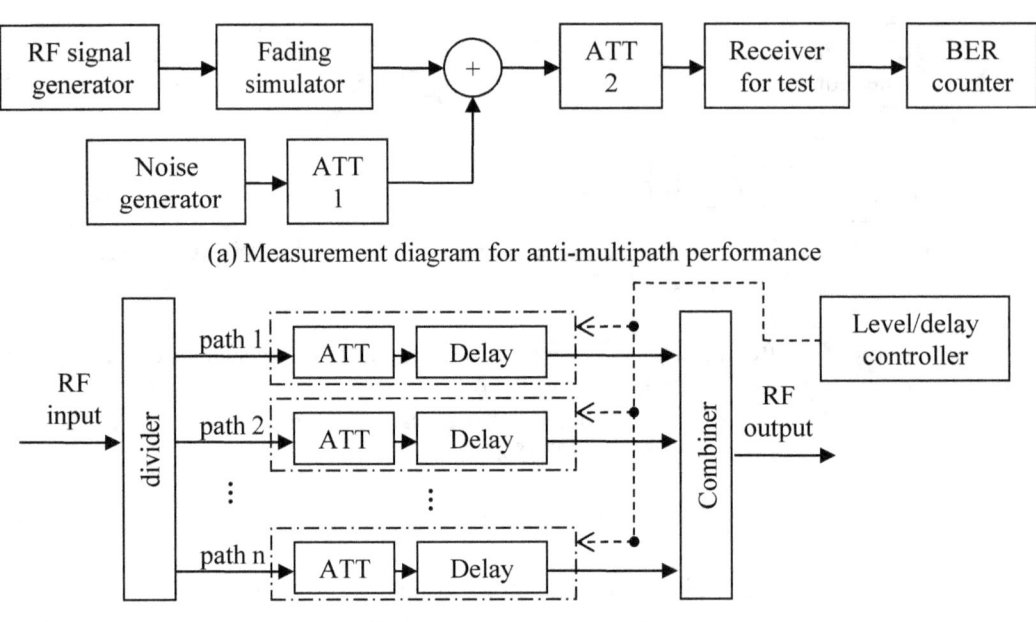

(a) Measurement diagram for anti-multipath performance

(b) Block diagram of "fading simulator"

Figure XII-4　An example of measurement diagram for receiving performance under multi-path condition

　　The report, ITU-R BT.2035 ANNEX 4 discusses on many multi-path ensembles (predefined parameter set on number of path and its delay, attenuation, frequency and phase).　Please refer this document for details of ensembles.

　　Recently, some types of signal generator have this fading simulator function inside of equipment.　It is a good idea to check the functions of signal generator, and get most applicable equipment.

(5) Receiving performance under interference condition[177]
The tolerance against interference is also another important performance of a receiver.　This parameter is used especially for channel planning.

---

[177] REPORT ITU-R BT.2035-2: Guidelines and techniques for the evaluation of digital terrestrial television broadcasting systems including assessment of their coverage areas

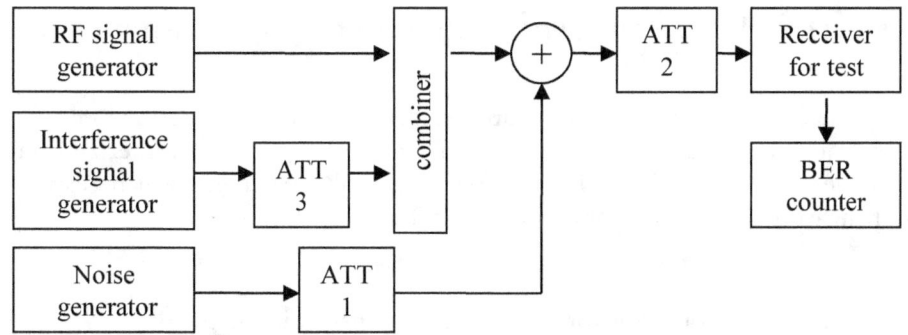

Figure XII-5 An example of measurement diagram for receiving performance under interference condition

In order to measure this parameter, a desired RF signal and an undesired RF signal are combined and fed to a receiver. The ratio of two signals is defined as the Desired to Undesired Ratio. The RF signal generator generates desired RF signal at a proper signal level. In the mean time, the undesired RF signal is set to 0 and gradually increased by adjusting the variable attenuator (ATT3) settings. By monitoring the receiver output and when the receiver output start to be distorted, record the desired RF signal level and the undesired RF signal level. This signal ratio is defined as protection ratio.

ITU R Rec. BT.1368-12 shows protection ratios of many cases for planning criteria. Table XII-2 shows an example of interference pattern written in ITU-R BT.1368-12 which is used for channel planning.

Table XII-2 An example of reception performance under interference[178]

Interference pattern	desired	undesired	6 MHz system (dB)	8MHz System (dB)
Co-channel	Analog (note 1)	Digital (note 4)	39 (note 5)	34 (note 5)
	Digital (note 2)	Analog (note 1)	5	5
	Digital (note 2)	Digital (note 4)	21	20
Lower adjacent channel	Analog (note 1)	Digital (note 4)	-6 (note 5)	-9 (note 5)
	Digital (note 2)	Analog (note 1)	-31	-32
	Digital (note 3)	Digital (note 4)	-26	-26
Upper adjacent channel	Analog (note 1)	Digital (note 4)	-6 (note 5)	-8 (note 5)
	Digital (note 2)	Analog (note 1)	-33	-38(note 6)
	Digital (note 3)	Digital (note 4)	-29	-29

(note 1) 6MHz analog system: M-NTSC,   8MHz analog system: G-PAL
(note 2) Digital system: ISDB-T, 64QAM, code rate 3/4
(note 3) Digital system: 64QAM, code rate 7/8
(note 4) Digital system: ISDB-T, transmission parameters are not defined
(note 5) In case of Tropospheric interference
(note 6) Digital system, ISDB-T 64QAM, code rate 2/3

---

[178] Summarized from ITU R BT.1368-12 Annex 3 for the protection ratio against interference of ISDB-T

XII.3 Overview of Measurement/Evaluation Items for Transmitter & Transmission Network

In this Section, it explains the overview of measurement/evaluation items classified into (2) Measurement / evaluation of signal quality and (3) Measurement / evaluation of digital transmitter and transmission site in Section XII.1.1 above. In Section XII.3.1, measurement/ evaluation items will be explained, followed by some examples of measurement configuration in Section XII.3.2.

XII.3.1 Measurement/evaluation items for transmitter & transmission network

The actual measurement items for transmitter/transmission network evaluation are shown in Table XII-3 below. For the measuring point, see Figure XII-6 below.

Table XII-3   An example of measurement items for transmitter/transmission network[179] [180]

Measurement item	Digital transmitter only (note 1)	Input signal Quality (note 2)	Signal quality through relay station (note 3)	Note
(General characteristics of transmitter)				
Frequency	X		X	
Output power	X		X	
Spurious domain emission	X		X	
Out of band emission	X		X	
Occupied bandwidth	X		X	
Power consumption	X			
(Input and output signal characteristics)				
Shoulder	X(note 1)		X(note 1)	Section 2-3 of the JEITA Handbook
MER	X	X	X	Section 2-3
BER; case 1 (note 4)	X			See Section 2-3 of the JEITA Handbook
BER; case 2 (note 5)		X	X	See Section 2-3 of the JEITA Handbook
END(note 6)	X	X	X	See Section 2-3 of the JEITA Handbook
Phase noise	X			
Amplitude frequency	X	X	X	

---

[179] REPORT ITU-R BT.2035-2: Guidelines and techniques for the evaluation of digital terrestrial television broadcasting systems including assessment of their coverage areas
[180] Original table shown in JEITA handbook, Methods of measurement for digital terrestrial transmission network

characteristics				
Delay profile		X		

note 1   Measurement point: Figure XII-6 (b) except "Shoulder". For "Shoulder", to avoid the influence of BPF, should measure at the input of BPF (indicated as(a) in Figure)
note 2   Measurement point: Figure XII-6 (c)
note 3   Measurement point: Figure XII-6(d)
note 4   Measurement method with PN signal
note 5   Simple measurement method using broadcast signal
note 6   For measurement of received signal END, it should be considered that receiving signal may imply noise component.

The measurement points of transmitters and transmission networks are illustrated in Figure XII-6.

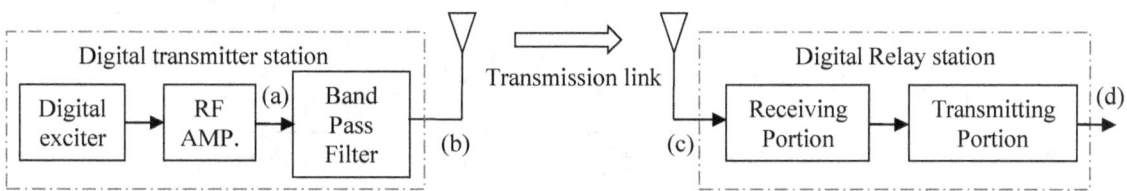

Figure XII-6   Measurement points of transmitter & transmission network

For, "Shoulder" measurement (to be discussed in XII.3.3), in order to avoid the influence of Band Pass Filter, the measurement point should be the input of BPF, as indicated in the Figure XII-6 (a). For other measurement items of digital transmitter, the measurement point should be the output of transmitter indicated as (b).

In case of transmission network chain measurement, point (c) will be usually used for received signal quality measurement. The point (d) is the measurement point of relay transmitter performances, which includes any of signal degradation generated at previous transmitters stage and transmission links.

As shown in Table XII-3, the main purposes of measurement are to evaluate digital transmitter and transmission network performances.

The measurement items are categorized in two groups:
(a) General characteristics of transmitter:
These measurement items are generally required for license application of transmitter stations. These items are common to analog transmitter.
(b) Signal quality of digital transmission:
These measurement items are necessary to verify/evaluate the performances of digital transmitter and digital transmission network.

XII.3.2   Overview of each measurement item

In this Section, the overview of several commonly used measurement items will be introduced. For the details of measurement, it is good idea to refer specialized texts for measurement as well[181] [182] [183] [184].

---

[181] REPORT ITU-R BT.2035-2: Guidelines and techniques for the evaluation of digital terrestrial television broadcasting systems

XII.3.3 Intermodulation (Shoulder)

Shoulder attenuation is defined as "ratio between signal and spectrum re-growth outside the channel" in the IEC standard[185]. This is the ratio between the level of signal spectrum and level of intermodulation products generated outside of useful OFDM spectrum (see Figure X-23).

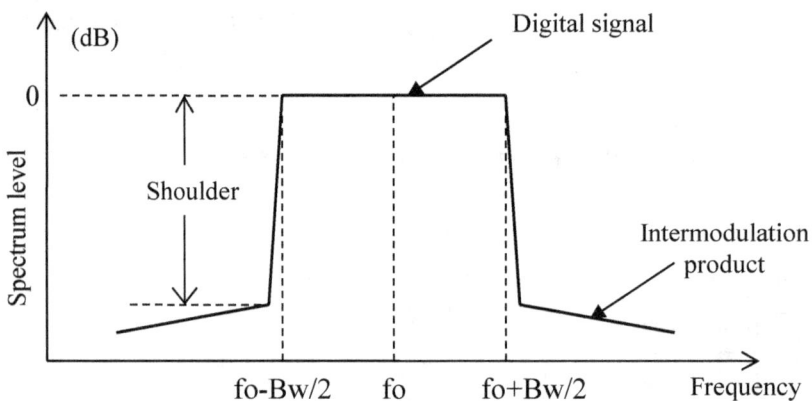

Figure XII-7 Definition of "shoulder"

This characteristic is very important to evaluate the linearity of power amplifier of digital transmitter which is critical to multi-carrier transmission system such as OFDM. To avoid the influence of Band Pass Filter (BPF) located at output portion of power amplifier, this items should be measured at point (a) in Figure XII-6. In IEC62273-1, the measurement points of RF spectrum are also defined.

XII.3.4 Modulation Error Ratio (MER)

MER is the measure of total degradation in the transmitted signal due to any degradation factors, such as thermal noise, phase noise, equipment degradation of transmitter and etc. This measure is defined by following formula:

---

including assessment of their coverage areas

[182] IEC 62273-1, Method of measurement for radio transmitter – performance characteristics for terrestrial digital television transmitters
[183] JEITA handbook, Methods of measurement for digital terrestrial transmitters
[184] JEITA handbook, Methods of measurement for digital terrestrial transmission network
[185] IEC 62273-1, Method of measurement for radio transmitter – performance characteristics for terrestrial digital television transmitters

$$MER = 10\log\left\{\frac{\sum_{j=1}^{N}\left(I_j^2 + Q_j^2\right)}{\sum_{j=1}^{N}\left(\delta I_j^2 + \delta Q_j^2\right)}\right\} \text{dB}$$

I: Inphase components
Q: Quadrature components
N: Number of carriers
j: Carrier index

Here, δ in formula means the signal variation component caused by noises, interferences and equipment degradations. MER is one of measures for the quality of digital modulated signal. The method of measurement of MER is simpler than BER measurement method, therefore, this method is used to check the signal quality at transmitter station and field measurement.

In case that more accurate data for signal quality should be necessary, END or ENF described later should be measured.

In addition, Constellation diagram is also effective as a means to analyze signal degradation factors. In ETSI TR 101 290 Measurement guidelines for DVB system, Chapter 9.18, the relationship between constellation diagram and several degradation factor of equipment are introduced, such as, "Carrier suppression", "Amplitude imbalance", "Quadrature error".

XII.3.5 Bit Error Rate (BER) (PN Method)

Bit Error Rate (BER) is the most fundamental and important measure of digital transmission. BER measurement use data whose contents are known. The data is sent from transmitter side and received by receiver. By comparing the received data with known data and by counting the number of unmatched bit, BER can be calculated. BER is defined as the ratio of (Number of erroneous bit)/(Number of total bit). Normally, PN (Pseudo-random Noise) pattern is used for known data. Therefore, during BER measurement, on air service should be interrupted. To avoid such service interruption, other methods described below are used.

XII.3.6 BER (Simple measurement method using broadcast signal)

Two measurement methods described below are proposed and used. These method are available even during "On air" periods.

(1) BER measurement using Null Packets
In digital terrestrial broadcasting, null packets are inserted into Transport Stream (TS) to adjust the transmission data rate[186]. Null packets are not used for information data transmission, therefore, data area of null packets can be used for BER measurement.

Figure XII-8 below shows a conceptual diagram of BER measurement using "Null Packets method."

---

[186] JEITA handbook, Methods of measurement for digital terrestrial transmission network

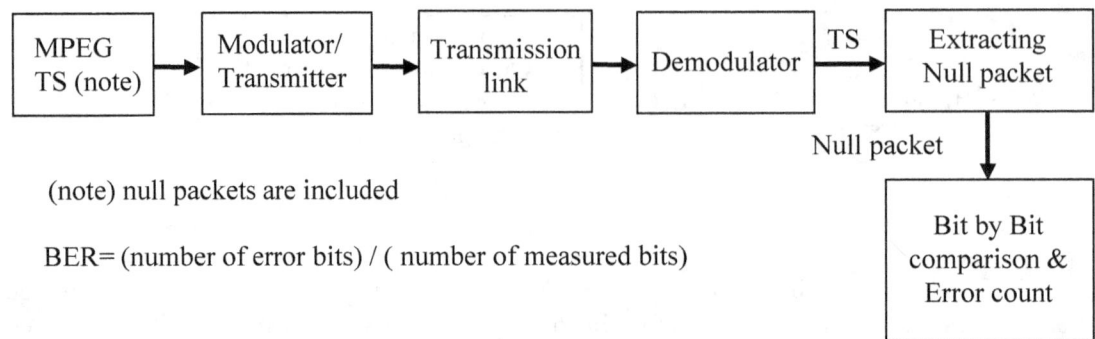

(note) null packets are included

BER= (number of error bits) / (number of measured bits)

Figure XII-8   Conceptual diagram of BER measurement using "Null Packets method"

The data area of null packets in TS should be replaced by known data, such as PN, then it is possible to count erroneous bits in null packets data area at receiver side.

This methods is available even in "On air" situation, but it should be noted that this method needs long term for measurement compared to PN method.   Because usable data capacity for BER measurement is less than PN method.

It is a good idea to estimate how long the measurement takes as usable data capacity depends on transmission and coding parameters.[187]

(2) Simplified method[188]

Figure XII-9 shows the measurement diagram of "Simplified method."   Both methods use the way to compare "data stream before error correction" and "data stream after error correction".   This method assumes that errors can be recovered by error correction mechanism and "data stream after error correction" contains no erroneous data bits.   Therefore, under low signal quality condition, the BER data measured by this method may not be reliable.   So it is necessary to check the measurement condition in which measured data are reliable or not.

These two types of simplified BER measurement method are introduced in ETSI TR 101 290 Measurement guidelines for DVB system.

---

[187] Usable data capacity depends on transmission parameter and coding parameter set.
[188] ETSI TR 101 290, Digital video broadcasting (DVB) Measurement guidelines for DVB System

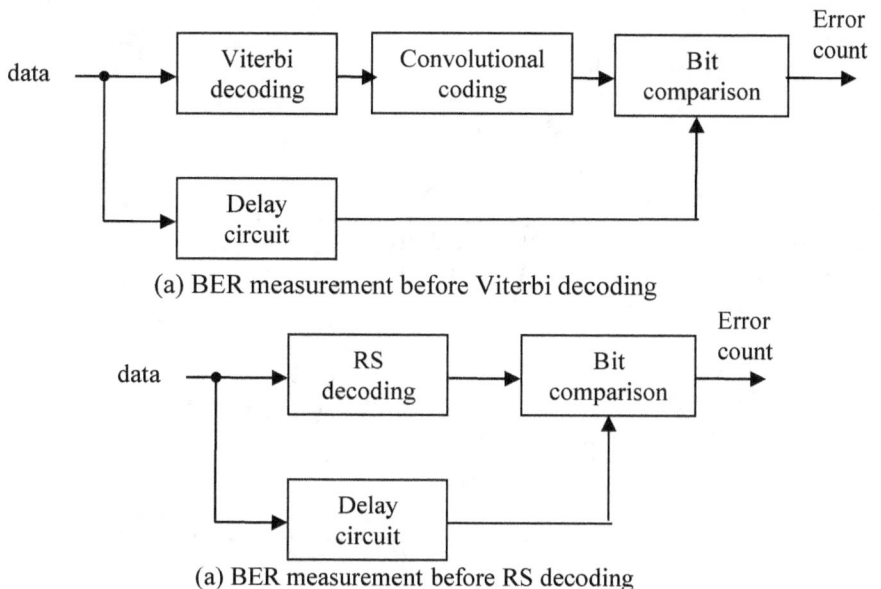

(a) BER measurement before Viterbi decoding

(a) BER measurement before RS decoding

Figure XII-9 Conceptual diagram of simplified BER measurement method

XII.3.7 Equivalent Noise Degradation (END) and Equivalent Noise Floor (ENF)
END is one of important measure to evaluate the digital terrestrial transmission network quality. In case of broadcast-wave relay network and IF signal relay network, discussed in previous Chapter, any signal degradations are accumulated in number of stages. END represents received signal quality by replacing any signal impairments to equivalent Gaussian noise level which gives the same signal quality impairments due to any kind of deteriorations such as interference, non-linearity, multi-path.

ENF[189] is another measure to evaluate the signal quality, in case that the degradation of signal quality is limited[190]. (In some case, term of "Equivalent C/N" is used.

In Figure XII-10, C/N at 2×10^{-4} of theoretical curve[191] is defined as "Required C/N(dB)", and C/N at 2×10^{-4} of measured curve is defined as "Superimposed C/N(dB)." END(dB) is defined in following formula as the difference between "Required C/N(dB) " and "Superimposed C/N(dB)". Please note required C/N is different for transmission system and of transmission parameter sets.

END (dB) = CN_{add}(dB) - CN_r(dB)
Here: CN_{add}(dB):   Superimposed C/N(dB)
      CN_r(dB):   Required C/N(dB)

---

[189] BBC R&D White Paper WHP105, Digital television services: equivalent noise floors and equivalent noise degradation
[190] JEITA handbook, Methods of measurement for digital terrestrial transmission network
[191] This curve shows calculated values without any degradation factors except thermal noise.

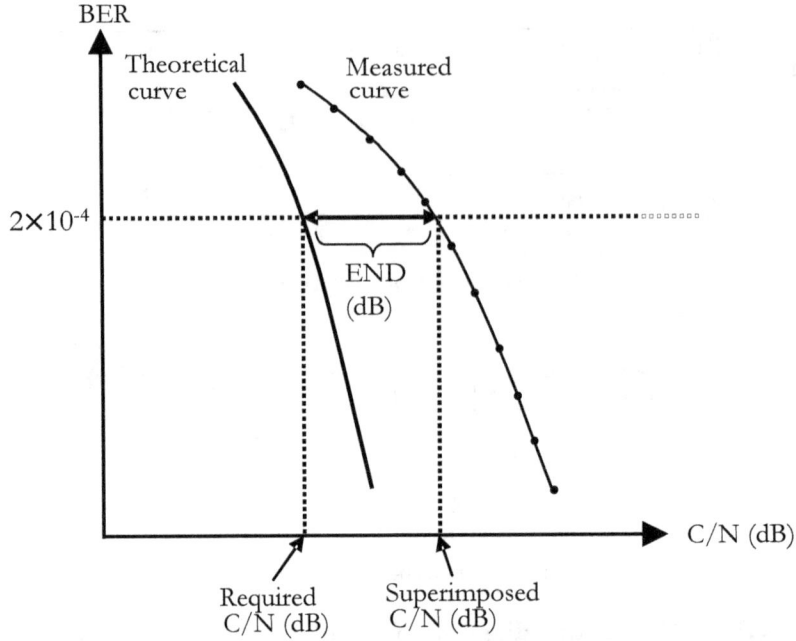

Figure XII-10   Definition of END

In case that the inherent degradation of OFDM demodulator which is used for measuring instrument is not negligible, END(dB) is given as following formula:

$$END[dB] = -10\log_{10}\left(10^{\frac{-CN_{add}}{10}} + 10^{\frac{-CN_{fix}}{10}}\right) - CN_r[dB]$$

Here, CN_{fix} (dB): inherent degradation of OFDM demodulator

ENF[192] is defined as the ratio of signal power (C) to the additionally superimposed Gaussian noise power (N) which gives the same BER as that brought by all of the deterioration causes. The relationship between END and ENF at BER of 2×10^{-4} is introduced in following formula:

$$ENF[dB] = -10\log_{10}\left(10^{\frac{-CN_r}{10}} + 10^{-\frac{CN_r + END}{10}}\right)$$

From above formulas, measurement of CN_{add} (dB) and CN_{fix} (dB) is necessary to calculate END(dB) and ENF(dB).

XII.3.8   Delay profile

---

[192] ENF is equal to the C/N (carrier to noise ratio), if there are no other degradation factors except thermal noise.   In case of ENF, all degradations including thermal noise are accumulated and this value is defined as "noise floor".

In general, delay profile represents the transmission path condition. Therefore, delay profile data is useful to investigate multi-path condition and the receiving situation in SFN operation.

Delay profile is defined[193] as follows:
Denoting the received signal as $r(t)$, the ideal ISDB-T signal as $s(t)$, and the impulse response of the transmission path as $h(t)$ in the time domain, the following relationship holds.

$$r(t) = \int_{-\infty}^{\infty} s(t) h^*(t-\tau) d\tau \quad \ldots (2)$$

(The asterisk "*" indicates a conjugated complex number.) Furthermore, denoting the respective frequency characteristics as $R(j\omega)$, $H(j\omega)$ and $S(j\omega)$, the following relation also holds.

$$R(j\omega) = H(j\omega) \cdot S(j\omega) \quad \ldots (3)$$

$H(j\omega)$ can be obtained from the pilot signal (SP or CP) in the received signal which are mapped in OFDM frame illustrated in Figure III-30 and III-32, and these original characteristics are already known. Then, $h(t)$ can be obtained by inverse Fourier transform of $H(j\omega)$. From $h(t)$, the delay profile $\tau(t)$ can be obtained by using the following equation.

$$\tau(t) = 10 \log_{10} |h(t+t_D)|^2 \quad [dB] \quad \ldots (4)$$

Here, the time t_D at which $|h(t)|^2$ is maximum is taken as the timing of the desired wave.

Please recall Figure III-29 in Chapter III.4.2. This Figure explained the relationship between the transmission path characteristics and the frequency characteristics of received signal. For ISDB-T system, scattered pilot signal is usable to measure the frequency characteristics of received signal, $H(j\omega)$. The transfer function of transmission path, $h(t)$, is obtained as Fourier transform of $H(j\omega)$.
An example of measured data is shown in next section.

XII.4 Some Examples of Measurement Data

In this Section, introduce some examples of measurement data.

(1) RF spectrum of ISDB-T signal
    RF spectrum is used for following purpose:
       (i) transmitter spectrum mask
       (ii) shoulder measurement
       (iii) received signal quality (especially verify the frequency characteristics)

Figure XII-11 shows an example of ISDB-T signal RF spectrum.

---

[193] JEITA handbook, Methods of measurement for digital terrestrial transmitters

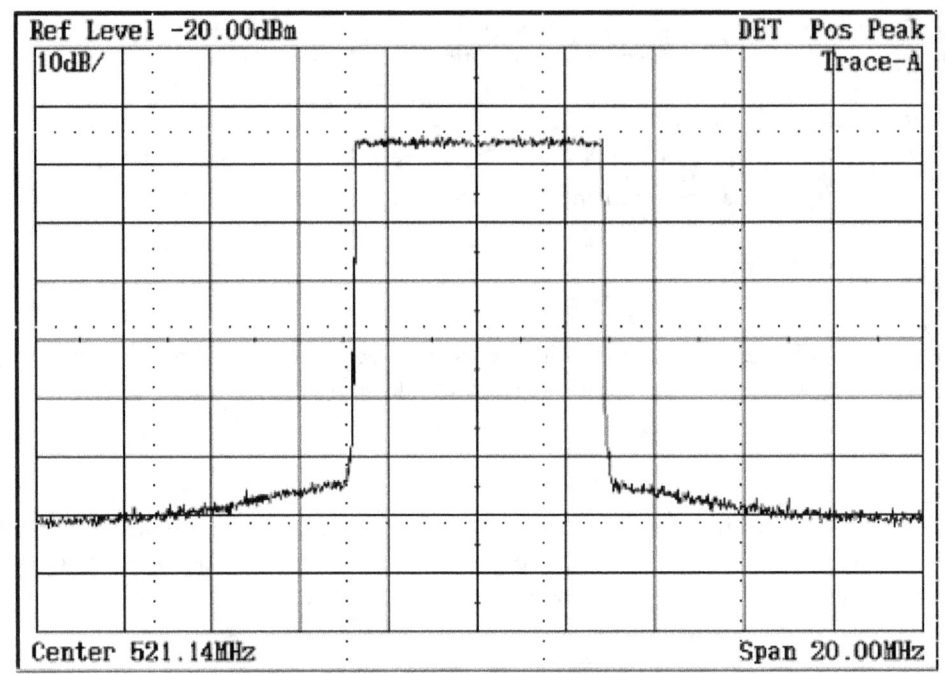

Figure XII-11 An example of RF spectrum of ISDB-T signal

(2) MER
The purpose of MER measurement is not only to measure the digital signal quality but also to analyze the signal degradation pattern (or situation).

There are many factors for signal quality degradation, such as, thermal noise, phase noise, imperfectness of modulator, etc.   It may be possible to estimate the cause of signal degradation by MER data.

END or ENF, explained in Section XII.3.7 before, only gives a degree of signal degradation, and are not possible to estimate the cause of degradation by END or ENF.

Some MER data are shown in Figure XII-12

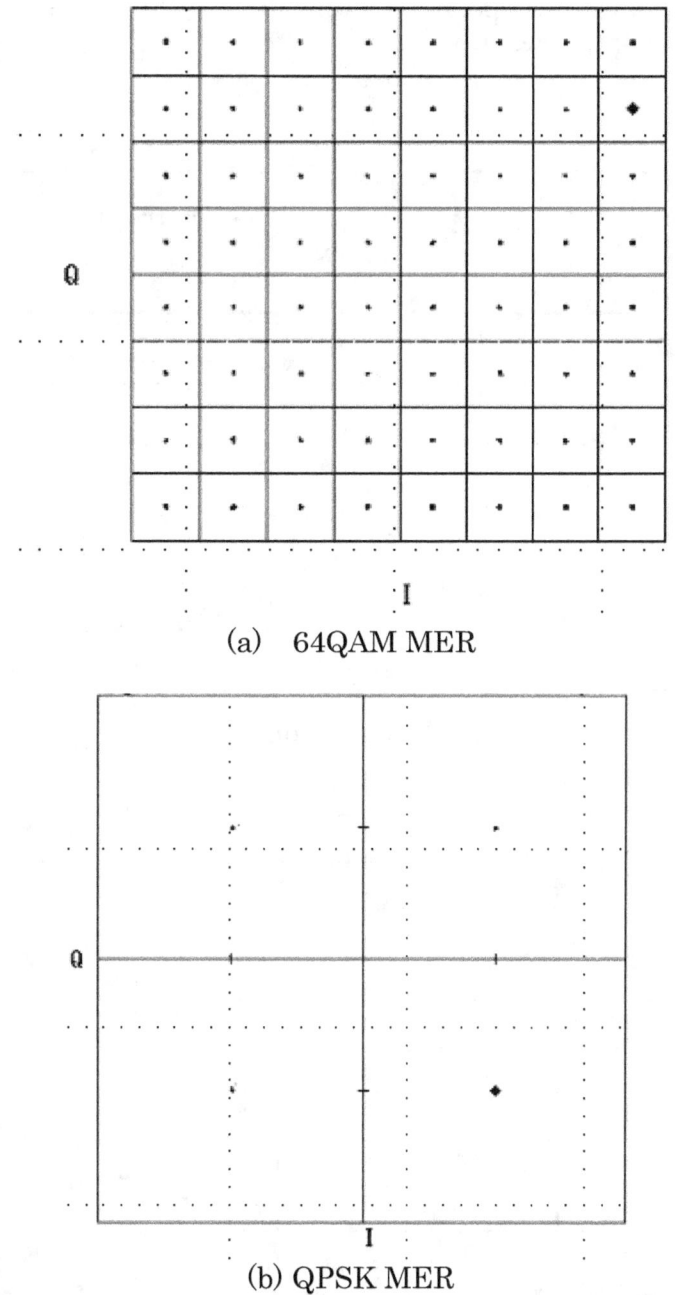

(a) 64QAM MER

(b) QPSK MER

Figure XII-12　Examples of MER data of ISDB-T signal

(3) Delay profile
As described in Section XII.3.8 above, Delay profile data shows the transmission characteristics.

So, this is one of dispensable data for evaluating the received signal quality in service area, especially in case of SFN operation.

Figure XII-13 (a) shows an example of delay profile of ISDB-T. In this case, single multi-path interference exists with about 2us delay. As shown in Figure, the main path signal and the multi-path signal are clearly distinguished.

Figure XII-13 (b) also shows the RF spectrum of same signal. The envelope ripple is observed. This ripple is caused by single multi-path interference.

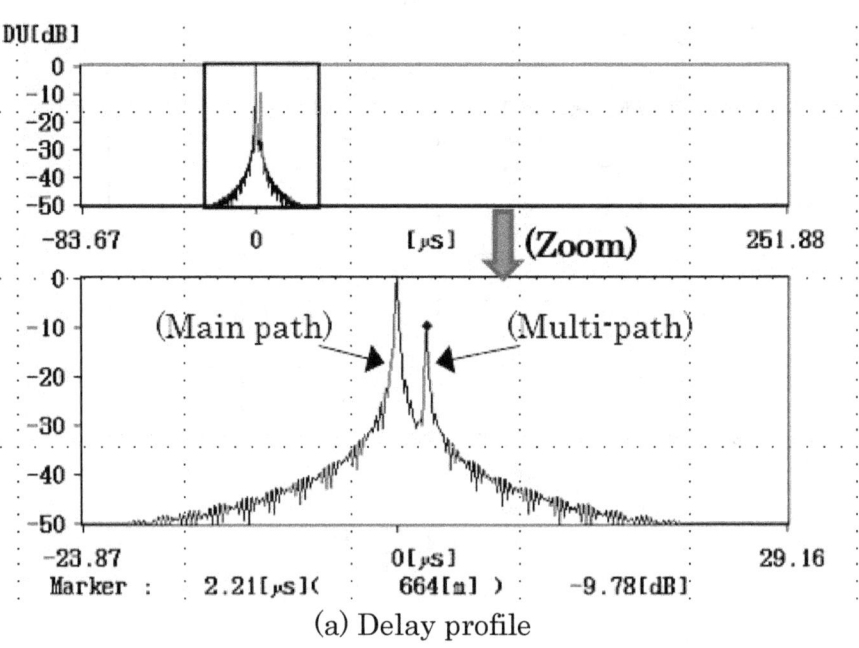

(a) Delay profile

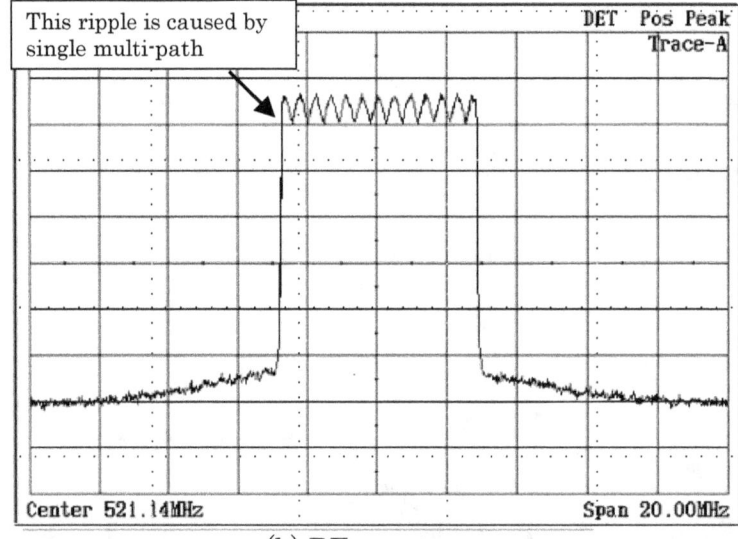

(b) RF spectrum

Figure XII-13   An example of RF spectrum and delay profile (single multi-path)

Chapter XIII. Frequency Allocation and Channel Planning

So far, we have discussed technical perspective on the digital terrestrial broadcast. This chapter discusses the digital terrestrial broadcast from a government's regulatory perspective. In order to introduce digital broadcast, governments needs to assign the frequency and license on the broadcaster stations. At the same time, in order to minimize interference and maximize frequency use efficiency, channel planning criteria needs be set based on the equipment's performance based on a broadcasting standard. This chapter will discuss how Japan handled these issues especially on the frequency allocation and channel planning from the government's perspective.

XIII.1 International Frequency Allocation

The fundamental law of international frequency coordination is the Constitution of the International Telecommunication Union (ITU) and the Convention of the International Telecommunication Union[194], that were elaborated with the object of facilitating peaceful relations, international cooperation among peoples and economic and social development by means of efficient telecommunication services. These documents specifies the purpose and organization of the ITU, and basic provisions on telecommunications and use of the Radio-Frequency Spectrum.

The provisions of both this Constitution and the Convention are further complemented[195] by the Radio Regulations (RR), which regulate the use of telecommunications. RR is a binding document on all Member States of the ITU, and defines on such as:
- the assignment and allocation of frequencies,
- coordination, notification and recording of frequency assignments,
- international coordination procedure on interferences,
- and provision for various radio services and stations.

The RR also defines the frequency allocation table, as an example shown on Table XIII-1, that defines what frequency is allocated for what services. For example, 470-694MHz is allocated for BROADCASTING in Region 1. There are footnote such as "5.149," that show the conditions accompanying the frequency use in the frequency allocation. This frequency allocation table is defined by regions. Region 1: Europe, Middle East, Russia and Africa. Region 2: The Americas. Region 3: Asia, Australia and Oceania. (See Figure XIII-1)

The updates of RR including frequency allocation table takes place in the World Radiocommunication Conferences (WRC), in which delegations from the administrations around the world get together to discuss on modifications. In case of Japan, the Ministry of Internal Affairs and Communications (MIC) is in charge of its coordination.

---

[194] International Telecommunications Union, Constitution and Convention, http://www.itu.int/en/history/Pages/ConstitutionAndConvention.aspx
[195] Constitution of the International Telecommunication Union, Article 4 – 3

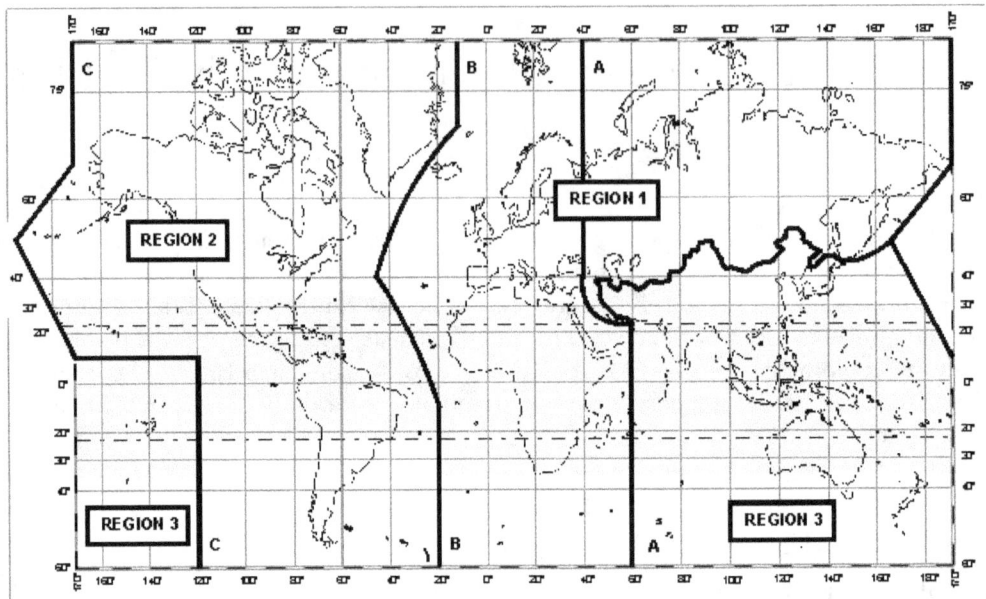

Figure XIII-1   ITU regions and areas[196]

Table XIII-1   Excerpt of frequency allocation table (part of 460-890 MHz)[197]

Region 1	Region 2	Region 3

470-694
BROADCASTING
 	
5.149 5.291A 5.294 5.296
5.300 5.304 5.306 5.311A
5.312
	470-512
	BROADCASTING
	Fixed
	Mobile
	5.292 5.293 5.295
	512-608
	BROADCASTING
	5.295 5.297
	608-614
	RADIO ASTRONOMY
	Mobile-satellite except
	    aeronautical mobile-satellite
	    (Earth-to-space)
	614-698
	BROADCASTING
	FIXED
		470-585
		FIXED
		MOBILE
		BROADCASTING
		
		5.291 5.298
		585-610
		FIXED
		MOBILE
		BROADCASTING
		RADIONAVIGATION
		5.149 5.305 5.306 5.307
		610-890
		FIXED
		MOBILE 5.313A 5.317A
		BROADCASTING

Aside from the clauses in the RR, there are several regional agreements[198] on specific services and frequencies. In relation to the terrestrial digital broadcast, there is GE06 Agreement. GE06 defines the procedure for administrations to bring broadcasting stations in use, and technical parameter and criteria for

---

[196] Quoted from ITU, Radio Regulations, Article 5.2
[197] Excerpt from ITU, Radio Regulations, Article 5, Section IV, 460-890 MHz
[198] ITU Radio communications bureau, FM / TV Regional Frequency Assignment Plans, http://www.itu.int/en/ITU-R/terrestrial/broadcast/Pages/FMTV.aspx

evaluating its conformity, applicable to region 1 and part of region 3 countries/areas[199]. In the GE06 framework, broadcasting stations need to be notified to ITU and registered on the Master International Frequency Register (MIFR) before bringing into use. Notification should list technical parameters of a broadcast station such as frequency, output power, antenna pattern, antenna height, longitude and latitude. Then, the notification will be evaluated by ITU based on the clauses especially on section II of annex 4 that evaluates interference level from the new station (or SFN networks). GE06 also covers protection ratio, spectrum masks, propagation model and other definition for calculating interference.

XIII.2 National Frequency Allocation

In Japan, there are two laws that regulate the broadcasting stations – Radio Law and Broadcasting Law. The purpose of the Radio Law is to promote public welfare by ensuring the equitable and efficient utilization of radio waves. The purpose of the Broadcasting Act is to regulate broadcasting so as to conform to public welfare and to achieve its sound development subject to the following principles: (i) To guarantee that broadcasting is disseminated to the greatest extent possible to the general public and that its benefits are achieved; (ii) To ensure freedom of expression through broadcasting by guaranteeing the impartiality, truth and autonomy of broadcasting; (iii) To enable broadcasting to contribute to the development of sound democracy by clarifying the responsibilities of the persons involved in broadcasting.

Radio Law regulates on the frequency use and provides provisions as follows: (excerpt of clauses mainly related to broadcasting)
- Licenses for Radio Stations
    - Licensing, validity, certificate, inspection
    - Frequency Assignment Plan – a list of available frequencies including mode of radio communications, purpose and requirement for using frequencies. Regarding broadcasting frequency, exclusively for broadcasting or else
    - The Plan for the Available Frequencies Allocated to Broadcasting (channel plan) – broadcasting licence application should be in conformity with the plan
- Radio Equipment
- Conformity Certification
- Radio Operators, Operation
- Supervision

Broadcast Law (Law No. 132 of 1950) regulates on the software side of the broadcasting and provides provisions as follows:
- General rules concerning the editing of broadcast programs
    - Editorial freedom of broadcast programs
    - Editing of the Broadcast Programs
        1. It shall not harm public safety or good morals.
        2. It shall be politically fair.
        3. Its reporting shall not distort the facts.
        4. It shall clarify the points at issue from as many angles as possible where there are conflicting opinions concerning an issue.
    - Deliberative Organ for Broadcast Programs - The broadcaster shall establish a deliberative organ

---

[199] Countries/Areas ratified GE06 is Region 1 (those parts of Region 1, as defined in No. 5.3 of the Radio Regulations, situated to the west of meridian 170° E and to the north of parallel 40° S, except the territories of Mongolia) and the Islamic Republic of Iran

in order to ensure the appropriateness of the broadcast programs.
- Japan Broadcasting Corporation (NHK)
  - Purpose, operation, organization, finance
  - Reception contract and reception fees
- Provisions depending on broadcaster classification
  - Approval/registration, operations, financial reporting
  - The dissemination plan for basic broadcasting
    - Broadcasting target region for simultaneous reception of the same broadcast program in each category of the Open University, NHK or domestic and international broadcasting in AM, FM, television and other types of broadcasting
    - Goals for the number of broadcasting networks for each broadcasting target region
  - Maintenance of equipment – the approved basic broadcaster shall maintain facility for basic broadcasting which conforms to the technical standards[200] stipulated in an ordinance of the Ministry of Internal Affairs and Communications (provision for basic broadcasters)

Having the international frequency allocation table set in RR, the Ministry updates its frequency assignment plan – Japanese national version of frequency allocation table. It also designates frequency exclusively allocated for the broadcasting purposes.

In addition to that, the Broadcasting Law mandates the minister to formulate the dissemination plan for basic broadcasting. This plan fixes the number of broadcasting networks in certain region, that is Tokyo, Chukyo and Kansai metropolitan area and other prefectures. For example in Kanto metropolitan area, the broadcasters are set to be NHK, NHK-Educational, Open University and 5 private networks.

Based on these frequency assignment plan and dissemination plan for basic broadcasting, the Ministry formulates the plan for the available frequencies allocated to broadcasting. This is the channel plan in the Japanese case. Table XIII-2 shows an excerpt of the plan as an example. This plan lists up broadcast networks and its transmitter site, channel number and antenna power of major broadcast stations. Small sized gap fillers are not included here.

Table XIII-2 Excerpt of the plan for the available frequencies allocated to broadcasting (NHK, Kanto metropolitan area)

Target area	Main station			Relay station		
	Transmitter site	Channel No.	Antenna power (kW)	Transmitter site	Channel No.	Antenna power (kW)
Kanto metropolitan area (This does not include Ibaragi, Tochigi and Gunma prefecture)	Tokyo	27	10	Chichibu	13	0.01
				Choushi	51	0.01
				Katsuura	34	0.01
				Togane	34	0.01
				Niijima	35	0.03
				Hachijyo	40	0.01
				Hiratsuka	19	0.1
				Odawara	19	0.01

---

[200] Ministry of International Affairs and Communications, Ministerial ordinance No. 87 (2011), "standard transmission method on digital television"

XIII.3 Parameters and Process of the Channel Planning

The channel plan is a result of calculation based on the field strength of the broadcast signal and protection ratio. The rule of MIC "License conformity evaluation criteria regarding Radio Law" set forth the evaluation procedure of the broadcast station's frequency assignment. The basis of this rule is the report from the Telecommunications Technology Council, "Technical requirement on the digital terrestrial television" published on May 1999. This report and resulting MIC's ordinance and rule[201] set forth the pre-requisites for calculation, link budget calculation and protection ratio and other parameters.

While the numbers discussed in this section is old as of late 1990's, the fundamentals how we develop channel planning remains the same. Therefore, this section choose the Japanese channel planning framework as an example to develop reader's understanding. For newer planning criteria of ISDB-T system, refer ITU-R BT.1368-12[202].

The conformity evaluation procedure formulated in the rule is summarized as follows:
1. The contour of the new assignment plan should be basically within the boundaries of designated broadcasting target region, with an exception when:
   i. The new assignment is for a gap filler
   ii. There is special reason from geographical or financial perspective
   iii. The frequency is available
   iv. There is no significant problem found with hearing from these broadcasters, even if interference to other broadcaster's installation plan exists
2. The contour should be calculated based on the ordinance on "the field strength calculation method."[203]
3. The transition signal is in conformity with the ministerial ordinance on "the standard transmission method on digital television"[204]
4. The antenna of transmitter should be horizontally polarized, with exceptions as follows:
   i. In order to avoid interference with broadcast stations in other areas
   ii. If there are existing stations with vertically polarization in the same area to assign an new station and in the same frequency
   iii. If the new assignment is for a gap filler and to avoid interference with other services
5. Protection ratio should be as follows: (summary)
   - Desired: Digital,   Undesired: Analog
     - Co-channel: 20dB
     - Upper adjacent: -24dB
     - Lower adjacent: -21dB
   - Desired: Digital,   Undesired: Digital
     - Co-channel: 28dB   (If applicant DTV station use SFN, evaluation does not necessarily based on this protection ratio. However, proper document can be requested in order to justify the conformity.)
     - Upper adjacent: -29dB
     - Lower adjacent: -26dB
6. Improvement by polarization should be calculated as follows:

---

[201] Ministry of Internal Affairs and Communications, Rule No. 67 (2001), License conformity evaluation regarding Radio Law, Annex 1, 2 Broadcasting Stations
[202] ITU-R BT.1368-10 (2013), Planning criteria, including protection ratios, for digital terrestrial television services in the VHF/UHF bands
[203] Ministry of Posts and Telecommunications (currently MIC), Ordinance No. 640 (1960), "field strength calculation method"
[204] Ministry of International Affairs and Communications, Ministerial ordinance No. 87 (2011), "standard transmission method on digital television"

- VHF
    - Difference (θ, deg): 0 to 26,   Improvement: 26 dB
    - Difference (θ, deg): 26 to 60,   Improvement: 16-((θ-26)×(12/34))dB
    - Difference (θ, deg): 60 to 180,   Improvement: 4 dB
- UHF
    - Difference (θ, deg): 0 to 20,   Improvement: 16 dB
    - Difference (θ, deg): 20 to 60,   Improvement: 16-((θ-20)×(16/40))dB
    - Difference (θ, deg): 60 to 180,   Improvement: 0 dB

7. In addition to the existing broadcast stations, future use should be incorporated.

XIII.3.1 Minimum usable field strength

The council report detailed the minimum usable field strength[205] calculations. The pre-requisites for reception used for the calculation was as follows:

(a) Reception type: fixed reception
(4-way splitter with booster as default)
(b) Receiving antenna height: 10m
(c) Receiving antenna: 14-element Yagi (8-10 dB in UHF band)
(d) Antenna directivity: compatible to ITU-R Rec.419-3
(e) Noise Figure(NF)
  - without booster amplifier: 7 dB
  - with booster amplifier: 3 dB
(f) Feeder loss
Assuming 20m without booster (loss=2 dB), and 30m with booster (loss=3 dB)

Based on these prerequisites, the council report showed examples of link budget for ISDB-T planning. (see Table XIII-3) As a result, 51dBuV/m is required for fixed reception. The report took the calculation with booster as more than 50% of viewers use boosters in Japan.

Table XIII-3   An example of the link budget for digital terrestrial TV broadcasting

Receiving condition	Fixed reception (with booster and 4-way splitter)		Fixed reception (without booster and splitter)	
Frequency (MHz)	470	770	470	770
Modulation	64QAM	64QAM	64QAM	64QAM
Coding rate of inner coder	7/8	7/8	7/8	7/8
Required C/N (after Viterbi decoding, BER=2x10-4) (dB)	22	22	22	22
Margin for equipment degradation (dB)	3	3	3	3
Required receiving C/N (dB)	25	25	25	25
Margin for multi-path (dB)	2	2	2	2
Margin for interference (dB)	1	1	1	1
Total noise Figure NF(dB)	3.3	3.3	7	7
Noise bandwidth B(kHz)	5600	5600	5600	5600

---

[205] There are several terms that have same meaning with "minimum usable field strength", including "minimum median field strength" or "required field strength." This book use minimum usable field strength thereafter.

Internal noise power of Receiver Nr(dBm)	-103.3	-103.3	-99.3	-99.3
External noise power of receiver (dBm)	-102.7	-108.1	-102.7	-108.1
Total noise power of receiver (dBm)	-100.0	-101.9	-97.7	-98.8
Minimum input voltage of receiver with termination(dBuV)	36.7	34.8	-39.0	-37.9
Receiving antenna gain (dB)	8	10	8	10
Receiving antenna effective length (dB)	-13.8	-18.1	-13.8	-18.1
Feeder loss (dB)	2	2	3	3
Field strength (dBuV/m)	50.5	50.9	53.9	55.0

Note 1; in case of "with booster", Booster NF=3dB, receiver NF=7dB, booster gain =27dB, feeder loss between booster and receiver =12dB

Note 2; In case of "with booster", "feeder loss" means the loss between antenna and booster

Note 3; in this table, margins for time availability and place availability are not included

Note 4; more than 50% on viewers use boosters, especially those situated in the fringe areas, boosters are popular.

In addition to the calculation result above, time availability and location availability were also incorporated. Margin of 9 dB was added assuming 99 % of time availability and 50 % of location availability.

As a result, the minimum usable field strength in a broadcast target region was set to be 60dBμV/m.

XIII.3.2 Field strength calculation

MIC Ordinance "field strength calculation method" set forth the field strength calculation method used for channel planning on the digital terrestrial television. While there are detailed conditions in the ordinance, essence of calculation method can be summarized as shown in this subsection. The boundary on a geographic map within which reception field level is above minimum usable field strength is called contour.

(1) VHF

$$E = E_0 + \theta + S$$

E: Field strength (dBμV/m)
E_0: Free space field strength (dBμV/m)
θ: Phase loss (dB)
S: Diffraction loss (dB)

$$E_0 = 20\log\left(\frac{7\sqrt{P_{ERP}'}}{d} \times 10^6\right)$$

d: Distance between transmission and reception site (m)
P_{ERP}: Effective radiated power (W)

$$\theta = 20\log\left(2\sin\left|\frac{2\pi h_1 h_2}{\lambda d}\right|\right) \quad \text{(If the path is line-of-sight)}$$

$$= 20\log\left(2\sin\left|\frac{2\pi h_1 h_2 f}{3\times 10^2 d}\right|\right)$$

h_1, h_2: Height of transmission and reception site (m)
λ: Wave length (m)
F: Transmission frequency (MHz)

$$S = 20\log\left(\sqrt{\frac{j}{\pi}}\int_x^\infty e^{-jt^2}dt\right)$$

$$x = \sqrt{\frac{\pi}{\lambda} \cdot \frac{d_1+d_2}{d_1 d_2}} H$$

d_1: Distance between transmission site and obstacles (mountains) (m)
d_2: Distance between obstacles (mountains) and reception site (m)
H: Height of obstacles (mountains) (m)

(2) UHF
Same formula as VHF applies. Regarding urban propagation within line-of-sight,

1) Adopt E0   if $d < \dfrac{4h_1h_2}{\lambda}$

2) Adopt E-6dB   if $d > \dfrac{4h_1h_2}{\lambda}$

XIII.3.3 Protection ratio

The Telecommunication Technology Council Report also studied on the protection ratio. The values bases on the test result of the first prototype receiver model in 1998. By incorporating the improvement of digital receiver performances, revised values for the digital receiver has been adopted in the MIC's rule shown in Table XIII-4. Currently, broadcaster's license application are evaluated based on this value.

Table XIII-4   Protection ratios the channel planning in Japan

Undesired	Item	D/U ratio according to the interference experiment[206]	Report of the Telecommunication Technology council	Current protection ratios[207]
Analog TV	Co-channel	25dB	30dB	20dB
	Lower adjacent	-33dB (If spurious -21dB)	-21dB	-21dB
	Upper adjacent	-35dB (If spurious -24dB)	-24dB	-24dB
Digital TV	Co-channel	23dB	28dB	28dB (note)
	Lower adjacent	-26dB	-26dB	-26dB
	Upper adjacent	-29dB	-29dB	-29dB

(note) If applicant DTV station use SFN, evaluation does not necessarily based on this protection ratio.

---

[206] The Digital Broadcasting System Committee Report, Telecommunications Technology Council (1999), Reference 2
[207] Ministry of Internal Affairs and Communications, Rule No. 67 (2001), License conformity evaluation criteria regarding Radio Law, Annex 1, 2 Broadcasting Stations

However, proper document can be requested in order to justify the conformity.

Let's see the relation between field strength and protection ratio based on the Figure XIII-2. Assume that transmitter A is transmitting a broadcast signal. As distance between transmission and reception site increases, the field strength becomes weaker. The reception area is defined as a area in which the field strength is larger than the desired field strength. In other words, viewers can expect good reception within the distance at site C.

The minimum usable field strength involves certain margins so that the receiver can satisfactory service in the presence of interference or noise from industrial and domestic equipment. If two transmitters operating in the same area and in the same frequency, the idea of protection ratio can be applied and treat undesired signal as a noise. For example, at the site C, this is a fringe of coverage area of transmitter A. Even if the signal from transmitter B reaches to site C and the signal level of transmitter B at site C is much smaller than that of transmitter A, it can be treated as a noise. For example if the protection ratio is 20 dB, it means if the undesired signal is 1/100 of desired signal it can be treated as a noise and receivers can safely receive the desired broadcast.

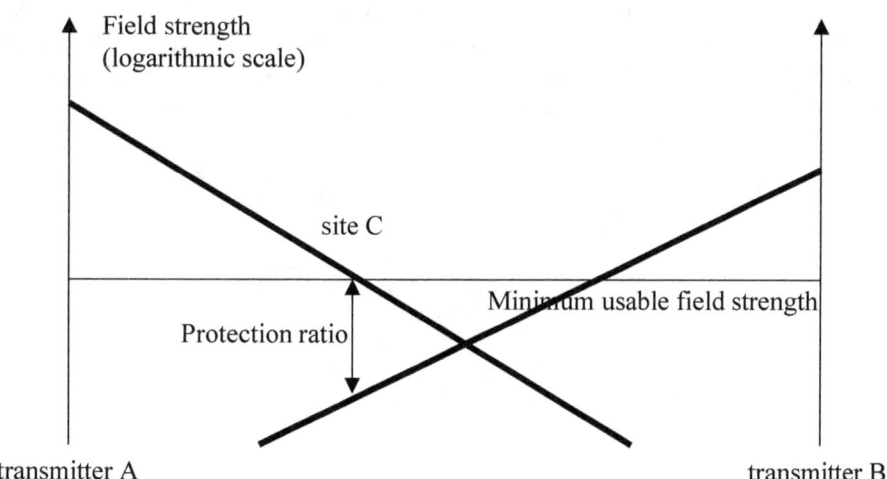

Figure XIII-2 Field strength and protection ratio

In calculation of D/U ratio, council report set out to use antenna directivity defined in the ITU-R BT.419-3 as shown in Figure XIII-3. If the antennas are looking directly towards wanted signal source, certain level of attenuation defined in the figure can be expected on the unwanted signals of which direction is different from the one of desired signal.

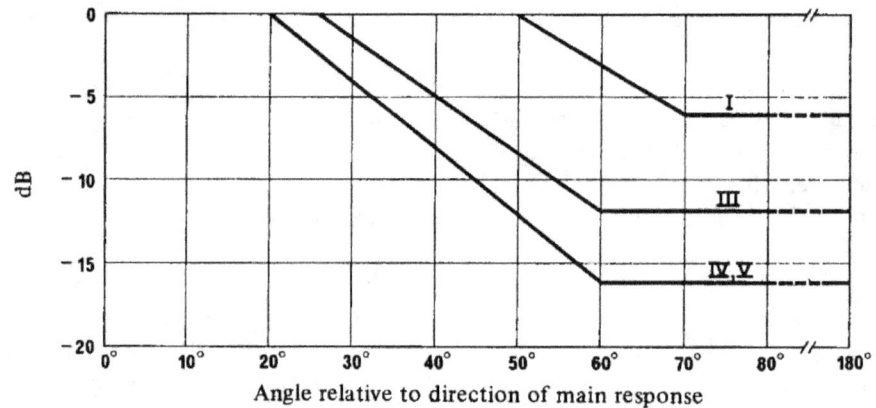

Figure XIII-3 Directivity of the receiving antennas defined in ITU-R BT.419-3

Recently, these coverage calculation and channel planning is done by computer simulation. Software for this purpose is readily available in the market. Generally if you input map and terrain information and also transmitter information, such as transmitter power, antenna pattern, gain, height, these software will automatically calculate coverage maps for you. Also, if you have database of the existing stations, the software will tell you if the new installation will cause interference toward existing coverage or not. It is a good idea to check the availability of these software in your area.

INDEX

1 pulse per second (pps), 213
16 states Quadrature Amplitude Modulation (16QAM), 70
64 states Quadrature Amplitude Modulation (64QAM), 70
8-level Vestigial SideBand (8VSB), 11
Advanced Audio Coding (AAC), 122
Advanced Television System Committee (ATSC), 1, 6
Analog transmitter system, 199
Associação Brasília de Normas Técnicas (ABNT), 37
Association of Radio Industries and Business (ARIB), 36
Asynchronous Transfer Mode (ATM), 7
Audio coding, 184
Auto-correlation circuit, 160
Auxiliary Channel (AC), 77, 80
Basic receivers, 188
B-frame, 113
Binary Phase Shift Keying (BPSK), 80
Bit Error Rate (BER), 68, 248
Bit interleaver, 64
Bit interleaving, 69
Broadcast Law, 265
Broadcast Markup Language (BML), 127
Broadcast TS (BTS), 51, 203
Byte interleaver, 64
Caption, 149
Carrier modulation, 42
Carrier synchronization, 161
Carrier to Noise ratio (C/N), 68
Cascading Style Sheets (CSS), 135
Channel coding, 49
Channel plan, 266
Cliff effect, 200
Clock rate, 55
Coded Orthogonal Frequency Division Multiplex (COFDM), 28
Color Look Up Table (CLUT), 137
Common Phase Error (CPE), 227
Composite Video Baseband Signal (CVBS), 108
Conditional Access Table (CAT), 97
Constellation, 10, 70
Continual Pilot (CP), 77
Contour, 267, 269
Convolutional coding, 67
Coupling loop interference, 208, 229
Cross-correlation method, 161
Cyclic Redundancy Check (CRC), 94, 96
Datacasting, 7, 127, 185
Delay adjustment, 53, 61, 62
Delay control device, 205
Delay profile, 258
Desired to Undesired Ratio (D/U), 251
Differential Binary Phase Shift Keying (DBPSK), 163
Differential Quadrature Phase Shift Keying (DQPSK), 70
Digital Storage Media Command and Control (DSM-CC), 151
Digital Terrestrial Multimedia Broadcast (DTMB), 11, 32
Digital Video Broadcasting-Terrestrial (DVB-T), 2, 6
Discrete Cosine Transform (DCT), 111, 114
Diversity reception, 222
Document Object Model (DOM), 129, 140
Dummy byte, 210
ECMAScript, 129, 139
Effective Radiated Power (ERP), 269
Electronic Program Guide (EPG), 237
Elementary Stream (ES), 90
Emergency information descriptor, 193
Emergency Warning Broadcast System (EWBS), 191
Emergency warning flag, 192
Encoder/Multiplexer (ENC/MUX), 233
Energy dispersal, 65
Entropy coding, 120
Equivalent C/N, 219
Equivalent Noise Degradation (END), 218, 255, 257
Equivalent Noise Floor (ENF), 255, 257

Error Correction, 25
Event synchronization, 131
F sync, 205
Fast Fourier Transform (FFT), 13, 45
FFT window, 158
Field strength, 268
Frame length, 42
Frequency allocation table, 263
Frequency capture range, 249
Frequency distortion, 79
Frequency interleaver, 64
GE06 Agreement, 264
Ginga, 141
Ginga-J, 141, 142
Ginga-NCL, 141, 145
Group of Picture (GOP), 113
Guard Interval (GI), 15, 83
Guard interval ratio, 42
H.264/AVC-10, 6
Hierarchical transmission, 19, 50
High Definition TV (HDTV), 6
Huffman coding, 117
IF relay network, 206
I-frame, 113
Integrated Service Digital Broadcasting-Terrestrial (ISDB-T), 2, 6
Inter Carrier Interference (ICI), 11, 227
Inter Symbol Interference (ISI), 13, 16, 221
Interleaver, 25
Intermodulation, 225, 254
International Telecommunication Union (ITU), 263
Inter-picture prediction, 119
Intra-picture prediction, 119
Invalid hierarchy packet, 210
Inverse Fast Fourier Transform (IFFT), 13, 45
I-Q coordinate, 69
I-Q modulation, 8
ISDB-T Information Packet (IIP), 210
Java DTV, 144
Java TV, 143
LUA Script, 148
Mapping, 69
Master International Frequency Register (MIFR), 265
Matrix Switcher (Sw'er), 235
Mode, 42
Model receiver, 52
Modified Discrete Cosine Transform (MDCT), 122, 123
Modulation Error Ratio (MER), 254
Monomedia, 132
Motion compensation, 111
MPEG-2, 6
Multi-Frequency Network (MFN), 201
Multipath interference, 79
Multi-path interference, 18
Multiplex frame, 55, 57
Multiplex frame pattern, 53
Multiplexing, 5
Narrow band reception, 174
Network Information Table (NIT), 97, 99
Network Synchronization Information (NSI), 210, 211
Non Uniform Mapping, 20
Null packet (Null TSP), 57, 60, 210
OFDM framing, 77
OFDM framing & modulation, 49
OFDM segment, 45
One-Seg, 173
Orthogonal Frequency Division Multiplex (OFDM), 11
Outer coder, 64
Packetized Elementary Stream (PES), 7, 90
Perceptual model, 123
P-frame, 113
Phase noise, 227
Plane model, 138
Presentation, 187
Program Association Table (PAT), 97
Program Clock Reference (PCR), 96, 179
Program Map Table (PMT), 97
Program Specific Information / Service Information (PSI/SI), 7, 89, 237
Protection ratio, 251, 270
Pseudo-Random Bit Sequence (PRBS), 65
Pseudorandom Noise (PN), 255
Punctured pattern, 67
Quadrature Amplitude Modulation (QAM), 11, 69
Quadrature Phase Shift Keying (QPSK), 11, 53, 70
Quality of Service (QoS), 200
Quantization, 111
Radio Law, 265
Radio Regulations (RR), 263
Reed-Solomon coding (RS coding), 64
Reference synchronization, 204
Remain property, 134
Re-multiplexer (RE-MUX), 57, 203, 237

Re-multiplexing, 52
Required C/N, 248
Required signal to noise ratio, 200
Sampling frequency, 51
Scattered Pilot (SP), 77, 78
Segmented OFDM, 23
Segmented OFDM Transmission, 20
Service Multiplexer (Service MUX), 237
Service multiplexing, 52
Set Top Box (STB), 156
Shoulder, 227, 254
simulcast, 233
Single Frequency Network (SFN), 18, 201
Slave synchronization, 204
Source coding, 5
Standard Definition TV (SDTV), 6
Strongest hierarchy, 53

Studio system, 235
Studio to Transmitter Link (STL), 218
Superimpose, 149
Synchronization Time Stamp (STS), 211, 213
System Time Clock (STC), 96
Time interleaver, 64
Transmission, 5
Transmission and Multiplexing Configuration Control (TMCC), 27, 77, 80
Transmitter to Transmitter Link (TTL), 218
Transport Stream Packet (TSP), 7, 90
TS re-multiplexing, 27, 48
Variable length coding, 111
Video coding, 183
Xlet, 142
Z-index, 137

(This page is left intentionally blank.)

ABOUT THE AUTHOR

Masayuki ITO

He is currently a Technical Officer and a Deputy Director of the Broadcasting Technology Division of the Ministry of Internal Affairs and Communications, Japan. He is in charge of developing a 4K and 8K technical roadmap for the satellite and terrestrial broadcasting.

He joined the ministry in 1999 and has been in charge of the frequency allocation policy coordination, international promotion of ISDB-T and the chief information officer at Maebashi City Government.

He received a M.S. in Computer Science in 1994 from the Graduate School of Engineering Science, Osaka University, and a M.P.A. in 2003 from Robert F. Wagner Graduate School of Public Service, New York University.

Yasuo TAKAHASHI

He is the President of YTC Planning and Consulting and technology consultant in the broadcasting engineering field. He had been working for the Toshiba corporation in the field of design and development of broadcast equipment. He served as a chief researcher in the Digital Broadcasting System laboratory in Toshiba, in the meantime, he engaged in development of the ISDB-T first transmission test motel as a proof of design concept of the ARIB STD-B31 standard. He later served as the chairman of DiBEG promoting ISDB-T technology for Latin American and South-East Asian countries.

He graduated from the Engineering Department at Kyoto University in 1967.

Member of IEC TC103 national committee, international expert of Japan of 19/NP (method of measurement for digital terrestrial transmitter) (2000-2012), Vice rapporteur of ITU-D SG2 Q11-3/2 (2011-2013)

James Rodney P. SANTIAGO

He is a graduate of Adamson University with a degree of Bachelors of Science in Electronics and Communications Engineering in 1996, a licensed Professional Electronics Engineer since 2012 and with a Master`s Degree in Electronics Engineering from Mapua University in 2017. He has been involved in the broadcast industry since 1998 and has 18 solid years of experience in this field of which majority has been rigorously spent on Digital Television, He is a leading Filipino Broadcast Engineer in the field of Digital Television Broadcast.

Currently, he is a Consultant for the Department of Information and Communication Technology (DICT) who's main task is to provide policy insights for the Digital TV migration and the Corporate Directions Inc., Japan for the promotional activity and channel plan for the NTC.

He has been responsible for the Philippines' adoption of ISDB-T thereby earning me the nickname "The Father of ISDB-T in the Philippines."

Major Awards he has received include:
- Outstanding Alumni for Engineering of Adamson University twice, in 2010 and 2012,
- International Telecommunications Union-AJ award for contribution in the dissemination of Digital Television Technology around the world (May 17, 2012),
- Broadcast Industry Excellence Award from the Asia Pacific Broadcasting Union (October 17, 2012),

- Citation from the Kapisanan ng mga Brodkaster sa Pilipinas in December, 2012 for the honor given to the Philippines,
- and Top 10 Outstanding Electronics Engineer by the Institute of Electronics and Engineers of the Philippines (December, 2014).

He is an instructor of Broadcast Engineering and Applied Acoustics in Mapua Institute of Technology[208] from 2001-2016 and was the Vice-President for External Affairs of the Institute of Electronics and Communications Engineers of the Philippines (2015-2016) and currently the Vice President-External for the Institute of Electronics Engineers-Quezon City chapter for year 2017-2018.

---

[208] http://drupal7.mapua.edu.ph/sites/default/files/eece/rodneysantiago.pdf

www.ingramcontent.com/pod-product-compliance
Lightning Source LLC
Chambersburg PA
CBHW081142180526
45170CB00006B/1889